Climate Change Impacts and Sustainability

Ecosystems of Tanzania

CABI CLIMATE CHANGE SERIES

Climate change is a major environmental challenge to the world today, with significant threats to ecosystems, food security, water resources and economic stability overall. In order to understand and research ways to alleviate the effects of climate change, scientists need access to information that not only provides an overview of and background to the field, but also keeps them up to date with the latest research findings.

This series addresses many topics relating to climate change, including strategies to develop sustainable systems that minimize impact on climate and/or mitigate the effects of human activity on climate change. Coverage will encompass all areas of environmental and agricultural sciences. Aimed at researchers, upper level students and policy makers, titles in the series provide international coverage of topics related to climate change, including both a synthesis of facts and discussions of future research perspectives and possible solutions.

Titles Available

1. **Climate Change and Crop Production**
   Edited by Matthew P. Reynolds
2. **Crop Stress Management and Climate Change**
   Edited by José L. Araus and Gustavo A. Slafer
3. **Temperature Adaptation in a Changing Climate: Nature at Risk**
   Edited by Kenneth Storey and Karen Tanino
4. **Plant Genetic Resources and Climate Change**
   Edited by Michael Jackson, Brian Ford-Lloyd and Martin Parry
5. **Climate Change Impact and Adaptation in Agricultural Systems**
   Edited by Jürg Fuhrer and Peter J. Gregory
6. **Livestock Production and Climate Change**
   Edited by Pradeep K. Malik, Raghavendra Bhatta, Junichi Takahashi, Richard A. Kohn and Cadaba S. Prasad
7. **Climate Change and Insect Pests**
   Edited by Christer Björkman and Pekka Niemelä
8. **Climate Change and Agricultural Water Management in Developing Countries**
   Edited by Chu Thai Hoanh, Vladimir Smakhtin and Robyn Johnston
9. **Climate Change Challenges and Adaptations at Farm-level: Case Studies from Asia and Africa**
   Edited by Naveen Prakash Singh, Cynthia. Bantilan, Kattarkandi Byjesh and Swamikannu Nedumaran
10. **Climate Change Impacts on Urban Pests**
    Edited by Partho Dhang
11. **Climate Change Impacts and Sustainability: Ecosystems of Tanzania**
    Edited by Pius.Z. Yanda, Claude G. Mung'ong'o and Edmund B. Mabhuye

Climate Change Impacts and Sustainability

Ecosystems of Tanzania

Edited by

Pius Z. Yanda, Claude G. Mung'ong'o and Edmund B. Mabhuye

CABI

CABI is a trading name of CAB International

CABI
Nosworthy Way
Wallingford
Oxfordshire OX10 8DE
UK

Tel: +44 (0)1491 832111
Fax: +44 (0)1491 833508
E-mail: info@cabi.org
Website: www.cabi.org

CABI
WeWork
One Lincoln St
24th Floor
Boston, MA 02111
USA

Tel: +1 (617)682-9015
E-mail: cabi-nao@cabi.org

A catalogue record for this book is available from the British Library, London, UK.

Library of Congress Cataloging-in-Publication Data

Names: Yanda, Pius Z, editor. | Mung'ong'o, C. G., editor. | Mabhuye, Edmund B., editor.
Title: Climate change impacts and sustainability : ecosystems of Tanzania / Pius Z Yanda, Claude G Mung'ong'o, Edmund B Mabhuye.
Other titles: Cabi climate change series.
Description: Boston, MA : CAB International, 2020. | Series: Cabi climate change series | Includes bibliographical references and index. | Summary: "Unique collection of chapters focussing on the climate change research carried out at the University of Dar es Salaam in Tanzania. The book covers a broad range of climate change impacts across a spectrum of systems"-- Provided by publisher.
Identifiers: LCCN 2020018110 (print) | LCCN 2020018111 (ebook) | ISBN 9781789242966 (hardback) | ISBN 9781789242973 (ebook) | ISBN 9781789242980 (epub)
Subjects: LCSH: Climatic changes--Tanzania. | Crops and climate--Tanzania. | Sustainable agriculture--Tanzania.
Classification: LCC QC903.2.T34 C56 2020 (print) | LCC QC903.2.T34 (ebook) | DDC 363.738/7409678--dc23
LC record available at https://lccn.loc.gov/2020018110
LC ebook record available at https://lccn.loc.gov/2020018111
References to Internet websites (URLs) were accurate at the time of writing.

ISBN-13: 978 1 78924 296 6 (hardback)
978 1 78924 297 3 (ePDF)
978 1 78924 298 0 (ePub)

Commissioning Editor: Ward Cooper
Editorial Assistant: Emma McCann
Production Editor: Marta Patiño

Typeset by SPi, Pondicherry, India
Printed and bound in the UK by Severn, Gloucester

Contents

Contributors vii

Preface ix

1 Introduction 1
Claude G. Mung'ong'o, Pius Z. Yanda and Edmund B. Mabhuye

PART I CLIMATE CHANGE AND AGROPASTORAL ECOSYSTEMS 7

2 Living and Responding to Climate Variability and Change among Coffee and Banana Farmers in the Highlands of Moshi Rural District, Tanzania 9
Pontian L. Temba, Noah M. Pauline and Patrick M. Ndaki

3 Cassava as an Adaptation Crop to Climate Variability and Change in Coastal Areas of Tanzania: A Case of the Mkuranga District 23
Naza A. Mbwambo and Emma T. Liwenga

4 Agroecosystems' Resilience and Social–Ecological Vulnerability Index to Climate Change in Kilimanjaro, Tanzania 34
Kelvine C. Shirima and Claude G. Mung'ong'o

5 Effects of Conservation Agriculture on Farmers' Livelihoods in the Face of Climate Change in Balaka District, Malawi 44
Josephine M. Zimba and Emma T. Liwenga

6 A Comparative Cost–Benefit Analysis of Mobile and Sedentary Pastoral Production Systems in Selected Villages in Northern Tanzania 59
Lucas E. Yamat and Claude G. Mung'ong'o

PART II CLIMATE CHANGE AND THE SOCIO-ECOLOGICAL SYSTEM 99

7 Locally Based Responses to Impacts of Climate Change in Pastoral Landscapes of Northern Tanzania 101
Edmund B. Mabhuye and Pius Z. Yanda

8 **Assessment of Socio-ecological Resilience of Agropastoralists to Climate Change and Variability Impacts in Bariadi District, Tanzania** 122
Eugen Cyrilo and Claude G. Mung'ong'o

9 **Natural Resource Use Conflicts in a Changing Climate: The Case of the Wetlands of Kilombero and Kilosa Districts in Tanzania** 153
Emma T. Liwenga and Florian Silangwa

10 **The Role of Ecosystem Services in Enhancing Climate Change Resilience of Local Communities: The Case of Ngarambe-Tapika Wildlife Management Area, Rufiji District, Tanzania** 169
Richard J.M. Katondo and Agnes M.S. Nyomora

11 **Effectiveness of Existing Climate Smart Agricultural Practices in Tehuledere District, North-eastern Ethiopia** 180
Adera S. Wassie and Noah M. Pauline

12 **Community Livelihoods and Ecosystem Integrity in Makere Forest Reserve, Western Tanzania** 194
Pius Z. Yanda, Edmund B. Mabhuye, Anselm R. Mwajombe and Namkunda Johnson

PART III KNOWLEDGE SYSTEMS AND CLIMATE CHANGE 215

13 **Weather Forecasting and Communication in the Upper Great Ruaha Catchment Area** 217
Anselm R. Mwajombe and Godwin A. Lema

PART IV CONCLUSION 235

14 **Lessons Learnt and the Way Forward for Research on Climate Change in Tanzania** 237
Pius Z. Yanda, Claude G. Mung'ong'o and Edmund B. Mabhuye

Index 245

Contributors

Eugen Cyrilo, Institute of Resource Assessment, University of Dar es Salaam, PO Box 35097, Dar es Salaam, Tanzania. E-mail: claude_gasper@yahoo.com
Namkunda Johnson, Institute of Resource Assessment, University of Dar es Salaam, PO Box 35097, Dar es Salaam, Tanzania. E-mail: kundakidee@gmail.com
Richard J.M. Katondo, College of Natural and Applied Sciences, University of Dar es Salaam, PO Box 35194, Dar es Salaam, Tanzania. E-mail: rjmkatondo@gmail.com
Godwin A. Lema, Department of Geography, College of Social Sciences, University of Dar es Salaam, PO Box 35091, Dar es Salaam, Tanzania. E-mail: godwinlema@gmail.com
Emma T. Liwenga, Institute of Resource Assessment, University of Dar es Salaam, PO Box 35097, Dar es Salaam, Tanzania. E-mail: liwenga99@gmail.com
Edmund B. Mabhuye, Institute of Resource Assessment, University of Dar es Salaam, PO Box 35097, Dar es Salaam, Tanzania. E-mail: edmund.mabhuye@gmail.com
Naza A. Mbwambo, Institute of Resource Assessment, University of Dar es Salaam, PO Box 35097, Dar es Salaam, Tanzania. E-mail: mbwambzd@yahoo.com
Claude G. Mung'ong'o, Institute of Resource Assessment, University of Dar es Salaam, PO Box 35097, Dar es Salaam, Tanzania. E-mail: claude_gasper@yahoo.com
Anselm R. Mwajombe, Centre for Climate Change Studies, University of Dar es Salaam, PO Box 33453, Dar es Salaam, Tanzania. E-mail: mwajombea@gmail.com
Patrick M. Ndaki, Centre for Climate Change Studies, University of Dar es Salaam, PO Box 33453, Dar es Salaam, Tanzania. E-mail: pattymzee@gmail.com
Agnes M.S. Nyomora, College of Natural and Applied Sciences, University of Dar es Salaam, PO Box 35194, Dar es Salaam, Tanzania. E-mail: agnesnyomora@gmail.com
Noah M. Pauline, Centre for Climate Change Studies, University of Dar es Salaam, PO Box 33453, Dar es Salaam, Tanzania. E-mail: noah.makula@gmail.com
Kelvine C. Shirima, Institute of Resource Assessment, University of Dar es Salaam, PO Box 35097, Dar es Salaam, Tanzania. E-mail: shirimamiro@gmail.com
Florian Silangwa, Institute of Resource Assessment, University of Dar es Salaam, PO Box 35097, Dar es Salaam, Tanzania. E-mail: silangwaf@yahoo.com
Pontian L. Temba, Centre for Climate Change Studies, University of Dar es Salaam, PO Box 33453, Dar es Salaam, Tanzania. E-mail: brponstemba@yahoo.com
Adera S. Wassie, Department of Plant Science, College of Agriculture, Wolaita Sodo University, PO Box138, Wolaita Sodo, Ethiopia. E-mail: aderasisay@gmail.com

Lucas E. Yamat, Institute of Resource Assessment, University of Dar es Salaam, PO Box 35097, Dar es Salaam, Tanzania. E-mail: lucasyamat@gmail.com

Pius Z. Yanda, Institute of Resource Assessment, University of Dar es Salaam, PO Box 35097, Dar es Salaam, Tanzania. E-mail: pyanda@gmail.com

Josephine M. Zimba, Institute of Resource Assessment, University of Dar es Salaam, PO Box 35097, Dar es Salaam, Tanzania. E-mail: liwenga99@gmail.com

Preface

The Centre for Climate Change Studies (CCCS) at the University of Dar es Salaam has been implementing a number of climate-change-related research projects focusing on climate change and tropical ecosystems. The overall project goal of these studies has been to generate scientific information for contribution to knowledge on climate change impacts in tropical regions with particular focus on Tanzania. Such information is also meant to inform policy and remedial action on the ground for the purpose of promoting adaptation to climate change and resilient livelihoods among communities not only in Tanzania, but also in other tropical regions globally.

Tropical ecosystems directly support the livelihoods of many people and much of a country's economy as a whole, providing goods and services such as food, water, medicine, building materials, fuel and numerous natural attractions that support tourism. The ecosystems also support ecosystem functions that mediate energy and material flow directly and control traits that may alter abiotic conditions (limiting resources, disturbance and microclimate). Recent decades of escalating climate change impacts on these systems and livelihoods worldwide and the vulnerability of ecosystem-dependent communities raise concerns about the consequences of ecosystem changes for human well-being. The pressure placed on ecosystems has been steadily growing as the human population has increased, the economy has expanded, and more ecosystem goods and services have been appropriated, traded and consumed.

This edited book presents the current state of research at the University of Dar es Salaam linking climate change and ecosystems' integrity in the tropics with particular focus on Tanzania. The focus of the book is the implications of climate change on various aspects of economic endeavour – from agriculture, wetland management to weather forecasting. The analyses concentrate on the potential impacts of climate change, focusing on changes in temperature and precipitation, as well as alternative adaptation strategies, such as changing crop types and cropping patterns, adopted by the local communities. Particular attention has been put on to impacts, vulnerability and resilience of ecosystems and communities to climate change outcomes with special reference to the impacts of extreme events such as droughts and flooding.

The chapters presented in this book provide a fairly detailed analysis of the impacts of climate change on various tropical ecosystems in Tanzania and their related aspects of economic endeavour – from agriculture, marine resources and wildlife to weather forecasting.

The analyses concentrate on real and potential impacts of climate change, focusing on changes in temperature and precipitation, as well as alternative adaptive capacity and resilience-enhancing strategies such as changing crop types and cropping patterns.

Pius Z. Yanda, Claude G. Mung'ong'o and Edmund B. Mabhuye

1 Introduction

Claude G. Mung'ong'o, Pius Z. Yanda* and Edmund B. Mabhuye*

Institute of Resource Assessment, University of Dar es Salaam, Dar es Salaam, Tanzania

An Overview

The increasing threat of climate change puts arduous demands on the human spirit towards a sustainable planet, 'as the sheer magnitude of the problem and the potential to challenge existing paradigms beyond the thresholds of historical practice make it one of the most complex and far reaching issues humans have ever encountered' (Engle, 2011). While discussions on possible responses to climate change in the international arena have more often than not concentrated on mitigation, there are similarly important implications regarding the ways ecosystems and societies at the local level will adapt to climate change. Researchers and practitioners are faced with the need to figure out how best to prepare 'for the expected and unexpected impacts of climate change' (Engle, 2011).

Tanzania hosts a variety of ecosystems, including mountains, drylands and wetlands, coastal and marine ecosystems, many of which are transboundary (Taylor *et al.*, 2011). As pointed out by Boon and Ahenkan (2011) for Ghana, the link between climate change, ecosystem services and livelihoods in African countries has been well established. For Tanzania these ecosystems directly support the livelihoods of many people and much of the country's economy as a whole, providing goods and services that include food, water, medicine, building materials, fuel and numerous natural attractions that support tourism. Recent decades of escalating climate change impacts on these systems and livelihoods worldwide and the vulnerability of ecosystem-dependent communities raise concerns about the consequences of ecosystem changes for human well-being.

In the case of Tanzania, the pressure placed on its ecosystems has been steadily growing as the human population has increased, the economy has expanded, and more ecosystem goods and services have been appropriated, traded and consumed. Looking at the total mainland population growth trends for the period 1957–2012, we see a fivefold increase (Wenban-Smith, 2015). At the current rate of growth of 3.2%, the population of 59.3 million is projected to increase more than fivefold by 2100, making Tanzania one of the ten highly populated countries in the world by the close of this century (Anyimadun, 2016). The relationship between population growth and economic development is complex (Agwanda and Amani, 2014), but it is undeniable that the tripling of the population since independence has placed pressure on the natural resource base as observed by Fang *et al.* (2018) for China.

Tanzania's focus in promoting an agro-industrial economy will most likely be based on the expansion, development and commercialization of agriculture. The Southern Agricultural Growth Corridor of Tanzania (SAGCOT) area and Lower Rufiji have been earmarked as high potential areas for promoting such agroprocessing industries.

*pyanda@gmail.com and edmund.mabhuye@gmail.com

The SAGCOT investments are, for example, already envisaged to increase export of processed foods and foreign earnings for national and local development. While doing so, based on the SAGCOT blueprint (2011), SAGCOT initiatives are intended to be sensitive to environmental issues by only focusing on agricultural land (around 5 million ha), leaving out protected areas, considering climate change impacts and envisaging how to reduce climate change vulnerability. No such precautionary measures have been taken for other sensitive but potential areas such as the Lower Rufiji.

The dynamics and main drivers of the change, the vulnerabilities and adaptation strategies being used by the communities have, however, not been clearly understood. Conventional approaches to understanding climate change have been limited to identifying and quantifying the potential long-term climate impacts on different ecosystems and economic sectors. While useful in depicting general trends and dynamic interactions between the atmosphere, biosphere, land, oceans and ice, this top-down, science-driven approach has failed to address the regional and local impacts of climate change and the local abilities to adapt to climate-induced changes. This impact-driven approach has since given way to a new generation of scholarship which utilizes bottom-up or vulnerability-driven approaches that assess past and present vulnerability, existing adaptation strategies, and how these might be modified by climate change. The research themes addressed in this book have focused on such impacts of climate change on ecosystems and communities in Tanzania.

Extreme events such as heat waves are among the most challenging aspects of climate change for societies. Climate models consistently project increases in temperature variability in tropical countries over the coming decades, with the Amazon as a particular hotspot of concern (Bathiany *et al.*, 2018). During the season with maximum insolation, temperature variability increases by ~15% per degree Celsius of global warming in Amazonia and Southern Africa and by up to 10% per degree Celsius in the Sahel, India and South-east Asia. Mechanisms include drying soils and shifts in atmospheric structure. Outside the tropics, temperature variability is projected to decrease on average because of a reduced meridional temperature gradient and sea ice loss. The countries that have contributed least to climate change, and are most vulnerable to extreme events, are projected to experience the strongest increase in variability. These changes would amplify the inequality associated with the impacts of a changing climate (Bathiany *et al.*, 2018).

As noted by Serdeczny *et al.* (2016), the repercussions of climate change are being felt in various ways throughout both natural and human systems in sub-Saharan Africa. According to these authors, climate change projections for this region point to a warming trend, particularly in the inland subtropics: (i) frequent occurrence of extreme heat events; (ii) increasing aridity; and (iii) changes in rainfall – with a particularly pronounced decline in Southern Africa and an increase in East Africa. The region could also experience as much as 1 m of sea level rise by the end of this century under a 4°C warming scenario. Particularly vulnerable to these climatic changes are the rain-fed agricultural systems on which the livelihoods of a large proportion of the region's population currently depend. Impacts across various socio-economic sectors are likely to amplify the overall effect but remain little understood.

Several local studies have been carried out on the impact of climate change on livelihoods and development, especially in developing countries. However, there is a general scarcity of literature that makes a comparative appraisal of the impacts of climate change across the various socio-ecological systems, including on agroecological-based livelihoods across the African continent. Dube *et al.* (2016) have attempted to address this gap by making a comparative analysis of the effects of climate change on agro-based livelihoods across the African continent, focusing on Eastern, Western, Southern Africa and the Sahel region. They have tried to provide a continental perspective on this issue in an attempt to inform current global climate change negotiations and response strategies both at global and national levels.

Despite this noble effort, there remains a dearth of literature that reviews and consolidates these findings to give an overall holistic picture about continental and subcontinental impacts in Africa, especially in relation to local agroecological-based livelihoods.

The collection of papers presented in this volume provides a fairly detailed analysis of the impacts of climate change on various tropical ecosystems in Tanzania and their related aspects of economic endeavour – from agriculture, marine resources and wildlife to weather forecasting. The analyses concentrate on real and potential impacts of climate change, focusing on changes in temperature and precipitation, as well as alternative adaptive capacity and resilience-enhancing strategies such as changing crop types and cropping patterns. Across all of the analyses, particular attention is paid to impacts, vulnerability and resilience of ecosystems and communities to climate change outcomes with special reference to the impacts of extreme events such as droughts and flooding.

The chapters in this book, taken as a whole, represent some of the early attempts at analysing the implications of climate change carried out at the Centre for Climate Change Studies (CCCS) of the University of Dar es Salaam. The chapters employ 'a bottom-up systems approach' (Arndt *et al.*, 2012) whereby the implications of climate change are evaluated based on analytical models from agriculture, wildlife management and weather forecasting systems. In order to deliver a comprehensive analysis, the authors of the chapters, by necessity, hail from multiple disciplines. This comprehensiveness of the key issues, multidisciplinarity, and structural approach allows for more robust insight into the potential implications of climate change. The approach also allows for experimentation with alternative policy options for achieving research and development objectives in the context of climate change.

Structure of the Book

After this introductory chapter the rest of the book is structured in four parts. Part I considers the issues of vulnerability and resilience to climate change of agropastoral systems that include cropping systems, pasture and grazing lands, and animal management. The many crops and livestock varieties kept in Tanzania are grown and reared in diverse climates, regions and soils. No matter the region, however, weather and climate factors such as temperature, precipitation, CO_2 concentrations and water availability directly impact the health and well-being of plants, pasture, rangelands and livestock. For any agricultural commodity, variation in yield between years is related to growing-season weather, which also influences insects, disease and weeds, which in turn affect agricultural production (CCSP, 2008).

In Chapter 2 Temba and colleagues tackle the issue of climate variability and change among the coffee and banana growers in the highlands of Moshi Rural District, Tanzania. The study focuses on uncovering the state of knowledge and strategies used by farmers to address the impacts of the climate variables by assessing people's perceptions to them in the study area. In addition, the chapter assesses the coping strategies employed by the smallholder farmers to improve production as well as the challenges that the smallholder farmers face in adapting to adverse impacts of climate variability and change.

In Chapter 3 Mbwambo and Liwenga investigate the role that is played by the root crop cassava in helping adaptation to climate variability and change in the coastal areas of Tanzania. Basing their study in the coastal district of Mkuranga in the Coast Region, these researchers' objective was to deepen understanding of the role of cassava as an adaptation crop to the changing climate, so as to promote cassava production and ensure household livelihood and food security in the coastal areas.

In Chapter 4 Shirima and Mung'ong'o report on research that investigates the agroecosystems' resilience to climate change on the footslopes of Mount Kilimanjaro, and attempts to develop a social–ecological vulnerability index for the area. The specific research objectives of this study include identifying and assessing the farming practices of the area in terms of their susceptibility to the impacts of climate change. It also examines the agroecosystems' natural resilience to the impacts of climate change.

Conservation agriculture (CA) thrives on the three major pillars of: (i) minimum or no tillage; (ii) permanent soil cover; and (iii) crop rotation. CA has been seen to be the alternative production method that can significantly not only improve soil quality, but also contribute to more economically viable farming systems that are environmentally friendly and climatically sustainable. In Chapter 5 Zimba and Liwenga assess whether the implementation of CA is incrementally improving the livelihoods of farmers in the Balaka District of Malawi, and thereby serving as an effective adaptation measure to drought in that part of Africa. The authors also assess the challenges and opportunities of CA as an adaptation strategy to climate change.

As far as livestock is concerned, Yamat and Mung'ong'o provide in Chapter 6 a comparative cost–benefit analysis of mobile and sedentary pastoralism in Tanzania. Using two cases from northern Tanzania the researchers try to answer the question whether sedentary livestock raising is more productive and utilizes fewer resources and space than the mobile pastoral system in the context of climate change (Niamir-Fuller, 1999; Hesse and MacGregor, 2006). This study attempts to update the debate in an effort to appreciate the social and economic benefits of mobile pastoralism against that of the sedentary system.

Part II of this volume deals with studies in the socio-ecological system based on the belief that 'society is formed and reformed in and through constant interaction with both its social and natural environments' (Baker, 2016). In this context, social environments are taken to be a function of the various individuals and groups that comprise them and the natural environments that sustain them. As elaborated by Baker (2016) 'natural environments can also be seen as both making certain forms of life and society possible, while in turn being modified by the forms of life and society that are actually created'.

In the Marxian conceptualization, nature is in a constantly changing dialectical relationship with humans. Just as human beings are embedded in the natural world, and are shaped by it, nature too is continuously shaped by our interactions with our environment (Marx, *Das Kapital*, 1890–1894, as cited in Baker, 2016). This relationship can be beneficial or harmful to humans, just as humans can interact in ways that support or disrupt natural ecocycles and systems, deplete natural resources and diminish plant and animal biodiversity. However, as emphasized by Baker (2016), 'humans are not the centre of this relationship, because a good, healthy, viable environment can exist without any human society'.

Chapters in this part of the book, therefore, tackle the issues of people's participation in development planning in the context of climate change, the socio-ecological resilience of agropastoralists to climate change and variability impacts. In Chapter 7, Mabhuye and Yanda present locally based responses to the impacts of climate change in pastoral landscapes of northern Tanzania. The main objective is to characterize the major changes in climatic conditions, their impacts and the response strategies undertaken by local communities to counteract the effects. In Chapter 8, Cyrilo and Mung'ong'o assess the socio-ecological resilience of agropastoralists to climate change and variability impacts in the Bariadi District, northwestern Tanzania.

Liwenga and Silangwa in Chapter 9 examine the contribution of climatic factors to the natural resource use conflict, based on a study conducted in Kilombero and Kilosa districts. Based on this assessment, it is evident that the impacts of changing climatic conditions on the availability of natural resources, coupled with factors such as population growth due to influx of people, weak governance and land tenure challenges, have led to increased competition over scarce natural resources. Further, the authors examine conflict resolution mechanisms in terms of adaptive capacity to address these issues of a changing climate. In Chapter 10, Katondo and Nyomora explore the links between ecosystem services and resilience of local communities to the impacts of climate change. The chapter aims to delineate forms of ecosystem services gained by local communities from protecting the ecosystem in the Ngarambe-Tapika Wildlife Management Area (WMA) in southern Tanzania. It also outlines the benefits accrued from conservation of the WMA ecosystem that would help the communities to adapt to climate change impacts. In Chapter 11, Wassie and Pauline evaluate the effectiveness of climate smart agricultural practices in Tehuledere District, north-eastern Ethiopia. The main objective of the chapter is to explore major contributions of climate smart agricultural practies in climate change adaptation. Finally, Part II ends with Chapter 12, where Yanda and others assess comunity livelihoods and ecosystem integrity in the Makere Forest Reserve, western Tanzania. They explore how communities are being pushed by climate and non-climate factors beyond their coping capacity and report on the eventual effects on forest resources management.

Part III deals with the analysis of knowledge systems and climate change. It explores the role for indigenous knowledge (IK) in climate change management in Tanzania. In the face of global climate change and its emerging challenges and unknowns, it is essential that decision makers base policies and actions on the best available knowledge. Biophysical and social sciences contribute significantly to the collective understanding of earth systems, social systems and their interactions. In recent years there has been a growing awareness that scientific knowledge alone is not adequate for solving climate crises, while the knowledge of local and indigenous people is increasingly recognized as an important source of climate knowledge and adaptation strategies (Mafongoya and Ajayi, 2017).

Chapter 13 in this part of the book explores weather forecasting and communication in a major catchment area in Tanzania. The chapter examines a case study of the different uses of IK in the seasonal prediction of climate, based on tree phenology, animal behaviour and astronomical observations, to facilitate decision making in managing and adapting to climate risks. Some IK indicators, such as tree phenology, are losing their value in the face of climate change; however, the integration and conservation of IK in scientific seasonal forecasting for more robust decision making is emphasized by Kijazi *et al.* (2013).

Lastly, Part IV concludes the book with Chapter 14 by drawing on the lessons learnt and charting the way forward for research on climate change in Tanzania.

References

Agwanda, A. and Amani, H. (2014) Population growth, structure and momentum in Tanzania. Tanzania Human Development Report (THDR) Background Paper No. 7/Economic and Social Research Foundation (ESRF) Discussion Paper 61.

Anyimadun, A. (2016) Politics and development in Tanzania: shifting the status quo. Research Paper, Chatham House Africa Programme, London.

Arndt, C., Chinowsky, P., Robinson, S., Strzepek, K., Tarp, F. and Thurlow, J. (2012) Economic development under climate change. *Review of Development Economics* 16(3), 369–377. doi:10.1111/j.1467-9361.2012.00668.x

Baker, S. (2016) Nature in the Anthropocene: political science meets ecology debates. Paper presented to European Consortium for Political Research (ECPR) Joint Session of Workshops, Workshop on Environmental Political Theory in the Anthropocene, Directed by John Barry and Manuel Arias Maldonado, Pisa, Italy, April 2016.

Bathiany, S., Dakos, V., Scheffer, M. and Lenton, T.M. (2008) Climate models predict increasing temperature variability in poor countries. *Science Advances* 4(5).

Boon, E. and Ahenkan, A. (2011) Assessing climate change impacts on ecosystem services and livelihoods in Ghana: case study of communities around Sui Forest Reserve. *Journal of Ecosystems and Ecography* S3, 001. doi:10.4172/2157-7625.S3-001

Dube, T., Moyo, P., Ncube, M. and Nyathi, D. (2016) The impact of climate change on agro-ecological based livelihoods in Africa: a review. *Journal of Sustainable Development* 9(1), 256–267.

Engle, N.L. (2011) Adaptive capacity and its assessment. *Global Environmental Change* 21, 647–656. doi:10.1016/j.gloenvcha.2011.01.019

Fang J., Yu, G., Liu, L., Hud, S. and Chapin III, F.S. (2018) Climate change, human impacts, and carbon sequestration in China. *Proceedings of the National Academy of Sciences* 115(16), 4015–4020.

Hesse, C. and MacGregor, J. (2006) Pastoralism: dry land's invisible asset? Developing a framework for assessing the value of pastoralism in East Africa. Paper no.42. International Institute for Environment and Development (IIED), London.

Kijazi, A.L., Chang'a, L.B., Liwenga, E.T., Kanemba, A. and Nindi, S.J. (2013) The use of indigenous knowledge in weather and climate prediction in Mahenge and Ismani wards, Tanzania. *Journal of Geography and Regional Planning* 6(7), 274–280. doi:10.5897/JGRP2013.0386

Mafongoya, P.L. and Ajayi, O.C. (2017) Indigenous knowledge and climate change: overview and basic propositions. In: Mafongoya, P.L. and Ajayi, O.C. (eds) *Indigenous Knowledge Systems and Climate Change Management in Africa*. CTA, Wageningen, the Netherlands, pp. 17–27.

Niamir-Fuller, M. (ed.) (1999) *Managing Mobility: the Legitimization of Transhumance*. ITDG, London/Food and Agriculture Organization of the United Nations (FAO), Rome.

Serdeczny, O., Adams, S., Baarsch, F., Coumou, D., Robinson, A., *et al.* (2016) Climate change impacts in Sub-Saharan Africa: from physical changes to their social repercussions. *Regional Environmental Change* 17(6), 1585–1600. doi:10.1007/s10113-015-0910-2

Southern Agricultural Growth Corridor of Tanzania (SAGCOT) (2011) Investment Blueprint. The Kilimo Kwanza Growth Corridors initiative. SAGCOT Centre, Dar es Salaam, Tanzania.

Taylor, A., Rubens, J., Masanja, M., Devisscher, T. and Jeans, H. (2011) Ecosystems, development, and climate adaptation: improving the knowledge base for policies, planning and management. Tanzania Study, Final Report. Stockholm Environment Institute (SEI), Stockholm and World Wide Fund for Nature (WWF), Oxford.

US Climate Change Science Program (CCSP) (2008) The effects of climate change on agriculture, biodiversity, land, and water resources in the United States. US Climate Change Science Program Synthesis and Assessment Product 4.3. CCSP, Washington, DC.

Wenban-Smith, H. (2015) Population growth, internal migration and urbanization in Tanzania, 1967–2012: Phase 2 (Final Report). International Growth Center, Working Paper C-40211-TZA-1, September. International Growth Center, London. Available at: https://www.theigc.org/wp-content/uploads/2015/09/Wenban-Smith-2015-Working-paper.pdf (accessed 12 December 2019).

Part I

Climate Change and Agropastoral Ecosystems

2 Living and Responding to Climate Variability and Change among Coffee and Banana Farmers in the Highlands of Moshi Rural District, Tanzania

Pontian L. Temba, Noah M. Pauline* and Patrick M. Ndaki

Centre for Climate Change Studies, University of Dar es Salaam, Dar es Salaam, Tanzania

Abstract

The study aimed at exploring perceived impacts of climate variability on coffee and banana farming and community responses in the highlands of Moshi Rural District. A socio-economic survey employing qualitative and quantitative research approaches was used. Data were collected using questionnaires, key informant interviews, focus group discussions as well as field observation. A total of 96 farmers were involved in the study. SPSS Statistics software package and Microsoft Excel were used for data processing and analysis. Findings showed that communities are knowledgeable about climate variability. Their knowledge is based on perceptions of the impacts already felt and attributed to climate variability, including unpredictable patterns of rainy seasons. Climate variability is associated with decrease in household food supply, unpredictable farming calendar and drying of water sources for irrigation and domestic use. Coffee yields showed a decreasing trend (at the rate of $R^2 = -0.494$) during the years 1990–2016. This was contrary to bananas, which indicated an increasing trend ($R^2 = 0.036$) of production during the same period. Communities were responding to impacts of climate variability in various ways, including intercropping, planting early maturing and drought-resistant varieties and gravity canal irrigation. Projected climate changes showed that the future was uncertain for farmers depending on rain-fed farming. Therefore, further research on viable options would help farmers adapt to current and future climatic stresses. Options may include intensified irrigation of crops and conservation farming which have the potential to increase banana and coffee production, thereby improving productivity and food security for communities.

Introduction

Climate variability and change continue to be serious challenges with devastating impacts on both social and ecological systems (IPCC, 2012, 2014a). Changes in frequency and severity of extreme climate events have significant consequences for human as well as natural systems (IPCC, 2014a). Climate variability already has substantial impacts on biological systems, smallholder communities and countries which depend on them especially developing ones (Thornton *et al.*, 2014). African countries are among the most vulnerable due to their economies being heavily dependent on climate-sensitive sectors, particularly smallholder rain-fed farming (IPCC, 2001; Niang *et al.*, 2014).

*noah.makula@gmail.com

In Tanzania, seasonal variations of rainfall and temperature continue to be reported in different parts of the country. These have noticeable impacts on livelihoods and the economy, particularly for the majority of the population in rural settings, due to not only weak adaptive capacity but also heavy dependence on rain-fed crop production (Yanda and Mubaya, 2011) The increasing intensity of droughts, floods and changes to growing seasons have significant effects on agricultural productivity, water supply and food security (Majule *et al.*, 2004). Likewise, climate variability seriously affects coffee and banana farmers in Tanzania, hence increasing food insecurity and decreasing household and community income among smallholder farming communities (Regassa *et al.*, 2010). Climate variability has contributed to crop pests (e.g. *Prostephanus truncatus*, *Bemisia tabaci*, grasshoppers, armyworms and mole rats) including diseases (such as coffee berry caused by *Colletotrichum kahawae*, a fungal plant pathogen that causes the green berries of the coffee to drop prematurely), which have been on the increase in the study area due to increasing temperatures and decreasing rainfall (URT, 2012).

In response to the adverse impacts of climate variability (e.g. unpredicted rainfall) on crop production and food security, communities have been adopting several adaptation strategies (Mwandosya, 2007). Some of the adaptation strategies adopted and introduced include small-scale irrigation farming, alternative off-farm income-generating activities and an indigenous knowledge system to curb harsh weather conditions (Mwandosya, 2007). Survival skills and coping mechanisms adopted have been used to reduce severity of the impacts of climate variability on people (Shemsanga *et al.*, 2010). Many coping mechanisms among farmers include actions that do not have formal systems recognized by agriculture agencies, such that their implications may have both negative and positive effects on coffee and banana farming (Low *et al.*, 2005).

Moshi Rural District, the geographical focus for this study, is adversely affected by the impacts of climate variability, particularly for the poor communities who mostly depend on coffee and banana as cash crops (Mwakalila, 2014). For example, the gradual increase in temperature is significantly reducing arabica coffee yields and quality (Craparo *et al.*, 2015). In addition, climate variability has led to an increase in banana weevils, mainly, and also banana parasitic nematodes (including *Radopholus similis* and *Helicotylenchus multicintus*) in the highlands due to the prolonged and recurrent dry spells (Gold *et al.*, 1999). Most studies cited in this chapter have addressed the impacts of climate variability specifically on lowland areas and these impacts have been generalized to the highland agroecological zone of Moshi Rural District (e.g. Ajuaye, 2010; Mwakalila, 2014; Mushy, 2016).

Therefore, this study sought to investigate implications of climate variability and change on coffee and banana farming in the highlands of Moshi Rural District, Tanzania. The study focused on uncovering the state of knowledge and strategies used to address the impacts of climate variability and change on coffee and banana farming. In so doing, the study started by assessing people's perceptions on the state of the climate and possible impacts of variability and/or change on coffee and banana production in the study area. In addition, an assessment was made on the coping strategies employed by coffee and banana smallholder farmers to improve production. Finally, a number of challenges facing smallholder farmers in adapting to adverse impacts of climate variability and change were identified before some recommendations could be made.

The vulnerability context of coffee and banana farming

Globally, climatic variability has been raising concerns about potential changes to crop yields. It is established that increasing atmospheric concentrations of greenhouse gases could lead to regional and global changes in temperature as well as precipitation (IPCC, 2007, 2014a). Climate variability is a short-term change or variation in temperature and rainfall and is caused by both natural as well as anthropogenic processes. It has contributed to changes in climate patterns such as increasing temperature, decreasing rainfall and increasing wind intensity; such patterns affect groundwater

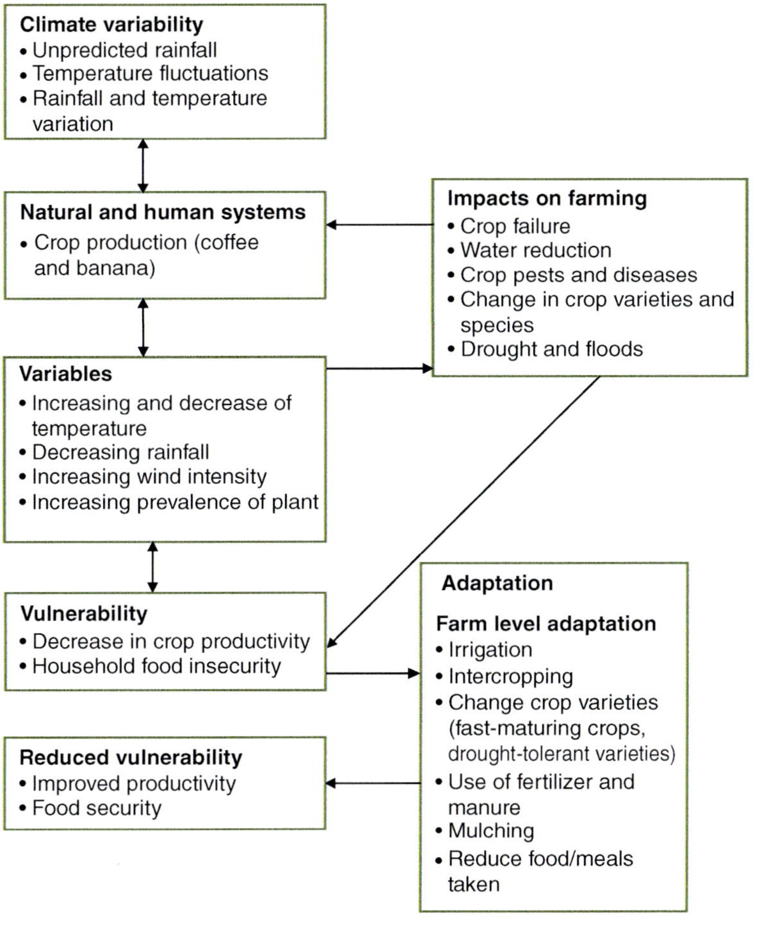

Fig. 2.1. Vulnerability and responses to climate change in banana and coffee farming. (Adapted from Srinivasan, 2004.)

in terms of its quantity and quality as well as its recharge, leading to damage caused to plants and poor crop production (Al-Gamal *et al.*, 2009). These changes in climate patterns have impacts on crop cultivation (Aggarwal *et al.*, 2010).

The connections between the variables in this study (the vulnerabilities and responses to climate change) are shown by the conceptual framework presented in Fig. 2.1.

Natural and human systems depend entirely on the climate of the area. Disturbances or changes to the climate result in alterations in the behaviour of natural and human systems which makes the people who live and depend on those systems for their survival vulnerable (Chapin *et al.*, 2000). Rain-fed agriculture is common in rural areas of Tanzania. Coffee and banana farming are being affected due to a decrease in rainfall as a result of climate variability in the Moshi Rural District, which has resulted in chronic food shortages and reduced people's incomes. Also, changes in the composition of some ecosystems as a result of increases in temperature affect coffee and banana farming in the highlands (Craparo *et al.*, 2015). As a result of the decline in revenues, gross domestic product (GDP) and poverty among people has become a tendency because agriculture is a vital economic activity employing millions of people particularly in rural areas (FAO, 2012).

Coffee is the world's most valuable tropical export crop, which has been threatened by climate variability. Higher temperatures, unpredictable rainfall, resilient pests and plant diseases are associated with climate variability, all affecting coffee production (Haggar and Schepp, 2012). Most studies show that temperature and rainfall conditions are considered important factors in determining coffee growing (Craparo, 2015). Many coffee growing regions are already suffering from such changing conditions and they are very likely to be affected in the near and long-term future (Marengo and Antonio, 2009). A study on the impact of climate change on coffee production in Kenya concluded that an increase in temperature of between 2.2°C and 2.4°C increases precipitation by 135–205 mm (CIAT, 2010).

Bananas are a major perennial multicycle food crop that enhances food security and it is mostly grown in warm temperate regions in Asia (e.g. in India and China), Africa and Latin America (Machovina and Feeley, 2013). A study conducted in Uganda and Burundi by Thornton and Cramer (2012) showed that water stress for long periods of time, low soil moisture and extended exposure to extreme temperatures (above 35°C) could reduce banana production. Also, the banana crop can be affected by climate variability, depending on the stage of the crop at the time of occurrence of climate extremes as well as frequency of extreme climate events within a given crop cycle (Nyombi, 2013).

Adaptation and mitigation is inevitable in order to cope with the impacts of climate variability on coffee and banana production. The prolonged drought causes immediate drying up of coffee and bananas and hence low crop production ensues. Effective use of irrigation, intercropping, change in crop varieties (fast-maturing crops, drought-tolerant varieties) and reduced food/meals taken can reduce risks associated with climate variability. For example, changes in farming practices may yield positive sustainable results in production (Howden *et al.*, 2007). Indigenous knowledge on various categories such as weather forecasting, crop management, water and soil fertility management is also vital for development of adaptation mechanisms (Molua, 2002). These mechanisms contribute to reduction of vulnerabilities by increasing productivity and food security of the community, hence reducing the impact of climate variability and change.

Materials and Methods

The study area

The study was carried out in Moshi Rural District, north-eastern Tanzania. The district is located between latitude 3°10' S and 3°48' S and longitude 37°15' E and 37°36' E (MDC, 2017). It is bordered by Siha and Hai Districts in the west, while Mwanga District lies to the south-east. In the north and north-east the district is bordered by Rombo District while to the south it is bordered by Simanjiro District, which is in Arusha Region (MDC, 2017). With 1529 km^2, the district is administratively divided into four divisions, 31 wards and 165 villages (MDC, 2010).

Moshi Rural District has an average daily temperature of 26°C. The highest temperatures occur in the months of February, March, April, September, October and November with mean maximum temperatures of around 31°C. The lowest temperatures are experienced in June, July, December and January averaging 15°C. The mean annual rainfall is 1520 mm (MDC, 2012).

The major economic activity is small-scale agriculture. Small-scale rain-fed crop production, mainly coffee and bananas, and zero grazing of animals form the major source of livelihood of the communities in the district. These activities are mostly affected by varying rainfall patterns and frequent droughts (URT, 1997). For the purpose of this study, three villages from three wards were randomly selected (see Fig. 2.2).

Data collection

Data collection included: (i) a literature review; (ii) a household questionnaire survey; (iii) key informant interviews; (iv) focus group discussions (FGDs); and (v) participant observations. Questionnaires and interviews were used to obtain both qualitative and quantitative data. A total of 96 households were involved in the study – 34, 32 and 30 in Samanga, Ruwa and Nduweni villages, respectively. Interviews were conducted with 12 key informants involving three ward executive officers, three village executive officers, three village chairpersons and three agricultural extension officers. FGDs were also carried out involving Village Environment Committee members and selected farmers. The data from interviews and FGDs supplemented data obtained from questionnaires. Observation was carried out to complement other methods such as questionnaires and interviews. Through observation it was possible to see efforts made by smallholder farmers to adapt to the adverse impacts of rainfall and temperature variability.

Secondary data sources for this investigation included documentation from locally available sources in the study area (e.g. village register books and district socio-economic profiles). In addition, climate data (rainfall and temperature) were obtained from the Tanzania Meteorological Agency (TMA) at Lyamungo Meteorological Station covering the period 1990–2016. Statistical Package for Social Sciences (SPSS) version 20.0 software was used to compute means, display frequency distributions and calculate percentages and for drawing histograms. Cross tabulation was used to obtain the relationship between variables. Finally, qualitative data were analysed using the structural-functional approach and content analysis.

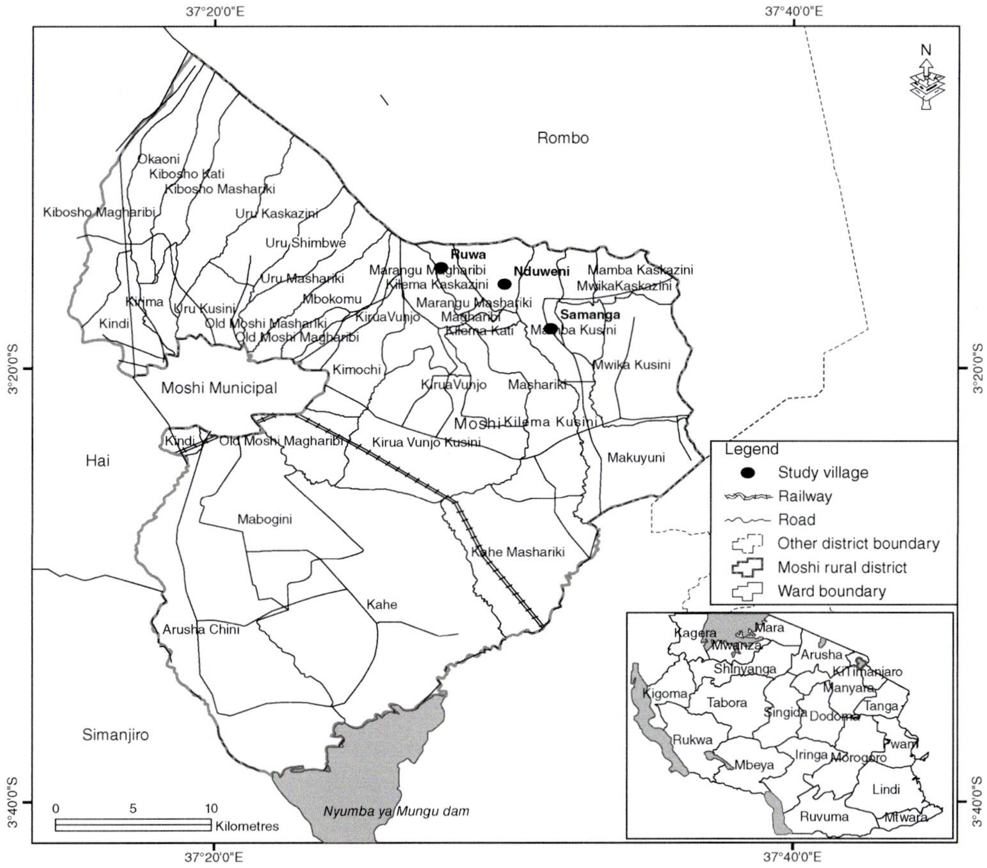

Fig. 2.2. Location of the study area. (Drawn by Cartographic Unit, University of Dar es Salaam, 2017.)

Results

Perceived climate variability

To elicit information regarding climate variability, farmers were asked about their perceptions on several weather elements using different instruments such as a questionnaire. Farmers had different perceptions about the impacts of climate variability on their coffee and banana farming. Findings on rainfall trends over the period 1990–2016 indicated that the majority (69%) of respondents perceived that rainfall was decreasing, while some (27%) noted that rainfall was fluctuating (Table 2.1). In addition, a small proportion (2%) of respondents pointed out that rainfall was increasing and only 1% perceived that there had been no change in the rainfall pattern (Table 2.1).

On the other hand, Table 2.1 also shows that most respondents (81%) perceived that temperatures were increasing. Another 18% perceived that they were fluctuating while only 1% pointed out that temperature was decreasing.

Rainfall and temperature variability in Moshi Rural District, 1990–2016

Records from the metrological station in the study area indicated that the average annual minimum and maximum temperature trends over the past 26 years have been increasing at the rates of R^2 = 0.818 and R^2 = 0.494, respectively (Fig. 2.3). This implies that there has been a general warming trend in the study area. Similarly, findings revealed that the annual trend for rainfall had been decreasing in the study area at the rate of R^2 = −0.181 annually, from 1990 to 2016, whereby 2012 was the driest year on record. It received an annual rainfall of only 26.4 mm. On the other hand, the year 2001 was found to be the wettest year in which an annual rainfall of 165.7 mm was recorded (Fig. 2.4).

Perceived impacts of climate variability and change on coffee and banana farming

The variation has greatly affected coffee and banana farming in the area. Findings indicated that the majority (92%) of respondents reported that there was low yield of crops, which made them shift from growing indigenous crops that required constant water to higher yielding crop varieties and shorter cycle crop varieties, for example, hybrid coffee and bananas (Table 2.2). Few respondents (8%) noted that there was no change in crop yield because sometimes they use irrigation. Also, the majority (93%) of respondents reported that some crops, which required adequate water were no longer being planted, while few (7%) pointed out that they still plant them because they use mulching and irrigation (Table 2.2).

Table 2.2 shows further that the majority (92%) of respondents reported that there was an outbreak of pests and deceases, while the rest (8%) noted that there was no outbreak of pests and deceases of crops because of shifting to hybrid crops. Moreover, findings revealed that the majority (91%) of

Table 2.1. Perceived rainfall and temperature trends (%). (From field survey, 2017.)

		Village			
Trend 1990–2016		Ruwa	Samanga	Nduweni	Total
Rainfall	Increasing	2	0	0	2
	Decreasing	10	30	29	69
	Fluctuating	21	5	1	27
	No change	0	0	1	1
Temperature	Increasing	18	33	30	81
	Decreasing	0	0	1	1
	Fluctuating	16	2	0	18

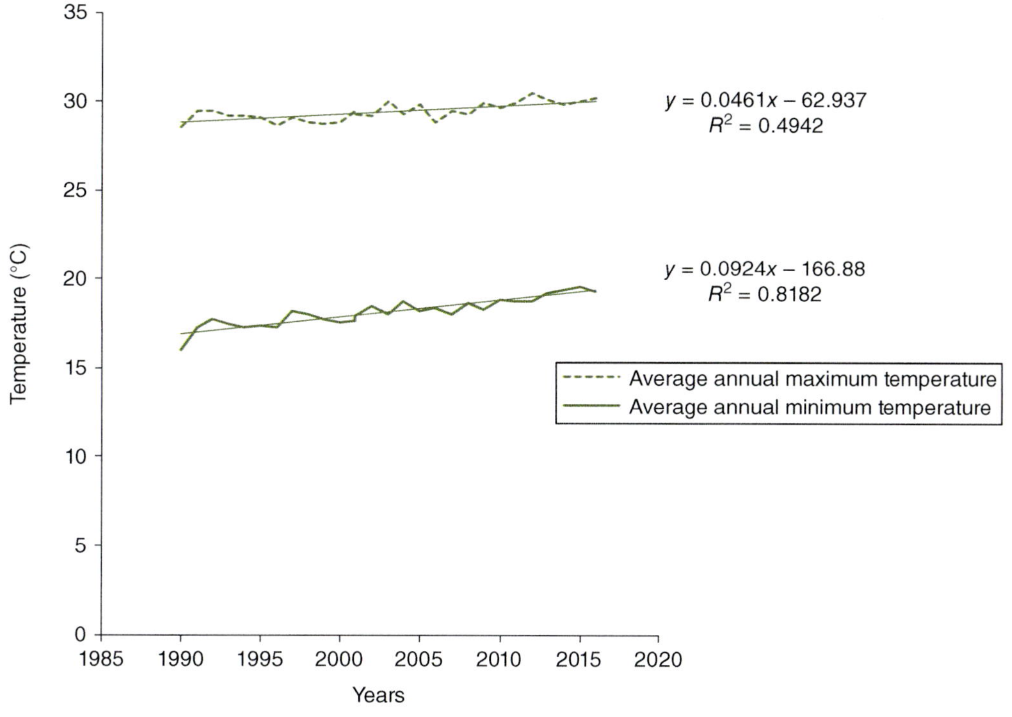

Fig. 2.3. Temperature trends from 1900 to 2016. (Data from the Tanzania Meteorological Agency (TMA) in 2016.)

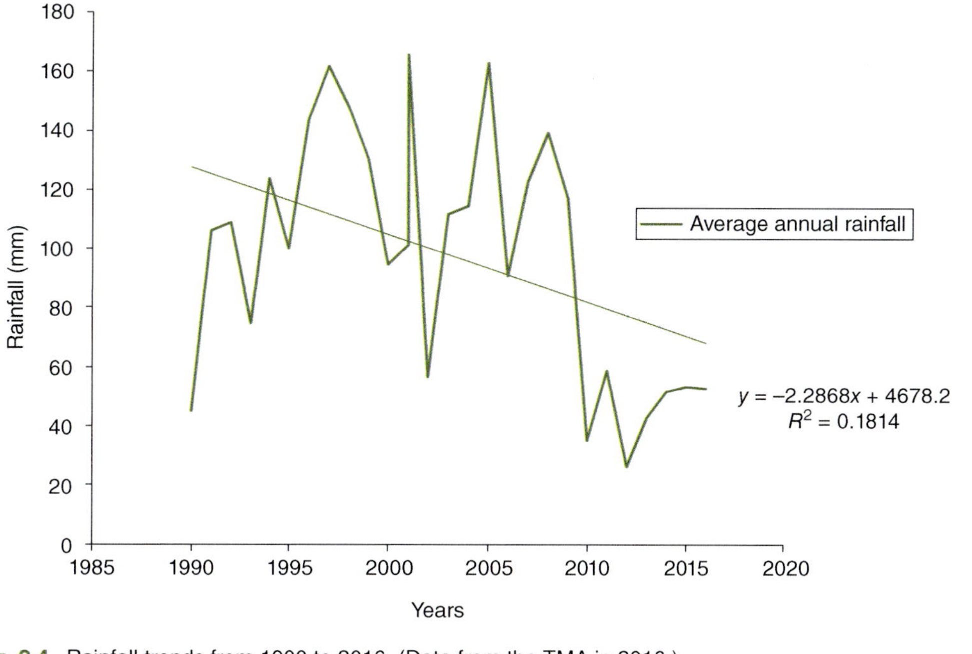

Fig. 2.4. Rainfall trends from 1900 to 2016. (Data from the TMA in 2016.)

Table 2.2. Responses on impact of climate variability on farming. (From field survey, 2017.)

Questions	Response	Frequency	Percentage
Is low yield an indicator of climate variability?	Yes	88	92
	No	8	8
	Total	96	100
Is the decrease of crops (i.e. number of crops grown) an indicator of climate variability?	Yes	89	93
	No	7	7
	Total	96	100
Are outbreaks of pests and diseases an indicator of climate variability?	Yes	88	92
	No	8	8
	Total	96	100
Are the dryng up of water sources an indicator of climate variability?	Yes	87	91
	No	9	9
	Total	96	100
Is reduction of pastures an indicator of climate variability?	Yes	83	87
	No	13	14
	Total	96	100

respondents noted that there was drying up of water sources such as rivers and streams. Besides, 87% of the respondents noted that there had been a reduction of pastures for animals, while the rest (14%) noted that there was no reduction of pastures at all.

Trend of banana and coffee production in the study area, 1990–2016

Findings indicated that there was a decreasing trend (at the rate of $R^2 = -0.494$) for coffee production from the year 1990 to 2016. But findings indicated an increasing trend ($R^2 = 0.036$) for banana production during the same period. Generally, there was a low rate of increase in banana production and a high rate of decrease of coffee production in the area. Data from District Agriculture and Livestock Development Officers (DALDOs) representing trends in crop production are summarized in Fig. 2.5.

Responses to impacts of climate variability on coffee and banana farming

Findings indicate that primary response measures to climate variability in order to improve production included: (i) use of drought-resistant varieties (54%); (ii) planting early maturing crop varieties (60%); (iii) intercropping (52%); (iv) irrigation (83%); and (v) 85% of respondents said use of manure (Table 2.3).

During the field survey, it was observed that the majority of households practise zero grazing and hence, they increase availability of manure. Thus, most people used manure as a response measure due to its availability. Moreover, farmers have started to diversify cash crops. For example, they were shifting from depending on coffee as a sole cash crop to production of avocado as a source of food and selling extra yields so as to improve their living standards. During key informant interviews and FGDs, it was disclosed that people started to grow grafted avocado in 2000. The grafted avocado grows faster and has higher production rates than the native type.

Challenges of adaptating to climate variability and change for coffee and banana farming

During the household survey, it was also necessary to get to know the challenges respondents experience in adapting to the adverse impacts of climate variability, particularly for coffee and banana farming. The challenges identified by respondents were:

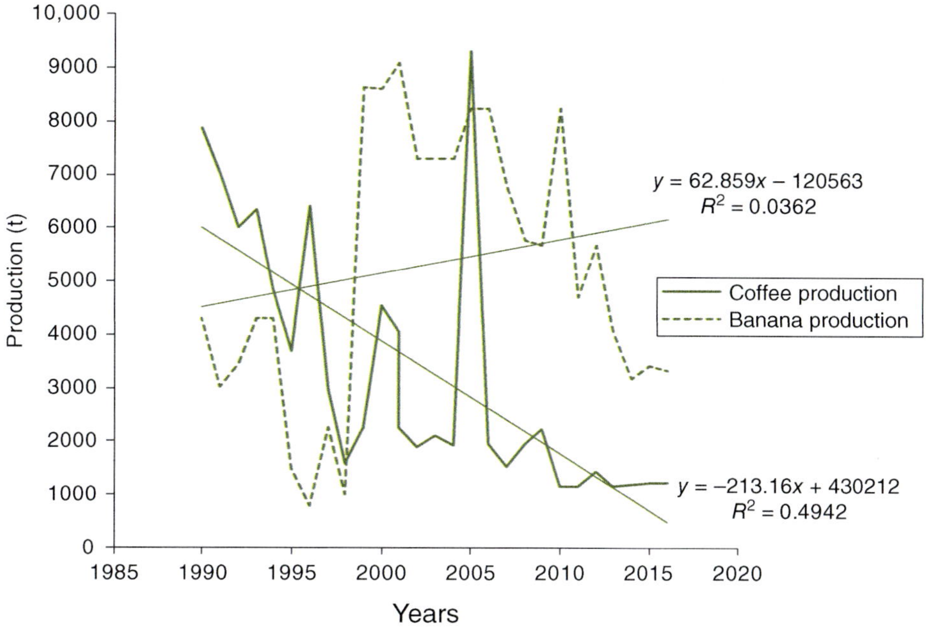

Fig. 2.5. Trends in coffee and banana production from 1990 to 2016. (From field survey, 2017.)

Table 2.3. Coping strategy to improve production in the face of climate variability and change. (From field survey, 2017.)

Coping method used	Response	Frequency	Percentage
Use of drought-tolerant varieties	Yes	52	54
	No	44	46
	Total	96	100
Use of early maturing varieties	Yes	58	60
	No	38	40
	Total	96	100
Intercropping	Yes	50	52
	No	46	48
	Total	96	100
Small-scale irrigation	Yes	80	83
	No	16	17
	Total	96	100
Use of manure	Yes	82	85
	No	14	15
	Total	96	100

(i) poor capital (13%); (ii) poor market coordination (26%); (iii) poor government support (22%); (iv) pests and diseases (22%); and (v) poverty and lack of clear information (17%) (Fig. 2.6). All these were reported as barriers to farmers' efforts to adapt to climate variability.

Discussion

Climate variability is already happening in the study area as revealed by the findings of this study. Communities perceived climate variability through observations of trends of climatic elements, especially changes in

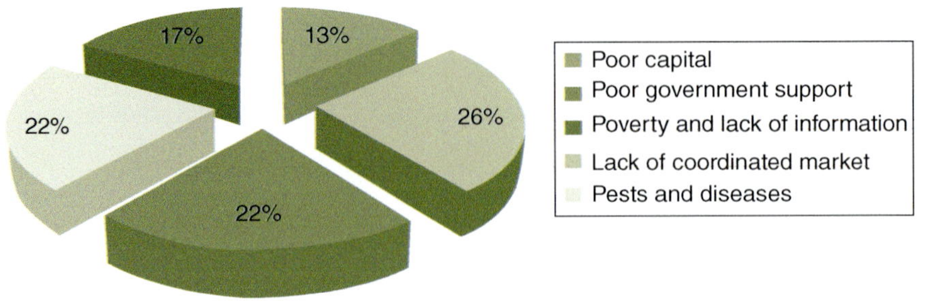

Fig. 2.6. Barriers to adaptation of coffee and banana farming. (From field survey, 2017.)

rainfall and temperature patterns. The study revealed that minimum temperatures had been increasing while, on the other hand, maximum temperatures had been increasing so there has been a general warming trend in the study area. The annual trend of rainfall had been decreasing in the study area as revealed by findings. Notable changes included drought, heavy rains accompanied by heavy winds, environmental destruction via soil erosion and local furrows for irrigation (*mifongo*) being destroyed by floods. There was also an increase in pest and disease infestation.

In recent years, there has been a noticeable increase in intensity of sunshine, a pattern which influences surface temperatures in the study area. Daily temperatures are perceived to have increased, leading to an increase in evening-to-midnight temperatures, which the respondents had never experienced before. Supported by the data from the TMA for the Moshi Rural District, it was revealed that both minimum and maximum temperatures had been increasing. As predicted by the Intergovernmental Panel on Climate Change (IPCC) (2015) the present study has shown that there has been a general warming trend in the study area. Hence crop distribution, reproduction and maturation would consequently be affected by pests and diseases caused by increases in temperatures.

The IPCC (2014b) projected that surface temperatures would rise over the 21st century, varying from place to place. This has been vindicated by the data from the present study, as well as earlier studies in other highland areas of East Africa (Kangalawe, 2009; Mndeme, 2016; Mushy, 2016). Researchers from other regions in sub-Saharan Africa have also pointed out similar findings. For example, Asante and Amuakwa-Mensah (2015) reported that the majority of farmers in Ghana and Zambia perceived that temperatures had increased in their areas while rainfall had declined. Likewise, Nyanga *et al.* (2011) showed that most smallholder farmers in their study areas perceived climate change as changes in temperature and rainfall.

Moreover, Arndt *et al.* (2012) commented that mean temperatures had increased by 1.87°C across Tanzania and thus crop production would decline by 6.25%. In due regard to Debaeke and Aboudrare (2004), increased evapotranspiration in the soil makes crops fail to reach maturity due to lack of enough moisture in the soil, leading to a shortage of crop production. In general, the respondents' perceptions on temperature trends seem to support meteorological data for the Kilimanjaro Region and studies on climate variability that showed that temperatures have been increasing.

The rainfall pattern from 1990 to 2016 as observed by the TMA (Tanzania Coffee Research Institute (TaCRI) station) shows that although overall the average rainfall was decreasing there is a great variation in the amount of rainfall from year to year. For example, in 2012 the recorded annual rainfall was only 26.4 mm while 2001 was found to be the wettest year in which the annual rainfall recorded was 165.7 mm at Lyamungo station. That was mainly attributed to extensive deforestation as a result of extraction of timber from Wachagga gardens for building materials (Kitalyi *et al.*, 2013). These variations have been confirmed by the study done by Mndeme (2016) in Kilimanjaro which reported that rainfall had become unpredictable, sometimes being heavier and sometimes declining.

According to Mngumi (2016), the majority of participants acknowledged tree cutting, especially in lowland areas, for making charcoal, burning bricks, firewood collection and building poles. Such trees are *Vachellia xanthophloea* and *Vachellia tortilis* (previously called and widely known as *Acacia xanthophloea* and *Acacia tortilis*, respectively), resulting in deforestation that leads to water shortage in the area.

Awareness and perception of climate variability help farmers to make decisions at the right time, so they can either change their practices to accommodate the changes or otherwise reduce the impact and their vulnerability to the changing climate – as observed elsewhere by Egbe *et al.* (2014) and Belay *et al.* (2017). However, communities also revealed that they depended on their own observations and knowledge as a source of climate information, which may not be precise and accurate sources thereby subjecting them to unforeseen climatic extremes.

The impacts of climate variability on coffee and banana farming are negative, but the impacts on these two types of farming can vary substantially (IPCC, 2015). Communities are already facing the vagaries of climate variability as noted in the findings from this study. The majority of respondents confirmed that there has been decreasing food availability at household level, a situation which threatens food security. Similar findings by the Tanzanian Coffee Board (TCB, 2010) reported in the northern highlands of Kenya that they had experienced climatic variations in the past 10 years, which consequently affected their crops. A study by Mbilinyi (2013) in Kondoa, Tanzania, pointed out that agricultural activities were affected due to climate variability whereby rainfall may in one season delay and alter the planting season or end before crops ripen, situations which have created problems in sustaining household livelihoods.

Conclusion

The study has revealed that most of the interviewed farmers had experienced climate variability, mainly increased temperature, increased rainfall variability and increased pest invasions in highland areas. The rainfall amount has been decreasing over time while temperatures have increased over time, which has an impact on coffee and banana farming. Communities are responding to impacts of climate variability in various ways, including intercropping, planting early maturing and drought-resistant varieties and canal irrigation using gravity. Projected climate changes show that the future is uncertain for farmers who depend on rain-fed farming. Therefore, further research on viable alternative options would help farmers adapt to current and future climatic stresses. Options may include irrigation of crops and conservation farming that has a potential to increasing banana and coffee production, thereby improving productivity and food security of farming communities. Also, this study has concluded that there are observable and reported challenges facing smallholder farmers in adapting to the impacts of climate variability and change in the study area. The study recommends that information systems should be developed that can help decision makers in devising sound policies to address these challenges, together with mitigation measures that are suitable and manageable in terms of physical and financial resources available to people in the study area.

Acknowledgements

Many thanks to Mwenge Catholic University (MWECAU), the staff of Moshi Rural District Council, and the various smallholder farmers who participated in the research that has led to the successful publication of this work.

References

Aggarwal, P.K., Baethegan, W.E., Cooper, P., Gommes, R., Lee, B., *et al.* (2010) Managing climatic risks to combat land degradation and enhance food security: key information needs. *Procedia Environmental Sciences* 1, 305–312.

Ajuaye, A. (2010) Analysis of farmers' adaptation to climatic change in Kilimanjaro region. PhD dissertation, Sokoine University of Agriculture. Morogoro, Tanzania.

Al-Gamal, S.A, Sokona, Y. and Dodo, A.K. (2009) Climatic changes and groundwater resources in Africa. *International Journal of Climate Change Strategies and Management* 1(2), 133–145.

Arndt, C.F., Strzepek, W.K. and Thurlow, J. (2012) Climate change, agriculture and food security in Tanzania. Policy Research Working Paper 6188. The World Bank, Development Research Group, Agriculture and Rural Development Team, Washington, DC.

Asante, F.A. and Amuakwa-Mensah, F. (2015) Climate change and variability in Ghana: stocktaking. *Climate* 3(1), 78–99.

Belay, A., Recha, J.W., Woldeamanuel, T. and Morton, J.F. (2017) Smallholder farmers' adaptation to climate change and determinants of their adaptation decisions in the Central Rift Valley of Ethiopia. *Agriculture and Food Security* 6(1), 24.

Centro Internacional de Agricultura Tropical (CIAT) (2010) *Climate Adaptation and Mitigation in the Kenyan Coffee Sector*. CIAT, Colombia, pp. 42–43.

Chapin, F.S., Zavaleta, E.S., Eviner, V.T., Naylor, R.L., Vitousek, P.M., *et al.* (2000) Consequences of changing biodiversity. *Nature* 405(6783), 234–242.

Craparo, A.C.W., Van Asten, P.J.A., Läderach, P., Jassogne, L.T.P. and Grab, S.W. (2015) Coffee arabica yields decline in Tanzania due to climate change: global implications. *Agricultural and Forest Meteorology* 207, 1–10.

Debaeke, P. and Aboudrare, A. (2004) Adaptation of crop management to water-limited environments. *European Journal of Agronomy* 21(4), 433–446.

Egbe, C.A., Yaro, M.A., Okon, A.E. and Bisong, F.E. (2014) Rural people perception to climate variability/change in Cross River State-Nigeria. *Journal of Sustainable Development* 7(2), 25.

Food and Agriculture Organization of the United Nations (FAO) (2012) *The State of Food Insecurity in the World: Economic Growth is Necessary but not Sufficient to Accelerate Reduction of Hunger and Malnutrition*. Food and Agriculture Organization of the United Nations, Rome, 65 pp.

Gold, C.S., Rukazambuga, N.D.T.M., Karamura, E.B. Nemeye, P. and Night, G. (1999) Recent advances in banana weevil biology, population dynamics and pest status with emphasis on East Africa. In: *Proceedings of a Workshop on Banana Integrated Pest Management (IPM)*, Nelspruit, South Africa, November 1998, pp. 23–28.

Haggar, J. and Schepp, K. (2012) Coffee and climate change: impacts and options for adaption in Brazil, Guatemala, Tanzania and Vietnam. Climate Change, Agriculture and Natural Resource. NRI Working Paper Series No.4. University of Greenwich, London.

Howden, S.M., Soussana, J.F., Tubiello, F.N., Chhetri, N., Dunlop, M. and Meinke, H. (2007) Adapting agriculture to climate change. *Proceedings of the National Academy of Sciences* 104(50), 19691–19696.

Intergovernmental Panel on Climate Change (IPCC) (2001) *Synthesis Report. A Contribution of Working Groups I, II, and III to the Third Assessment Report of the IPCC*. [Watson, R.T. and the Core Writing Team (eds)]. Cambridge University Press, Cambridge, 398 pp.

Intergovernmental Panel on Climate Change (IPCC) (2007) *Climate Change 2007: Impacts, Adaptation and Vulnerability. Contribution of Working Group II to the Fourth Assessment Report of the Intergovernmental Panel on Climate Change*. [Parry, M.L., Canziani, O.F., Palutikof, J.P., van der Linden, P.J. and Hanson, C.E. (eds)]. Cambridge University Press, Cambridge.

Intergovernmental Panel on Climate Change (IPCC) (2012) *Managing the Risks of Extreme Events and Disasters to Advance Climate Change Adaptation. A Special Report of Working Groups I and II of the Intergovernmental Panel on Climate Change*. [Field, C.B., Barros, V., Stocker, T.F., Qin, D., Dokken, D.J., *et al.* (eds)]. Cambridge University Press, Cambridge, 582 pp.

Intergovernmental Panel on Climate Change (IPCC) (2014a) *Climate Change 2014: Impacts, Adaptation, and Vulnerability. Part A: Global and Sectoral Aspects' Contribution of Working Group II to the Fifth Assessment Report of the Intergovernmental Panel on Climate Change*. Cambridge University Press, Cambridge, 1132 pp.

Intergovernmental Panel on Climate Change (IPCC) (2014b) *Climate Change 2014: Synthesis Report. Contribution of Working Groups I, II and III to the Fifth Assessment Report of the the Intergovernmental Panel on Climate Change*. [Core Writing Team, Pachauri, R.K. and Meyer, L.A. (eds)]. IPCC, Geneva, Switzerland, 151 pp.

Intergovernmental Panel on Climate Change (IPCC) (2015) *Meeting Report of the IPCC Expert Meeting on Climate Change, Food, and World*

Meteorological Organization. IPPC, Geneva, Switzerland, 68 pp.

Kangalawe, R.Y.M. (2009) Impact of climate change on human health: example of highland malaria – Mbeya Region. Study report submitted to the Division of Environment, Vice President's Office, Dar es Salaam, Tanzania.

Kitalyi, A., Wambugu, R.O.C. and Kimaro, D. (2013) *FAO Characterisation of Global Heritage Agroforestry Systems in Tanzania and Kenya. Agroforestry and Development Alternatives (AFOREDA), Tanzania*. Food and Agriculture Organization of the United Nations (FAO), Rome.

Low, P.J., Perry, A.L., Ellis, J.R. and Reynolds, J.D. (2005) Climate change and distribution shifts in marine fishes. *Science* 308(5730), 1912–1915.

Machovina, B. and Feeley, K.J. (2013) Climate change driven shifts in the extent and location of areas suitable for export banana production. *Ecological Economics* 95, 83–95.

Majule, A.E., Mbonile, M.J. and Campbell, D.J. (2004) Ecological gradients as a framework for analysis of land use change. *Michigan State University, LUCID Working Paper*, Series No. 45. Michigan State University, East Lansing, Michigan.

Marengo, J.A. and Antonie, S. (2009) Assessments of moisture fluxes east of the Andes in South America in a global warming scenario. *International Journal of Climatology* 29(10), 1395–1414.

Mbilinyi, D. (2013) The role of indigenous knowledge systems in adaptation to the effects of climate change variability in Kondoa District, Dodoma Region, Tanzania. University of Dar es Salaam, Dar es Salaam, Tanzania.

Mndeme, D.F. (2016) Adaptation strategies to climate variability and climate change impacts on food security among smallholder farmers in Moshi Rural District, Kilimanjaro Region, Tanzania. Department of Agroecology, Faculty of Science and Technology, Aarhus University, Denmark.

Mngumi, J.W. (2016) Perceptions of climate change, environmental variability and the role of agricultural adaptation strategies by small-scale farmers in Africa: the case of Mwanga District in northern Tanzania. PhD thesis, University of Glasgow, Glasgow, UK.

Molua, E.L. (2002) Climate variability, vulnerability and effectiveness of farm-level adaptation options: the challenges and implications for food security in southwestern Cameroon. *Environment and Development Economics* 7(3), 529–545.

Moshi District Council (MDC) (2010) *Socio-economic Profile Briefs*. District Executive Director, Moshi.

Moshi District Council (MDC) (2012) *Socio-economic Profile Briefs*. District Executive Director, Moshi.

Moshi District Council (MDC) (2017) *Socio-economic Profile Briefs*. District Executive Director, Moshi.

Mushy, R.B. (2016) Impact of climate change adaptation strategies on smallholder farmers' livelihoods in Moshi, Tanzania. PhD thesis, University of Dar es Salaam, Dar es Salaam, Tanzania.

Mwakalila, S. (2014) Climate change impacts and adaptation strategies in Kilimanjaro transect in Tanzania. *Journal of Environmental Science and Water Resources* 3(1), 007–014.

Mwandosya, M.J. (2007) Tourism, human welfare and climate concerns: the case of Tanzania. Presented by Professor Mark J. Mwandosya (MP), Minister of State, Vice President's Office (Environment), to the International Conference on Secure and Sustainable Living: Social and Economic Benefit of Weather, Climate Change and Water Services, Madrid, Spain.

Niang, I., Ruppel, O.C., Abdrabo, M.A., Essel, A., Lennard, C., *et al.* (2014) *Climate Change, Impacts, Adaptation, and Vulnerability. Part B: Regional Aspects. Contribution of Working Group II to the Fifth Assessment Report of the Intergovernmental Panel on Climate Change*. Cambridge University Press, Cambridge.

Nyanga, P.H., Johnsen, F.H., Aune, J.B. and Kalinda, T.H. (2011) Smallholder farmers' perceptions of climate change and conservation agriculture: evidence from Zambia. *Journal of Sustainable Development* 4(4), 73.

Nyombi, K. (2013) Towards sustainable highland banana production in Uganda: opportunities and challenges. *African Journal of Food, Agriculture, Nutrition and Development* 13(2), 7544–7561.

Regassa, S., Givey, C. and Castillo, G. (2010) The rain doesn't come on time anymore: poverty, vulnerability, and climate variability in Ethiopia. *Climate Change and Resilience* 6(1), 90–134.

Shemsanga, C., Omambia, A.N. and Gui, Y. (2010) The cost of climate change in Tanzania: impacts and adaptations. *Journal of American Science* 6.

Srinivasan, A. (2004) Local knowledge for facilitating adaptation to climate change in Asia and Pacific: policy implications. Institute for Global Environmental Strategies (IGES)-CP Working Paper No. 002. IGES, Kanagawa, Japan.

Tanzanian Coffee Board (TCB) (2010) *Tanzanian Coffee Industry Development Strategy, Kilimanjaro, New Cooperative Initiative. Strategy 2011–16*. TCB/Tanzania Coffee Research Institute (TaCRI), Moshi, Tanzania, 35 pp.

Thornton, P. and Cramer, L. (eds) (2012) Impacts of climate change on the agricultural and aquatic systems and natural resources within the CGIAR's mandate. CCAFS Working Paper 23. CGIAR Research Program on Climate Change, Agriculture and Food Security (CCAFS), Copenhagen.

Thornton, P.K., Ericksen, P.J., Herrero, M. and Challinor, A.J. (2014) Climate variability and vulnerability to climate change: a review. *Global Change Biology* 20(11), 3313–3328.

United Republic of Tanzania (URT) (1997) *Agriculture and Livestock Policy 1997*. Ministry of Agriculture and Co-operative Development, Government of Tanzania, Dar es Salaam, Tanzania.

United Republic of Tanzania (URT) (2012) *National Sample Census of Agriculture 2007/2008, Regional Report* (Volume V). Ministry of Agriculture, Food Security and Cooperatives, Government of Tanzania, Dar es Salaam, Tanzania.

Yanda, P.Z. and Mubaya, C.P. (2011) *Managing a Changing Climate in Africa: Local Level Vulnerabilities and Adaptation Experiences*. African Books Collective, Oxford.

3 Cassava as an Adaptation Crop to Climate Variability and Change in Coastal Areas of Tanzania: A Case of the Mkuranga District

Naza A. Mbwambo and Emma T. Liwenga*

Institute of Resource Assessment, University of Dar es Salaam, Dar es Salaam, Tanzania

Abstract

This study was carried out in two villages, Kizapala and Kazole, of the Mkuranga District, in the Coast Region of Tanzania. The objective of the study was to establish the role of cassava as an adaptation crop to the changing climate and household food security. Primary data were obtained using household questionnaires and different participatory rural appraisal (PRA) techniques which included focus group discussions (FGDs), key informants and expert meetings. Secondary data were collected through a literature review, whereas temperature and rainfall data from 1984 to 2014 was obtained from the Tanzania Meteorological Agency (TMA). In each village, a sample size of 10% of all households was interviewed. Findings showed that 96% of respondents from Kazole village and 90% from Kizapala linked climate change with major climatic extreme events such as prolonged droughts and occasional abnormal floods. Analysis of temperature data for the last 30 years (1984–2014) revealed that temperature had significantly risen by a correlation coefficient of R^2 = 0.4936 for maximum and R^2 = 0.777 for minimum temperature. The field survey results closely correlated with findings from the analysis of TMA rainfall and temperature data. Findings revealed a decline in crop production which resulted in food shortages and livelihood insecurity in the study villages. The respondents in both villages consider cassava as a crop that is least affected by climate and environmental extremes, thus serves to ensure food availability and security in their households. As a result, growing cassava should be considered as an adaptation strategy to climate change and variability now and in the future. Improving cassava production, processing, marketing and value chain infrastructures is, therefore, crucial for enhancing sustainable adaptation in the district.

Introduction

Climate variability and agriculture have tremendous effects on each other. Climate variability in the form of higher temperature, reduced rainfall and increased rainfall reduces crop yield and threatens food security in low income and agricultural-based economies (IPCC, 2007). Studies by Agrawala *et al.* (2003) and Boko *et al.* (2007) have predicted significant impacts from climate variability and change on African agriculture, although estimates of impacts vary widely between authors (Challinor and Wheeler, 2008). Given the negative impacts of climate variability on economic livelihood and food security in much of the developing world, helping farmers better adapt to this variability is a central concern of development.

*liwenga99@gmail.com

Cassava (*Manihot esculenta*) is an important crop grown in the coastal areas and upcountry as a perennial crop. Its roots serve as an important source of carbohydrates to over 800 million people in the world. It is a major staple food for more than 500 million people in Africa, and is renowned for its drought tolerance and hardiness in stressful environments (Mbanzibwa *et al.*, 2011). Cassava has the largest production volume compared with other agricultural products in Tanzania. Its production is 6.6 million MT, which is 80% more than maize, Tanzania's second crop (3.65 million MT). It is a traditional food crop in Mkuranga District, and one of the main staple foods. Within this district the main staples are cassava (30%), maize (30%) and rice (40%). Cassava is not only treated as a food crop but also as a cash crop.

Some studies have quantified the impacts or responses of cassava to climate variability and have found cassava to be the least affected crop when compared with other major staples such as maize, sorghum and millets. A study done by Lobell *et al.* (2008) found that cassava production in Africa would moderately benefit from climate variability and change by 2030 with an average increase in yield of 1.1% in production from 2000. Schlenker and Lobell (2010) found a decrease in production of 8% for cassava by mid-century, compared with much more severe impacts for maize (–22%), sorghum (–17%) and millets (–17%).

Due to its ability to withstand environmental shocks, many communities in different areas have adopted cassava to ensure food and livelihood security for their households. Despite the fact that this crop has proved to be resilient to climate change impacts and therefore supports food and livelihood security, little is understood about the role this crop plays in enhancing resilience of communities to the impacts of climate change and variability. Not much has been documented regarding the relevance of this crop in climate change adaptation in coastal agroecosystems, which have also faced climate change challenges. This study aimed to bridge this gap by analysing the role of cassava as an adaptation crop to the changing climate in Mkuranga District.

The main objective of the study was to deepen our understanding of the role of cassava as an adaptation crop to the changing climate, so as to promote cassava production and ensure household livelihood and food security especially in the coastal areas of Tanzania. The study specifically aimed to: (i) evaluate the local people's perceptions of climate variability and change in the study area; (ii) establish the climatic trend, particularly over time and its implications to crop production in the study area; and (iii) assess the position of cassava as an enhancer of a household's food and livelihood security.

Materials and Methods

The study area

Mkuranga District is located between latitude 7°7'0" S and 39° 12' 0" E and is 30–40 km away from the city of Dar es Salaam. It is a relatively small district, covering 2432 km^2 out of the total area of 33,539 km^2 covered by the Coast Region. The population size of Mkuranga District is 222,921, made of 108,024 males and 114,897 females (NBS, 2012). The district has about 90 km^2 of coastline, extending from Temeke in Dar es Salaam to the Rufiji (MDC, 2002).

Mkuranga District is warm and humid, with temperatures of around 28–30°C through most of the year, dipping just below 25°C in the coolest months. The district has two rainy seasons, in November–December (short rains or *Vuli*) and in March–June (long rains or *Masika*). On average, the annual rainfall is between 800 mm/year and 1000 mm/year, while the temperature is 28°C.

Agriculture is the principal economic activity, with over 90% of the households engaged in farming. The most common food crops grown are cassava, rice and legumes, and the major cash crops are cashew nuts, coconut, pineapple and citrus. Of these, cassava accounts for 33.7% of the produce and it stands as the highest-ranking staple food in the area. Other economic activities in the district are forest products. According to the 2012 census, the average income per household was TSh600,000 (about US$250)/year (NBS, 2012). The average household size was five people per family. The data show that on average one person earned less than US$1/day.

The study villages

The study was conducted in two villages, Kizapala and Kazole, in Mkamba and Vikindu administrative wards of Mkuranga District, respectively. The criteria of ward selection were based on existing agroecological variations, and the variation in social–economic and livelihood activities between the two wards in the district. In this regard, Mkamba Ward was selected on the basis that it has the highest cassava production in the Mkuranga District, while Vikindu Ward was selected because it is relatively low in terms of cassava production. From each ward, one village which performed the best in terms of cassava production, but was also vulnerable to the impacts of climate variability and change, was selected. The villages were selected with the consultation of district level stakeholders such as District Agriculture and Crop Officers, as well as environmental and other extension officers.

Methods of data collection

Primary data

According to Kothari (2004), primary data are those which are collected afresh and for the first time from the field. They are original in nature and directly related to the problem or the topic of the research study, so the degree of accuracy is very high. In this study, both quantitative and qualitative sets of questions were used to capture information from both the local people and the experts in the field. Primary data were collected through household questionnaire surveys, focus group discussions (FGDs), together with key and/or expert informants' interviews.

Secondary data

Major sources of secondary data in this study included the Tanzania Meteorological Agency (TMA) for rainfall and temperature data, local authorities in the study area for cassava production data, and the Ministry of Agriculture and Food Security for cassava production data at the national level. Some data were also accessed via the Internet. Research reports prepared by scholars were accessed from public records and statistics at the University of Dar es Salaam libraries. Other secondary data sources were historical documents on crop production from private individuals and organizations involved in cassava production and processing in the country.

Data analysis

Descriptive statistical analysis was used for household survey data. Qualitative data collected were edited and classified based on specific objectives. Similarly, quantitative data were also edited, coded and analysed using Statistical Package for Social Sciences (SPSS) computer software SPSS version 20.

Results and Discussion

Both descriptive and analytical methods were used to analyse the data collected from various sources both quantitatively and qualitatively.

Climate change perception of the respondents

Perception is the process of attaining awareness or understanding of sensory information. What one perceives is a result of the interplay between past experiences and one's culture. In studying adaptation to climate change in agriculture, understanding farmers' perception is very important. Gbetibouo (2009) revealed that farmers' ability to perceive climate change is a key precondition for their choice to adapt. In the context of knowledge on climate variability and change, the majority of the respondents in both villages perceived that they were aware about the change in climate. Results indicated that 96% of respondents from Kazole village and 90% from Kizapala understood climate change in different ways (for details of the ways in which these respondents perceived climate change see Table 3.1), while only 6% from Kazole and 10% from Kizapala appeared to be unaware about climate change and/or variability.

The study had 77.4% literate informants, the majority of them (about 71%) having received at least primary school education. There was also age diversity within the respondents, the youngest being 18 years old and the oldest being 90 years old. Most (80.0%) of the respondents were aged between 30 and 90 years old, therefore information gathered from respondents with regard to climate change perception might have been influenced by the age, character, experience and literacy level. From the discussion it was found that around the 1960s, when the majority of these elders were still young and aware of the environment around them, rains used to be heavy, frequent and throughout the year, and agricultural production was consistent.

Considering rainfall and temperature, the majority of people perceived that April used to be the peak of the rainfall season. People could not leave their houses for several hours a day. Moreover, everyone was busy with farm activities and food was abundant. Nowadays, April is one of the driest months in the region and rains are no longer predictable. The elders further claimed that many of the observed changes started around the 1970s, and things started to get a little worse in the1980s, and effects intensified in the 1990s. In both villages, it was reported during FGDs that temperatures have been increasing steadily since the 1990s, such that cold and hot months are no longer predictable. Observed major climate change events in the area in the past 30 years are shown in Table 3.2.

Table 3.1. Respondents' perception of climate change. (From household survey, 2015.)

	Kazole village		Kizapala village	
Climate perceptions	Frequency	Percentage	Frequency	Percentage
Increase in temperature	15	30	15	30
Temperature decrease	2	4	2	4
Rainfall increase	4	8	3	6
Rainfall decrease and changing patterns	26	52	24	48
Disappearance of some flora and fauna	0	0	4	8
Other (strong winds, diseases)	3	6	2	4
Total	50	100	50	100

Table 3.2. Respondents' observed major climate change events in the area in the past 30 years. (From household survey, 2015.)

	Kazole village		Kizapala village	
Other climatic events	Frequency	Percentage	Frequency	Percentage
Prolonged droughts	39	78	40	80
Floods	3	6	10	20
Climate-related land slides	3	6	0	0
Deaths of livestock and people	2	4	0	0
Other (strong winds, diseases)	3	6	0	0
Total	50	100	50	100

Climatic trends in the study area

Temperature trends over the past 30 years

Results from the majority of respondents clarified that formerly temperature patterns and levels used to be steady, and it was known that cool conditions would start in around April and persist until late September. Over 70% of respondents in Kizapala village and about 80% of respondents from Kazole reported during FGDs that temperatures have been increasing steadily since the 1990s, such that cold and hot months are no longer predictable in the villages; they associated these changes with climate change and variability.

Data for annual temperatures were obtained at the nearest weather station to Mkuranga District from the TMA, Dar es Salaam weather station, located at coordinates: 6°87' S, 39°20' E. The analysis of the mean maximum and mean minimum temperature data from TMA for the last 30 years (1984–2014) showed that temperatures had significantly risen by correlation coefficients of R^2 = 0.4936 for maximum temperature (Fig. 3.1) and R^2 = 0.777 for minimum temperature (Fig. 3.2). The general view of many respondents from both study villages was that temperatures had been steadily increasing since the 1990s so the meteorological data is in line with this. These results also agree with other reports on the general

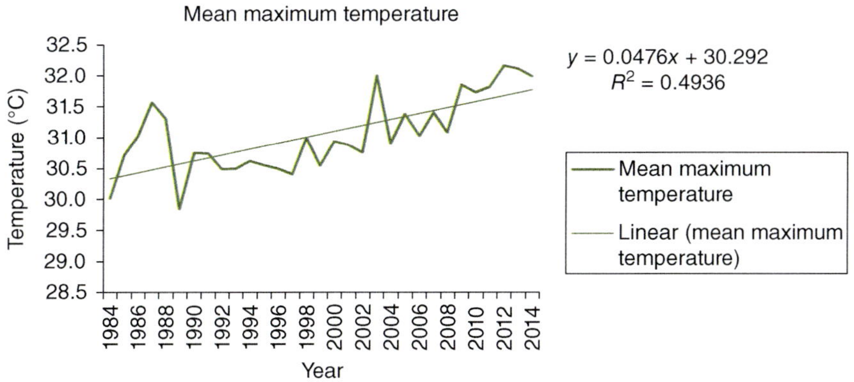

Fig. 3.1. Average mean maximum temperature trend for 30 years (1984–2014) in Dar es Salaam. (Data from the Tanzania Meteorological Agency (TMA) in 2016.)

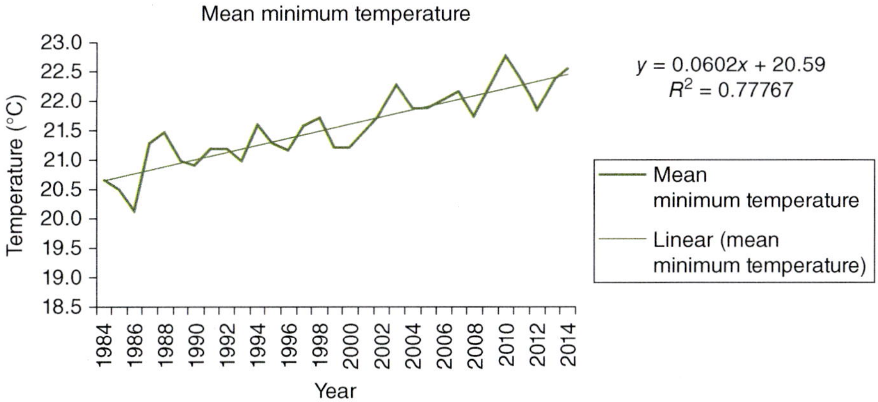

Fig. 3.2. Average mean minimum temperature trend for 30 years (1984–2014) in Dar es Salaam. (Data from the TMA in 2016.)

climatic trend in Tanzania to the effect that mean annual temperatures have increased by 1.0°C since 1960, at an average rate of 0.23°C per decade (URT, 2014).

Rainfall trend in the study area over the past 30 years

Analytical results in Fig. 3.3 indicate that 72% of respondents in Kizapala village and 60% in Kazole confirmed that rainfall amounts have been decreasing, while 22% in Kizapala and 34% in Kazole considered rainfall patterns had changed tremendously over time. Only 2.0% of respondents in Kizapala believed that rainfall had been increasing but not decreasing. The rest of the respondents (6% in Kazole and 4% in Kizapala) believed that there had been no change at all in rainfall trends. It was further reported by key informants that rainfall distribution had also changed on both spatial and temporal scales.

This study also made an analysis of rainfall data from the TMA recorded from 1980 to 2014. The findings from the household survey and FGDs are supported by rainfall data from the TMA which shows a decrease in rainfall trends from the 1990s to 2013 (Fig. 3.4). Although rainfall data for the years 1982, 1983 and 1985 are missing, the general picture in Fig. 3.4 is of both rainfall

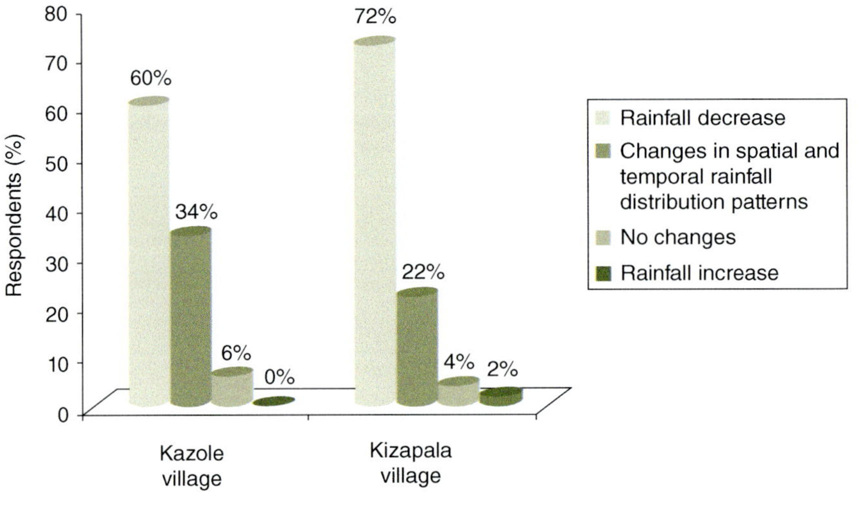

Fig. 3.3. Respondents' observed changes in rainfall trends in the study villages. (From household survey, 2015.)

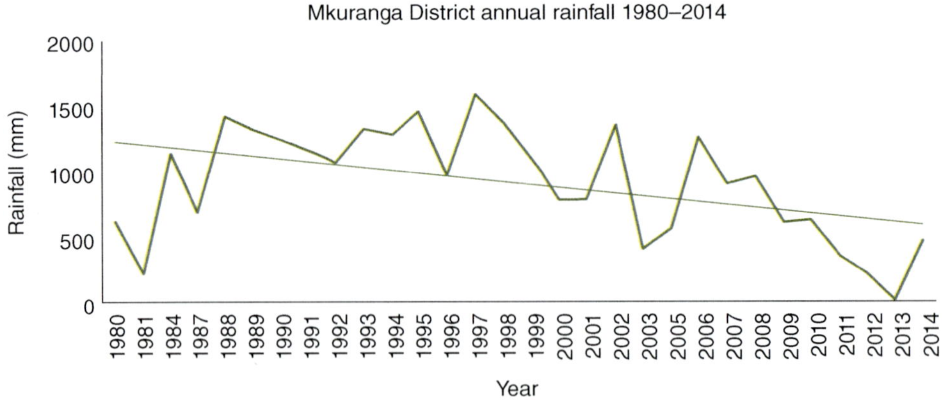

Fig. 3.4. Trend of rainfall in the Mkuranga District over 34 years (1980–2014). (Data from the TMA in 2016.)

variability from year to year and a clearly decreasing trend of rainfall amounts over the period. For example, the years 1981, 1996, 2003 and 2013 had the lowest amounts of rainfall, with 2013 being the driest year over the past 30 years. Accordingly, farmers explained that in the year 1986/87 there were very serious droughts in Kizapala village that left the soil cracked, crops dried and coconut palms dried up, and this resulted in very low crop production that affected the local community's livelihood as well as food security. On the other hand, the year 1988 had very heavy rains such that it was not possible to harvest. The years 1987/88 and 1997/98 had the highest amount of rainfall, and heavy rains were also reported in 2014 after the extreme droughts of 2013.

FGD respondents in all villages claimed that the amount of rainfall had been decreasing gradually since the 1990s causing a decrease in agricultural production and water scarcity in the study area. In Kizapala village participants mentioned that the drying of Kidogoli, Mwena and Nyakenge ponds (that used to have plenty of water throughout the year and whose waters were very useful for agricultural production) was a living example of climate change impacts in the village.

Impacts of climate variability and change on households' food status

More than 78% and 84% of respondents in Kizapala and Kazole, respectively, felt their agriculture industry was being affected due to climate variability and change. The climatic instability has resulted in decreased crop production, productivity and quality of the harvests. Other challenges included the increased costs of production, pests and disease control, soil degradation due to either too much heating of the land or leaching as a result of too much rain. Food security and crop productivity are strongly linked in the study area given the fact that more than 85.7% of respondents in Kazole and 100% in Kizapala depend on agricultural production to manage household food security and their livelihood. Food security in the study area is directly proportional to agricultural production, and any deficit in harvesting means inadequate food for the household. However, through FGD respondents, key informants and experts, it was reported that there was a correspondingly large array of possible autonomous adaptation options to climate variability and change. These included production of drought-resistant varieties of both cash and food crops, and performing non-farm activities that had a role in securing people's livelihoods.

Cassava and households' food and livelihood security

In this study, one of the specific objectives was to assess the position of cassava production in ensuring food security in the era of climate change across the study villages. In so doing, a question was set to enquire which crop among those produced in the study villages was of greatest importance for household food needs and as an adaptation crop to climate change. Results showed that, while 88.7% of respondents in both villages are involved in producing several crops such as maize, millet, paddy, sweet potatoes and cassava, 80.0% of the respondents in Kizapala village and 54.0% in Kazole depend on cassava as their main food crop. Other crops produced include paddy (produced by 12.0% of all respondents in Kizapala and 20.0% in Kazole), and maize (produced by 8.0% in Kizapala village and 20.0% in Kazole village), as shown in Fig. 3.5.

The results above suggest that while respondents can have a wide range of choices of crops to produce depending on the circumstances of the household, cassava is the number one priority. According to participants in the study villages, cassava: (i) is largely traditional and less expensive than other crops; (ii) is tolerant to climate and environmental extremes and poor soils; (iii) can be harvested at any time of the year; (iv) can produce many tubers even without application of fertilizer or pesticide; and (v) can be intercropped together with other crops. Cassava tubers are also fairly resistant to damage from insects or animals. According to El-Sharkawy (2006), cassava seems to respond fairly well to the projected 2030 climate in contrast to other staples of dietary importance in Africa. Fig. 3.6 shows the trend of selected crop production in the Mkuranga District for the period 2006–2014.

Cassava consumption in the study area

Figure 3.7 shows the main uses of cassava in the study villages. The majority of respondents consume cassava as food (78.0% in Kazole and 58.0% in Kizapala), with a considerable number of them producing cassava for earning income through selling it to

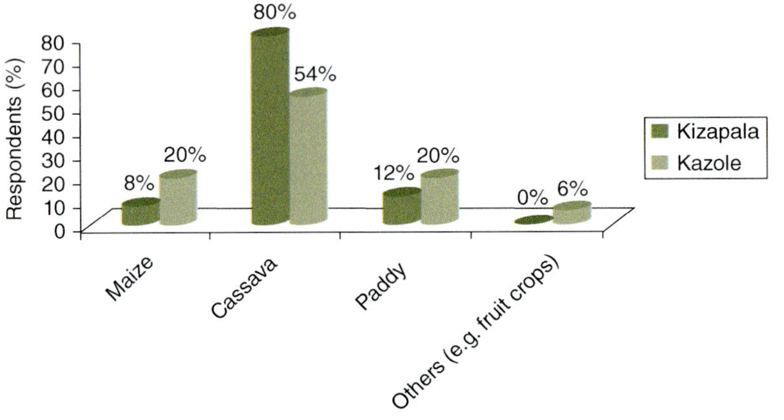

Fig. 3.5. The main food crops produced in study villages in the Mkuranga District. (From household survey, 2015.)

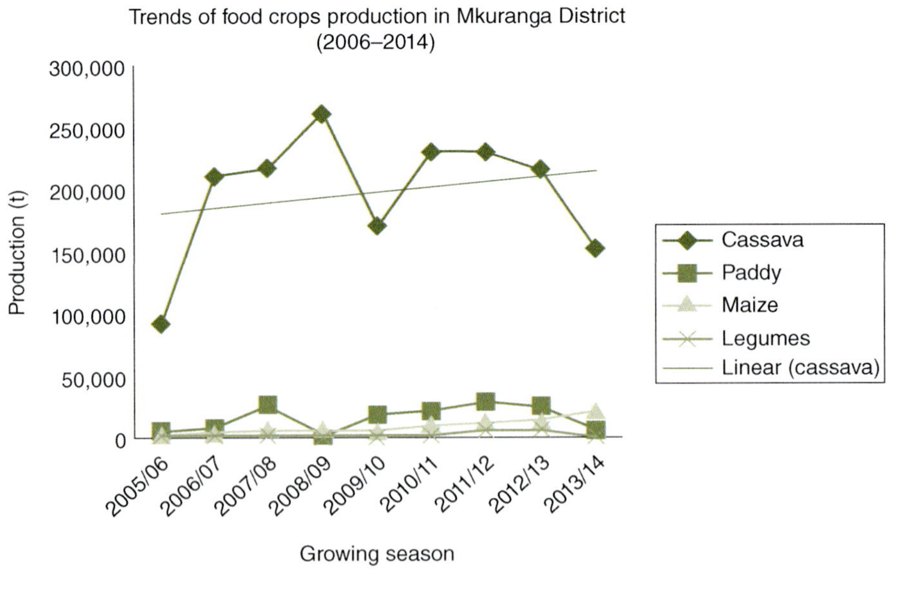

Fig. 3.6. Trend of selected crop production in the Mkuranga District for the period 2006–2014. (From Department of Agriculture, Cooperatives and Irrigation, Mkuranga District, 2013.)

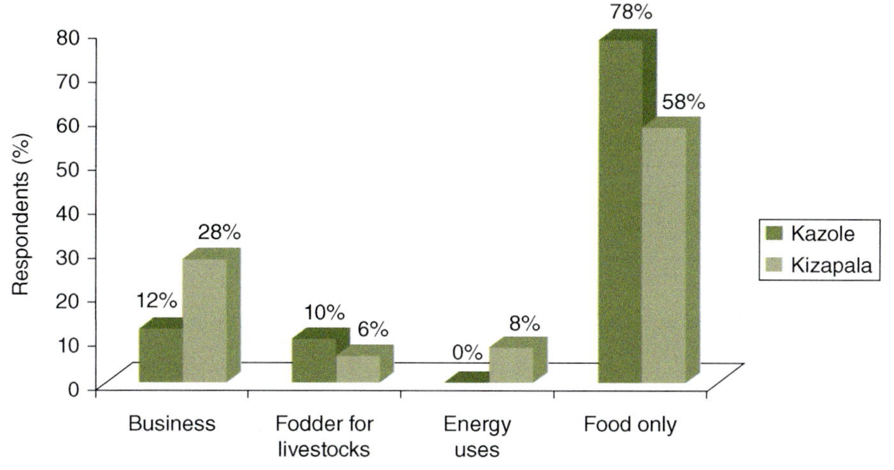

Fig. 3.7. Main uses of cassava in the study area. (From household survey, 2015.)

vendors (28.0% in Kizapala and 12.0% in Kazole). Findings show that cassava is also used as fodder (10.0% in Kazole and 6.0% in Kizapala) in different ways. Dried leaves are mixed in the diets in the proportions required (drying leaves reduces hydrogen cyanide (HCN) content to non-toxic levels). Peeled roots are normally fed to pigs and poultry while unpeeled roots are fed to cattle at low levels. This observation is also supported by Mngulwi (1983) who showed that the peel was rich in protein, fat and ash for animals. Cassava stems are used as a source of household energy (8.0% in Kizapala). However, findings from FGD respondents in all villages indicated that people grow cassava primarily for food needs. But since cassava can also serve as an income-earning crop, people produce surplus cassava so that they can sell it to earn money for other food and non-food needs. Furthermore, respondents maintained that cassava could be consumed as food in a variety of ways as presented in Table 3.3.

Generally, major uses of cassava include snacks, raw food, stiff porridge (*ugali*, *udaga*, *bada*, *msigenge*, etc.), as starch, simply boiled, and as a vegetable (*kisamvu*). It is commonly eaten by people of all ages and can be consumed at any time for breakfast, lunch or dinner. Cassava can also be mixed with other foods such as yams, sweet potatoes and round potatoes, while cassava flour can be mixed with maize/sorghum/millet flour for *ugali* or snacks.

Conclusion

Findings of this study have shown that local people in the two study villages understand climate change in different ways, depending on experience, age and literacy level. They are aware that rains are unpredictable, increasing and sometimes decreasing and causing floods or droughts, respectively. They are also aware of rising temperatures, occurrence of strong winds and disappearance of some flora and fauna. Both direct and indirect impacts of climate variability and change have resulted in decreasing crop production. Food security in the study area is directly proportional to agricultural production, and any deficit in harvests means inadequate food for households.

Observed trends of rainfall and temperatures in meteorological data obtained in some of the stations in Mkuranga District concur with local people's perceptions that rainfall amount has been decreasing gradually, and temperatures have been rising progressively since the 1990s. This changing weather has resulted in water scarcity and a continuous reduction in agricultural production in the study area. Local communities in

Table 3.3. Various ways in which cassava is consumed as food in the study area. (From focus group discussions (FGDs), key informant interviews and household surveys, 2015; photographs by N.A. Mbwambo, 2015.)

Ways of eating cassava	Processing/preparation
Eaten raw; sweet cassava can be eaten raw and tastes slightly sweet	Cassava tubers/roots are washed and their backs stripped off and eaten fresh. Informants stated that only sweet cassava types can be eaten raw, for example, *cosmas cassava* species (as shown in photo) is the most common cassava, which is produced for the purpose of raw consumption. It was reported that there is a high demand for cosmas cassava from towns, especially Dar es Salaam. Farmers reported that many women and men cassava vendors usually come to the villages looking for this type of cassava to sell in towns.
Ugali (boiled cassava flour ground into a paste)	The dark brown outer skin of cassava roots is removed manually using knives. The peeled cassava is then sliced into smaller pieces, and dried in the sun or over a fire. Dried cassava can either be preserved for future use or ground into cassava flour using a traditional mortar and pestle or by milling machines. Cassava can also be ground before drying. The finely ground cassava flour is stirred with boiling water to make *ugali* which is eaten together with other dishes such as salad, vegetables, meat, fish or spices. *Ugali* made from a mixture of cassava and maize flour was also observed in the study area. Women in Kazole peeling cassava in preparation for making cassava flour. A woman in Kazole spreading handmade chips of peeled cassava in the sun to dry before grinding it into cassava flour. Machine-made cassava chips put in the sun to dry ready for milling in Kizapala village.
Simply boiled cassava	Peeled cassava is often cooked and served with a bit of salt. FGD respondents said that boiled cassava is preferred for breakfast and is very common in the study villages because its preparation is easy and it does not require other ingredients (so is less costly).
Fried baked/ grilled	Peeled cassava tubers are fried up in boiling cooking oil in a similar way to cooking potato chips. Cassava chips can also be baked on charcoal or over a less smoky fire. Fried and baked cassava chips are sold in local markets to individuals in the same way as potato chips.
Cassava vegetable (locally known as *kisamvu*)	Cassava leaves are eaten as vegetables. Informants noted that there are many cooking styles, including: (i) boiling the leaves and then crushing them into a paste; and (ii) cooking the leaves with oil, peanuts or fish. FGD participants said that cassava leaves can be dried, pounded into powder and stored easily for use during dry seasons. *Kisamvu* is also sold in markets, and is well known even in towns and cities such as Dar es Salaam, making cassava a commercially viable crop.

the villages admit to have contributed to the climate variability and change through their human activities, while natural conditions have also played a part.

The findings also showed that respondents grew crops such as maize, paddy, legumes, sweet potatoes and cassava for their livelihood and food security. However, the majority of respondents depended on cassava as their staple food, followed by paddy and maize. Cassava holds great potential as a cash crop through the sale of roots, leaves, processed products and planting materials. It remains widely accepted, despite the increasing range of choices of crops produced in the study area mainly because cassava is tolerant to harsh climates and environmental extremes, the crop can be harvested at any time of the year and it can be used both for food and for earning money.[1] As cassava is able to withstand predicted impacts posed by climate variability and change, it is suitable to be an adaptation crop now and in the future, not only in coastal areas, but also across the whole of Tanzania in general.

Note

[1] The opening up of a possible lucrative market for cassava in China will surely boost the commercial value of this crop.

References

Agrawala, S.A., Moehner, A., Hemp, M., van Aalst, S., Hitz, J., *et al.* (2003) Development and climate change in Tanzania: focus on Kilimanjaro. Organization for Economic Co-operation and Development, Paris.

Boko, M., Niang, A., Nyong, A., Vogel, C., Githeko, M., *et al.* (2007) Africa. In: *Climate Change 2007: Impacts, Adaptation and Vulnerability. Contribution of Working Group II to the Fourth Assessment Report of the Intergovernmental Panel on Climate Change*. [Parry, M.L., Canziani, O.F., Palutikof, J.P., van der Linden, P.J. and Hanson, C.E. (eds)]. Cambridge University Press, Cambridge, pp. 433–467.

Challinor, A.J. and Wheeler, T.R. (2008) Use of a crop model ensemble to quantify CO_2 stimulation of water-stressed and well-watered crops. *Agricultural and Forest Meteorology* 148, 1062–1077. doi:10.1016/j.agrformet.2008.02.006

El-Sharkawy, M.A. (2006) International research on cassava photosynthesis, productivity, ecophysiology, and responses to environmental stresses in the tropics. *Photosynthetica* 44, 481–512.

Gbetibouo, G.A. (2009) Understanding farmers' perceptions and adaptations to climate change and variability. International Food Policy Research Institute (IFPRI) discussion paper 00849. IFPRI, Washington, DC.

Intergovernmental Panel on Climate Change (IPCC) (2007) *Climate Change 2007: Impacts, Adaptation and Vulnerability. Contribution of Working Group II to the Fourth Assessment Report of the Intergovernmental Panel on Climate Change*. [Parry, M.L., Canziani, O.F., Palutikof, J.P., van der Linden, P.J. and Hanson, C.E. (eds)]. Cambridge University Press, Cambridge.

Kothari, C.R. (2004) *Research Methodology: Methods and Techniques*, 2nd edn. New Age International Publishers, New Delhi.

Lobell, D.B., Burke, M.B., Tebaldi, C., Mastrandrea, M.D., Falcon, W.P. and Naylor, R.L. (2008) Prioritizing climate change adaptation needs for food security in 2030. *Science* 319, 607–610. doi:10.1126/science.1152339

Mbanzibwa, D.R., Tian, Y.P., Tugume, A.K., Mukasa, S.B., Tairo, F., *et al.* (2011) Genetically distinct strains of cassava brown streak virus in the Lake Victoria Basin and the Indian Ocean coastal area of East Africa. *Archives of Virology* 154(2), 353–359.

Mkuranga District Council (MDC) (2002) *Integrated Coastal Management Action Plan*. The Mkuranga District Council, Mkuranga, Tanzania.

Mngulwi, K.G.J. (1983) The digestibility and net energy gain from cassava crops. MSc thesis, University of Dar es Salaam, Tanzania.

National Bureau of Statistics (NBS) (2012) *Population and Housing Census: Population Projection*. President's Office, Planning and Privatization, Dar es Salaam, Tanzania.

Schlenker, W. and Lobell, D.B. (2010) Robust negative impacts of climate change on African agriculture. *Environmental Research Letters*. Available at: https://iopscience.iop.org/article/10.1088/1748-9326/5/1/014010/meta (accessed 11 December 2019).

United Republic of Tanzania (URT) (2014) *Tanzania Agricultural Climate Resilience Plan (ACRP) 2014–2029*. Vice President's Office, Dar es Salaam, Tanzania.

4 Agroecosystems' Resilience and Social–Ecological Vulnerability Index to Climate Change in Kilimanjaro, Tanzania

Kelvine C. Shirima and Claude G. Mung'ong'o*

Institute of Resource Assessment, University of Dar es Salaam, Dar es Salaam, Tanzania

Abstract

The concept of resilience has gained momentum during the current climate change era. Resilience is said to be the measure of the amount of change the system can undergo while still retaining the same controls on function and structure. Taking into account the effects of changing climate, the term resilience has been used to assess the vulnerability of social–ecological systems. Most agroecosystem studies have focused on dryland ecosystems and this prompted the need to shift concern on to mountainous ecosystems whose susceptibility to climate change is not adequately addressed. This chapter assesses the resilience of maize-coffee-banana agroecosystems on the southern slope of Mount Kilimanjaro in Tanzania. Also, it assesses agronomic practices and the social–economic status of farmers and computes a social–ecological vulnerability index for the ecosystem. The study depicts variation of agronomic practices with altitude due to microclimatic differences, terrain and soil characteristics that determine the type of crops and their farming system which have both positive and negative implications. Climatic shocks (e.g. drought frequency, floods and below average rains) were found to have an impact on agricultural yield. Social–economic indicators (e.g. the number of household dependants, social safety nets, off-farm contribution, possession of land title, usage of wood for cooking energy and access to extension services) have also shown a significant influence on household vulnerability to changing climate which may later affect the agroecosystem productivity as these parameters are associated with the natural environment. Indicators chosen for the vulnerability index depict slight variations of vulnerability altitude wise, except for the mid-lower zone which appears to be more vulnerable.

Introduction

Agroecosystems, which are controlled by humans and include various communities of plants, animals, their biophysical environment and the interactions between them, can be considered as social–ecological systems (Gomiero *et al.*, 2006). In most contemporary agroecosystems, the native ecosystem has been replaced and has been dominated by humans from time immemorial (van Aperdoorn *et al.*, 2011). Although globally modern agroecosystems are seen as the epitome of non-resilience with their monocultures and energy-intensive farming practices (Holling and Meffe, 1996), they are highly resilient at farm-field level. During the course of this century the resilience of many ecosystems is likely to be exceeded by an unprecedented combination of change in climate and other global change drivers (especially land use change and overexploitation), if greenhouse gas emissions and other changes continue at or above current rates

*claude_gasper@yahoo.com

(IPCC, 2007). The primary purpose of assessing resilience is to identify vulnerabilities in social–ecological systems so that action can be taken to create a more sustainable future for people and the land (Berkes *et al.*, 2003).

Tanzania's economy depends on agriculture, which accounts for more than one-quarter of gross domestic product (GDP), provides 85% of exports and employs about 80% of the workforce (WFB, 2013). The three most important crops are maize, coffee and cotton with maize being a major food staple, coffee a major cash crop grown in large plantations (and contributing significantly to the gross national income), while cotton is another cash crop grown largely by smallholder farmers (Agrawala *et al.*, 2003). An interesting observation is that in Kilimanjaro, which is considered to be the main coffee producing region in Tanzania, cash income from coffee appears to be a very small share of total cash income among coffee producing households (a mere 8.7% of total cash income of coffee producers) (Sarris *et al.*, 2006).

Wider institutional changes, in addition to internal changes in subsistence farming households, have contributed to the decline of coffee and the rise of maize and rice as the principal crops and Kilimanjaro subsistence farmers now produce only about 5000 t/year of coffee, which is less than half the yearly amount produced, on average, between the 1950s and the early 1990s (Maghimbi, 2007). The Kilimanjaro Region is also important in terms of food crops such as maize, bananas, beans, rice and millet. Since the 1970s, generational fragmentation of subsistence farms in Kilimanjaro has increased as the population has grown. In such situations, potential coffee farmers (i.e. the sons of older subsistence farmers) are likely to abandon coffee farming in lieu of other activities and to remain on their parents' tiny farms only as a last resort (Maghimbi, 2007).

A number of studies conducted recently in Tanzania have recognized that climate change and variability is happening and is coupled with significant impacts on natural resources, including agriculture which is the main source of livelihood in rural areas (Agrawala *et al.*, 2003). The specific research objectives of this study include: (i) identifying the farming practices and assessing their susceptibility to the impacts of climate change; (ii) examining the socio-economic status of farmers and assessing their capacity to adapt to the impacts of climate change; and (iii) examining the agroecosystems' natural resilience and assessing their susceptibility to the impacts of climate change.

Materials and Methods

The study area

The Kilimanjaro Region is one of Tanzania's 30 administrative regions situated in north-eastern Tanzania. According to the 2012 national census, the region had a population of 1,640,087 inhabitants (URT, 2013). The seasonal rainfall distribution in particular greatly influences agricultural practices. There are two rainy seasons interspaced by two dry spells. The major rainy season occurs from early March to the end of May (*Masika*) while the minor one begins in late September and ends in early November (*Vuli*). There is a marked variation in the amount of rainfall according to altitude and the direction of the slope in the mountainous areas. The mean annual rainfall varies from 500 mm in the lowlands to over 2000 mm in the mountainous areas (over 1600 m above sea level (masl)) (URT, 1998). The hot season lasts from October to March with high humidity and temperatures as high as 40°C in the lowlands. In the mountainous areas, temperatures range from about 15°C to 30°C. The soils in the region also vary according to terrain. Alluvial soils in the various zones can be potentially used for irrigated agriculture (URT, 1998).

This study was carried out along a specific 21.7 km-long and 2 km-wide transect located in the southern part of Mount Kilimanjaro between the villages of Kisangesangeni and Makunduchi/Kirua Vunjo at 3°28′0″ S to 3°16′0″ S and 37°30′0″ E to 37°26′0″ E in Moshi Rural District (Fig. 4.1). Based on altitude this area was divided into four zones: (i) the lower zone (about 700–900 masl); (ii) the mid-lower zone (about

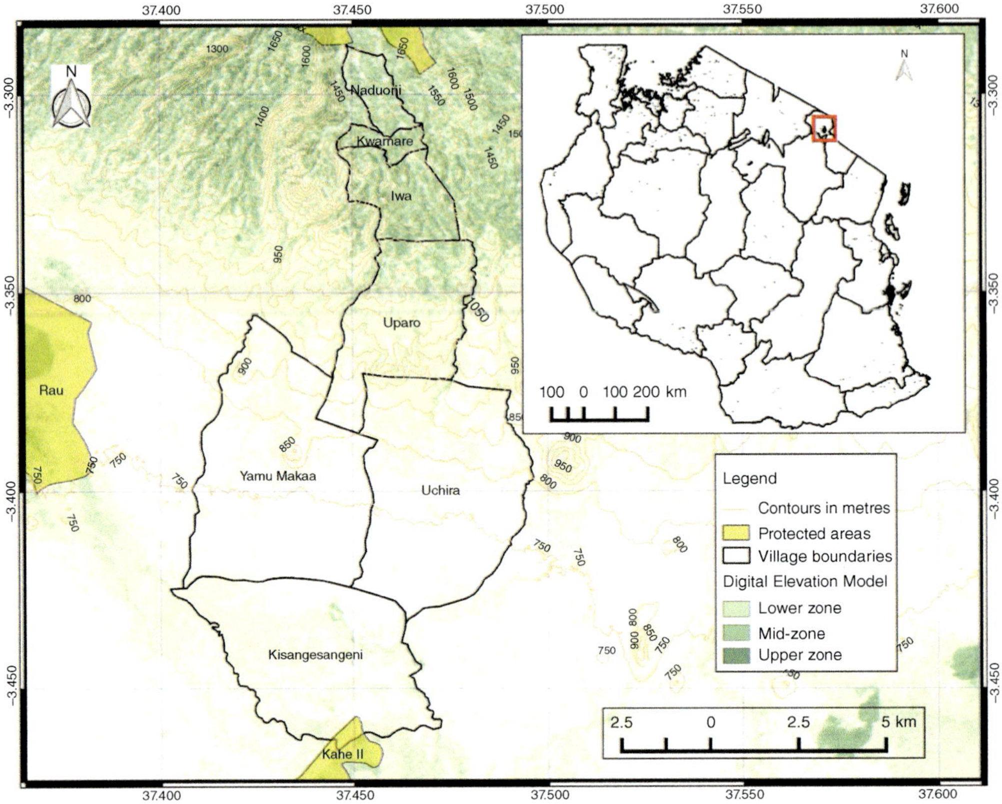

Fig. 4.1. Map of the study area. (From Shirimamiro Geographic Information System (GIS) Laboratory.)

1000–1300 masl); (iii) the mid-upper zone (about 1300–1500 masl); and (iv) the upper zone (1500–1600 masl). The upper and mid-upper zones were dominated by banana and coffee crops while the crops in the mid-lower zone were banana, coffee and maize. The lower zone was predominantly maize.

This study employed several sampling strategies. First, proportionate sampling was done to select respondent villages on a particular zone. Second, lists of farmers were purposively selected from each village across the transect. Third, the names from the lists were then randomly selected to get representative households across the transect. In total, six villages were sampled along the transect and 400 households were covered.

The villages included:

- Nduoni (in the upper zone where 50 households were randomly sampled out of 625 households);
- Iwa (in the mid-upper zone where 100 out of 528 households were sampled);
- Uparo (in the mid-lower zone where 100 out of 747 households were sampled);
- Yamu Makaa (in the lower zone where 50 out of 777 households were sampled);
- Uchira (in the lower zone where 50 out of 600 households were sampled); and
- Kisangesangeni (in the lower zone where 50 out of 530 households were sampled).

Sampling intensity differed according to the size of the zone and number of villages it contained. Four methods of data collection were involved, namely: (i) household questionnaires; (ii) focus group discussions (FGDs); (iii) key informant interviews; and (iv) direct transect walks/field observation.

The construction of the vulnerability index consisted of several steps. The first was the selection of the study area, which consists of several regions. In each region a set of indicators were selected for each of the three components of vulnerability (exposure, sensitivity and adaptive capacity). For each component of vulnerability, the collected data were then arranged in the form of a rectangular matrix with rows representing regions and columns representing the indicators as presented in Table 4.1.

It was assumed that there would be *M* regions/districts and *K* indicators collected. Letting X_{ij} be the value of the indicator *j* corresponding to region *i* then the table would have *M* rows and *K* columns. In this case, indicators would be in different units and scales. The methodology used in the United Nations Development Programme (UNDP)'s Human Development Index (HDI) (UNDP, 2006) was then followed to normalize them in order to obtain figures, which were free from the units, and also to standardize their values, so that they all lay between 0 and 1. There are two functional relationships here, vulnerability increases or vulnerability decreases, with increase or decrease in the value of the indicator. Hence, the higher the value of the indicator (e.g. change in maximum temperature or in annual rainfall or diurnal variation in temperature) the greater was the vulnerability of the area. In this case it can be said that the variables have an (up or raised) ↑ functional relationship with vulnerability, and the normalization is done using the in formula in Eqn 4.1.

NB: The formula is for normalization of variables with functional relation (FR) up ↑.

$$x_{ij} = \frac{X_{ij} - Min\{X_{ij}\}}{Max_i\{X_{ij}\} - Min_i\{X_{ij}\}} \quad \text{(Eqn 4.1)}$$

Table 4.1. Matrix used for computation of the vulnerability index. (From ICRISAT, 2013.)

Region/ district	Indicator			
	1	2	*j*	*K*
1	X_{11}	X_{12}	X_{1j}	X_{1K}
i	X_{i1}	X_{i2}	X_{ij}	Z_{iK}
M	X_{M1}	X_{M2}	X_{Mj}	X_{MK}

where:
x_{ij} = normalized scores for the variables having ↑ FR
$Max_i\{X_{ij}\}$ = maximum value of normalized scores for the variables having ↑ FR
$Min_i\{X_{ij}\}$ = minimum value of normalized scores for the variables having ↑FR.

The normalization of variables with FR (down) ↓ (e.g. literacy rate of the community) is computed by the formula in Eqn 4.2.

$$y_{ij} = \frac{Max_i\{X_{ij}\} - X_{ij}}{Max_i\{X_{ij}\} - Min_i\{X_{ij}\}} \quad \text{(Eqn 4.2)}$$

where:
y_{ij} = variables with FR down ↓
$Max_i\{X_{ij}\}$ = maximum value of normalized scores for the variables having ↓ FR
$Min_i\{X_{ij}\}$ = minimum value of normalized scores for the variables having ↓ FR.

After computing the normalized scores, the index is constructed by giving either equal weights to all indicators/components or unequal weights. In this case the indicators will be given equal weight and the vulnerability index will be calculated using the formula in Eqn 4.3.

$$VI = \frac{\sum_j x_{ij} + \sum_j y_{ij}}{K} \quad \text{(Eqn 4.3)}$$

where:
VI = vulnerability index
Σ = summation
j = indicator
x_{ij} = normalized scores for the variables having ↑ FR
y_{ij} = normalized scores for the variables having ↓ FR
K = number of indicators involved.

Results and Discussion

Livelihood strategies for smallholder farmers

Smallholder farmers in the study area had embarked on a variety of living strategies, which primarily relied on small-scale agriculture.

The study found that 95% of the indigenous people depended on subsistence farming as their main source of livelihood. Secondary activities included petty trading, livestock keeping and short-term wage labour. Crops grown in the area varied with altitude and the micro-climate as already stated ealier (see under heading 'The study area'). Livestock keeping in the upper zone was based on zero grazing with two to four cattle being kept per household while in the lower zone their grazing was based on cattle routing (a free-range system of grazing cattle) with between four and 20 cattle per household.

The upper and mid-upper zone generated their income from selling bananas, coffee and avocado while the lower and mid-lower zones sold their surplus maize, sorghum, groundnuts and livestock. However, some households had enrolled in social safety networks to improve their income. For example, in upper and mid-upper zones 9% of the households had enrolled in Savings and Credit Cooperatives (SACCOs) while in the lower and mid-lower zones 14% of households belonged to such SACCOs from which they received soft loans for petty trading and purchase of agricultural inputs. Generally, some of the farmers in the study area (19%) belonged to different farmers' associations from which they were assisted in marketing their crops, particularly coffee. Access to communication systems for farmers was promising as 87% of households had one or two operating mobile phones and 90% of them had an operating radio for accessing information.

Farmland and plot sizes

The major land use of the study area is crop production with major crops being maize and bananas. The head of the household is the owner of the land in 99% of the respondent households in the area. The study also noted that 85% of the household possessed their own pieces of land while 12% of them borrowed between 1 and 2 acres of land for cultivation. Ownership of a title deed to the land was a challenge with 77% of the households having no documents for their land at all; 75% of such households inherited the land from their forefathers. Some 10% had a letter of offer[1] concerning the right of land occupancy and only 2% had legal documents for their land parcels.

Land possession in the upper and mid-upper zones varied from less than 1 acre (10% of households) to 5 and 6 acres (14% of households). Further, it was found that 29% of the households in the two zones had 2–3 acres, 19% had 1–2 acres, while 15% had 3–4 acres. The upper and mid-upper zones were dominated by agroforestry farming.

In the lower and mid-lower zones 35% of households possessed 1–2 acres of land, 28% possessed 2–3 acres while 16% had 3–4 acres. A negligible percentage (2%) had between 12 and 21 acres of land in the lower and mid-lower zones. The slope of land parcels varied with altitude in which 60% of the lower and mid-lower zone households had their farmland on flat terrain and 39% of them having slightly inclined farm plots.

It was also found that 35% of the upper and mid-upper zone households had plots located on flat terrain while 65% of household farmland was located on slightly inclined terrain. There was little or non-existent erosion of farmland in these zones due to the agroforestry system they practised. In fact, 53% of the respondents reported to have not experienced erosion at all in their life, while 45% had experienced a little erosion in the upper and mid-upper zones. In the lower and mid-lower zones, a little erosion was also reported by 57% of households while only 6% of them reported severe erosion of their farmland.

Cropping and farm management systems

Cropping and management systems of farmland and the crops cultivated showed little variation across the transect (as noted in the preceding paragraphs). It was found that more than half of farmers (51%) prepared their farmland by using tractors while 48% used traditional hand hoes. The study found that intercropping was common in the home gardens in the upper and mid-upper zones, with coffee, bananas and a little maize being grown during rainy seasons. Even though in the mid-upper and upper zones farmers consider banana as their staple crop, most of them also depended on maize because it can be stored for a long time without damage. For farms located in the mid-lower and lower zones monocropping was common on the land away from home on which maize was grown. Such farms were larger than the home gardens and most farmers leased land from other people for prices ranging from TSh20,000/acre to TSh50,000/acre, depending on the condition of the plot.

However, during the short rains (*Vuli*) the majority of farmers cultivated small plots of less than 2 acres because these rains were not enough for maize. During this time most of them cultivated beans. While in the past mixed farming was practised in the lower and mid-lower zones, monocropping has become common (56% of households) among farmers presumably due to climate change. In the past they were reportedly mixing sunflower and maize, or beans and maize. However, although some 46% of the respondents were trying to maintain mixed farming of maize and beans on irrigated farms, the rest reported that they had abandoned mixed farming 8 years ago.

Farmers admitted that the use of industrial fertilizers had been increasing recently following the decrease in yields. It was reported in Iwa village, for example, that urea, NPK and *Minjingu mazao* (phosphate) fertilizers were increasingly being used in the lower zone and mid-lower zone farms where the farmers of Iwa village cultivated. Without fertilizer they said they could not get enough yields per farm plot due to exhaustion of the soil, which is caused by lack of fallowing. The rate of application of *Minjingu mazao* fertilizer was 25 kg/acre. The government has encouraged use of this fertilizer for growing maize but it is also said to require a lot of water for it to function well.

It was further established that other fertilizers such as sulfate have been applied since 1976. NPK was reported to be in use since the 1980s. However, farmers claimed that sulfate was detrimental to the soil and they had no option but to use farmyard manure. It was reported that livestock keepers had decreased due to the decline of pastures and therefore there was reduced availability of farmyard manure. It was reported, for example, that applying farmyard manure rather than industrial fertilizer was so expensive that it was not economical. While it could cost up to TSh120,000 to apply farmyard manure to 1 acre of maize, it could cost only TSh50,000 to apply with industrial fertilizer to the same acre. Given the economy of most farmers they always opted for industrial fertilizer, which is not effective in nourishing the soil with natural fertility. Urea and sulfates were used for plant growth while NPK and *Minjingu mazao* were used at the planting stage.

Farmers used different soil and water conservation strategies to ensure productivity at field level. According to the Uchira villagers, water conservation infrastructures such as ridges are commonly used in the lower parts of farms where few farmers practise irrigation agriculture. The way tractors are used is another conservation strategy; the tractor is orientated so that it cultivates across farms to avoid disturbing the gravitational draining ridges, which could exacerbate the surface runoff during heavy downpours. Orientation of 1 m-high terraces was practised in the past and it was helpful but currently they are not efficient due to the increasing incidence of floods on the farms. These 0.5 m-long terraces are made in a zigzag pattern depending on the farm orientation and this helps conserve water but if the rains exceed the optimum level these terraces are washed away.

Farmers reported during FGDs that due to the increase in incidence of floods over time some conservation strategies that they adoped have been found not to be effective. For example, in the past farmers used bags to protect against detrimental surface runoffs on their farms but they abandoned the use of bags because of increased incidence of floods. Also, the main water channels are haphazard and this has exacerbated the floods into their farmlands. However, it was found that while 74% of farmers have not practised any soil water conservation in the past 10 years, currently 27% have adopted terracing while 35% of farmers have dug 1 m-deep and 2 m-wide infiltration ditches across the farms.

Infiltration ditches are commonly used in home gardens, which are located on steep slopes, for several purposes including: (i) to reduce surface runoff which may destroy soil fertility; (ii) to provide ventilation and reduce heating in the soils so as to reduce the risk of Panama disease in bananas; and (iii) to control slow water infiltration into the soils and maintain the soil water retention capacity. Fallowing is practised in the lower zone by only 11% of the respondent households. The study noted that 69% of farmers did not practise mulching while 73% had no idea about conservation agriculture.

Climate extremes, impacts and responses

By the time of this investigation, farmers in the study area were experiencing climate stresses more than they had done in the past decades and this was leading to losses of crops at the farm level. Whereas the farmers were able to report droughts occurring at 10 year intervals (i.e. 1974, 1984, 1994 and 2004) the current dry spells were becoming common climate extremes that occurred very unpredictably. Hence, the study found that 59% of households had never received any early warnings concerning climate extremes. Further, even those who admitted they had received warnings said that the information provided was not precise enough to guide adaptation action.

In the mid-upper and upper zone, it was found that 97% of the farms had been affected by climatic events. Of these, 74% were affected by drought and 11% were affected by below average rains, which caused decline in crop yields. Some 62% of farmers in the mid-upper zone and 33% of farmers in the upper zone claimed loss of entire crops because of drought. To get round this, 58% of farmers said the only option they had was to buy food and 15% responded that they had to rely on savings to do that.

In the mid-lower and lower zones 90% of the farmers reported drought and floods as climatic extremes, which resulted in crop and animal losses. It was further noted that 70% of the farmers in the two zones had encountered losses due to drought while 20% of them encountered losses due to floods. It was noted also that 44% of mid-lower and lower zone households lost entire crops while 42% experienced decline in crop yields due to these extremes. As a result, 30% of the households in these zones depended on government food handouts during such extreme events while 38% of them bought food from their own savings.

Response by changing farm management practices after climate extremes was not much taken into account by farmers. The data show, for example, that only 46% had changed in one way or another after a drought. Changes included: (i) planting early-maturing maize seed varieties; (ii) planting drought-resistant crop varieties (e.g. sorghum and cassava); (iii) timing early planting; (iv) decreasing the amount of land under cultivation; and (v) changing fertilizer and soil water conservation strategies. Of all the respondents across the transect, 51% had not changed their farm management practices in any way after a drought. Such farmers blamed lack of capital and information as the culprit.

Non-climatic challenges and responses

Apart from climatic challenges, the farmers in Kilimanjaro faced other non-climatic challenges. These challenges included: (i) pre-harvest and postharvest losses; (ii) market access; and (iii) lack of inputs and other farm support services such as extension services. The common devastating pre-harvest losses came from pests. It was reported that the most common pest in maize in the lower zone was the African armyworms (*Spodoptera exempta*), which have been increased by the deforestation of nearby Kahe Forest in the 2000s. According to the farmers, before this period the pests were very minimal because the forest acted as a host for many pests, including the African armyworm. According to the Ward Extension Officer, different efforts had been made to forecast the outbreak of the African armyworm but the exercises did not yield positive results. Other pre-harvest pests included maize stalk borer (*Busseola fusca*) and locusts (*Locusta migratoria*). Others were rodents and vervet monkeys (*Chlorocebus pygerythrus*).

Postharvest pests in maize grains included 'scania' (*Teretriosoma nigrescens*) but the use of tanks to store maize grains had been an effective storage system to avoid these postharvest pests. Unfortunately, most of the farmers were unable to buy these drums for storage of grains. Mole rats (*Heterocephalus glaber*) were said to be the most common pre-harvest rodent for bananas. The most common pest in coffee included antestia (*Antestiopsis orbitalis*), coffee berry borer (*Hypothenemus hampei*) and coffee stem borer (*Xylotrechus quadripes*).

Meanwhile, farmers mentioned crop diseases as one of the non-climatic challenges. Common diseases in coffee were reported to be coffee leaf rust. Common banana diseases included Panama disease and cigar and rot diseases. Poor farm care was reported to be the cause of Panama disease in bananas. Pests and diseases were generally reported to be on the increase over time, so much so that people had formulated a saying following decreased crop yields, that: 'our forefathers have gone with their farms'.

In the upper and mid-upper zone, for environmental reasons the government had prohibited the application of industrial chemicals against pests and diseases in coffee and therefore diseases like coffee berry disease (CBD), rust and leaf minor were reported to be on the increase. In the case of access to coffee markets, the farmers reported that in the past two decades, coffee had a very good market and could generate enough income for the household, but currently prices had declined significantly and some of them had started to abandon coffee production altogether. Among the reasons mentioned by farmers for instability of coffee prices was the improved status of Brazilian coffee.

The market for maize was also a challenging part of livelihood due to bulk production of maize in other regions of Tanzania coupled with corruption in the cross-border market to Kenya which would have provided a good price for their surplus maize. Access to extension services had also been problematic. Most of the farmers claimed to buy counterfeit seeds and insecticides, and used poor farm management and storage systems. Data from this study further showed that 47% of the farming households had no access to extension services.

Calculating the vulnerability index

According to Chamber (1983) vulnerability had two sides: (i) an external side of risks and shocks to which an individual or household was subject to climate change; and (ii) an internal one which was defencelessness, lacking the means to cope without damaging loss. Blaikie *et al.* (1994) defined vulnerability as the characteristics of a person or group in terms of their capacity to anticipate, cope with, resist and recover from the impacts of natural hazards. Vulnerability could be viewed along a continuum from resilience to susceptibility.

Table 4.2 shows the scores as percentages for different variables across the four altitude zones of the southern slope of Kilamanjaro. These scores were used to calculate the vulnerability indices for each zone and an overall vulnerability index for the study area shown in Table 4.3. For example, Table 4.2 shows the maximum value for drought frequency is 78% for the mid-upper zone and the minimum value is in the upper zone at 34%. Therefore, the normalization for the up FR can be calculated as shown below derived from the formula in Eqn 4.1:

$$x_{ij} = \frac{X_{ij-34}}{78-34} = \frac{X_{ij-34}}{44}$$

So the normalization of drought frequency for the lower zone in this case will be:

$$x_{ij} = \frac{67-34}{44} = 0.750$$

Table 4.2. Scores (%) for different variables across four altitude zones of the southern slope of Kilamanjaro.

Variable	FR[a]	Zone			
		Upper	Mid-upper	Mid-lower	Lower
Drought frequency	↑	34	78	49	67
Household dependants	↑	22	39	55	47
Social safety nets	↓	38	55	30	39
Off-farm contribution	↓	38	33	30	38
Possession of land title	↓	6	2	3	6
Wood as cooking energy	↑	94	97	99	91
Access to extension services	↓	40	54	59	55
Fallowing practice	↓	2	16	11	11
Farm acreage	↓	30	36	16	22
Response to drought cases	↓	20	56	52	43

[a]FR, Functional relationship.

Table 4.3. Normalized scores with vulnerability indices for each zone and overall vulnerability index.

Variable	FR[a]	Zone			
		Upper	Mid-upper	Mid-lower	Lower
Drought frequency	↑	0	1	0.341	0.750
Household dependants	↑	0	0.515	1	0.758
Social safety nets	↓	0.680	0	1	0.640
Off-farm contribution	↓	0	0.625	1	0
Possession of land title	↓	0	1	0.750	0
Wood as cooking energy	↑	0.375	0.750	1	0
Access to extension services	↓	1	0.263	0	0.211
Fallowing practice	↓	1	0	0.357	0.357
Farm acreage	↓	0.300	0	1	0.700
Response to drought cases	↓	1	0	0.111	0.361
Vulnerability indices		**0.436**	**0.415**	**0.656**	**0.377**
Overall vulnerability index = 0.471					

[a]FR, Functional relationship.

The normalization of down FR can be calculated by referring to the formula in Eqn 4.2. In this case normalization will be calculated as follows (using the example of social safety nets):

$$y_{ij} = \frac{55 - x_{ij}}{55 - 30} = \frac{55 - x_{ij}}{25}$$

Therefore, the normalization of social safety net variables for the upper zone can be calculated as:

$$y_{ij} = \frac{55 - 38}{25} = 0.680$$

After the normalization of variables, the vulnerability index can be calculated with reference to formula in Eqn 4.3 above. For example, the vulnerability index of the lower zone can be calculated as:

$$VI = \frac{1.508 + 2.269}{10} = 0.377$$

Conclusion

The variation in microclimate and agroecosystem across the topographical landscape of the southern part of Mount Kilimanjaro has shown difference in resilience and exposure to climate change impacts. The upper and mid-upper zone agroecosystem exhibit a typical mountainous cool climate with wet loamy volcanic soils. This highland microclimate has gradually been changing over time due to climate change. The *Vuli* rains were reported to have almost disappeared in all the zones posing a threat to stable food and cash crop production. The change has discouraged banana production while the coffee crop production challenges have led to it being abandoned in some zones. This has posed a threat to sustainable livelihoods because coffee was used as an income-generating crop.

The limited parcel of farmland in the study area is viewed as a challenge to sustainable agroecosystem resilience in which a constant periodic soil disturbance is not an option. Sustainable livelihood challenges coupled with poor extension services for farmers have created another loophole to climate extreme exposure. The non-climatic stresses existing in the study area seem to be in parallel to the current change in climate and therefore have exacerbated vulnerability of the agroecosystem to the anticipated climate change impacts. Hence, the vulnerability index computation shows the mid-lower zone to be the most vulnerable based on the selected indicators at 0.656; the upper zone is the second most vulnerable at 0.436, followed by the mid-upper zone at 0.415. The lower zone has proved to be the least vulnerable at 0.377. However, the overall index of 0.471 shows the vulnerability of the agroecosystem on the southern slope of Mount Kilimanjaro to be at risk of future climate change impacts.

Note

[1] A letter of offer is a document that land owners have before having the title deed to the land.

References

Agrawala, S., Moehner, A., Hemp, A., van Aalst, M., Hitz, S., *et al.* (2003) *Development and Climate Change in Tanzania: Focus on Mount Kilimanjaro*. Organisation for Economic Co-operation and Development (OECD), Paris.

Berkes, F., Colding, J. and Folke, C. (eds) (2003) *Navigating the Social–Ecological Systems: Building Resilience for Complexity and Change*. Cambridge University Press, Cambridge.

Blaikie, P., Cannon, T., Davis, I. and Wisner, B. (1994) *At Risk: Natural Hazards, People's Vulnerability and Disasters*. Routledge, London.

Chamber, R. (1983) *Rural Development: Putting the Last First*. Longman, New York.

Gomiero, T., Giampietro, M. and Mayumi, K. (2006) Facing complexity on agro-ecosystems: a new approach to farming system analysis. *International Journal of Agricultural Resources, Governance and Ecology* 5(2–3), 116–144.

Holling, C.S. and Meffe, G.K. (1996) On command-and-control, and the pathology of natural resource management. *Conservation Biology* 10, 328–337.

Intergovernmental Panel on Climate Change (IPCC) (2007) *Climate Change 2007: Impacts, Adaptation and Vulnerability. Contribution of Working Group II to the Fourth Assessment Report of the Intergovernmental Panel on Climate Change*. [Parry, M.L., Canziani, O.F., Palutikof, J.P., van der Linden, P.J. and Hanson, C.E. (eds)]. Cambridge University Press, Cambridge.

International Crops Research Institute for the Semi-Arid Tropics (ICRISAT) (2013) Quantitative Assessment of Vulnerability to Climate Change. Available at: www.icrisat.org (accessed 16 June 2013).

Maghimbi, S. (2007) Recent changes in crop patterns in the Kilimanjaro Region of Tanzania: the decline of coffee and the rise of maize and rice. *African Study Monographs* 35, 73–83.

Sarris, A.H., Savastano, S. and Christiaensen, L. J.M. (2006) The role of agriculture in reducing poverty in Tanzania: a household perspective from rural Kilimanjaro and Ruvuma. Paper presented at International Association of Agricultural Economists 2006 Annual Meeting, 12–18 August 2006, Queensland, Australia, article no. 25573.

United Nations Development Programme (UNDP) (2006) *Human Development Report*. UNDP, New York.

United Republic of Tanzania (URT) (1998) *Kilimanjaro Region: Social Economic Profile*. Planning Commission, Dar es Salaam, Tanzania.

United Republic of Tanzania (URT) (2013)s *2012 Population and Housing Census: Population Distribution by Administrative Areas*. National Bureau of Statistics (NBS), Dar es Salaam, Tanzania.

van Apeldoorn, D.F., Kok, K., Sonneveld, M.P.W. and Veldkamp, T. (2011) Panarchy rules: rethinking resilience of agro-ecosystems: evidence from Dutch dairy farming. *Ecology and Society* 16(1).

World Fact Book (WFB) (2013) Tanzania Economy: Business Addresses. Available at: https://theodora.com/wfbcurrent/tanzania/tanzania_economy.html (accessed 12 January 2014).

5 Effects of Conservation Agriculture on Farmers' Livelihoods in the Face of Climate Change in Balaka District, Malawi

Josephine M. Zimba and Emma T. Liwenga*

Institute of Resource Assessment, University of Dar es Salaam, Dar es Salaam, Tanzania

Abstract

Conservation agriculture (CA) has been highly promoted due to its potential to ensure high crop yields even in the face of changes in climatic factors. However, the actual benefits associated with CA are not only focused on food security but are also site specific. This study sought to understand the benefits of CA in improving livelihoods in a changing climate in Hanjahanja and Sawali sections of Bazale Extension Planning Area in Balaka District. Specifically, it analysed CA's contribution to farmers' livelihoods and also the challenges and opportunities of CA in climate change adaptation. Data was collected through household surveys (n = 153), key informant interviews (n = 9), focus group discussions and field observations. The study found that due to CA adoption, the majority of the farmers in both Hanjahanja and Sawali sections had realized positive livelihood outcomes, mainly through improved food security and increased incomes. Despite the similarity, Hanjahanja farmers reported decrease in yields in seasons marred by floods. However, farmers faced several challenges due to CA adoption, which included high labour demands, rainfall variability and lack of inputs. Even so, improvement in soil moisture, soil erosion control, improved food security, presence of several institutions and enabling environment offered more opportunities of CA in adapting to climate change. CA, therefore, improves the livelihoods of the farmers except in times of floods. Hence, deliberate policies by the government to promote adoption of CA are required to take advantage of the benefits of CA. Research should also be done on how best to reduce the negative effects of CA on farmers' livelihoods.

Introduction

Climate change is now considered one of the biggest challenges facing humankind. To date, climate change has had effects on hydrological systems, terrestrial biological systems, marine and freshwater biological systems (IPCC, 2007), most of which are on the negative side. These effects have had repercussions on agricultural production, consequently affecting farmers, whose livelihoods depend on agriculture (Julie *et al.*, 2014). This situation is not expected to positively change in the future; rather, most of the projected future impacts are negative. This calls for adaptation that is more efficient to even cater for the future effects of climate change.

Conservation agriculture (CA) is widely promoted as an adaptation strategy to climate change. The Food and Agriculture Organization of the United Nations (FAO) defines CA as 'an approach to managing agroecosystems for improved and sustained

*liwenga99@gmail.com

productivity, increase profit and food security while preserving and enhancing the resource base and the environment' (FAO, 2015). It is a package of three intertwined principles, namely: (i) minimum tillage; (ii) permanent soil cover; and (iii) crop rotations and associations. According to Grabowski *et al.* (2013), CA improves soil quality, which results in increased crop yields, even during intermittent dry spells. As such, it is promoted to ensure food security under negatively related climate changes that increase food insecurity. However, the outcomes of CA implementation vary from place to place.

Several studies have been conducted to understand the effects of CA on crop yields. A study by Kamtimaleka (2009) revealed that there was a significant difference in the yield realized by adopters and non-adopters, with adopters realizing more crop yields than non-adopters. This suggests that CA leads to increase in crop yields. However, such a study focuses on the crop yield and not the livelihoods of the farmers, in general. For the studies that address the relationship between CA and farmers' livelihoods, the results are mixed. Milder *et al.* (2011) found the relationship between CA and the livelihood outcomes for some farmers to be positive unlike others where the relationship between CA and the same livelihood outcomes was negative.

For instance, CA seemed to have different effects on labour requirements for men and women, thereby affecting their livelihoods differently (Milder *et al.*, 2011). However, a study by Nyanga (2011) revealed that most farmers, even the adopters, did not perceive CA as an adaptation strategy to climate change. This clearly shows how varied and, sometimes, contradicting outcomes of CA practices can be. As such, the adaptation benefits associated with CA are not only site specific but also individual or personal in nature too and they ought to be understood to avoid hasty conclusions. Therefore, it is important that the specific effects of CA on the livelihoods of the farmers are studied in order to understand its contribution to adaptation of the farmers to climate change.

This study, therefore, aimed to assess whether the implementation of CA is incrementally improving the livelihoods of farmers in Balaka District, thereby serving as an effective adaptation measure to drought. It also assessed the challenges and opportunities of CA as an adaptation measure to climate change. Most importantly, the outcome of this study informs policy makers on the possible strategies of how best to deal with challenges that hinder the maximum realization of the potential adaptation-related benefits associated with CA adoption. The study also informs stakeholders on how best to take advantage of opportunities associated with CA in order to improve livelihoods and assist farmers adapt to climate change.

Conceptual framework

This study adopts the definition of a livelihood as stipulated by Chambers and Conway (1991) as comprising 'people, their capabilities and their means of living, including food, income and assets'. A livelihood is considered sustainable when it maintains or enhances the assets on which people depend, is resilient from stress and shocks and provides for future generations (Chambers and Conway, 1991). Therefore, the study used the UK Department for International Development's (DFID) Sustainable Livelihoods Framework in order to analyse and understand the role of CA in improving livelihoods of the people in Balaka District. This framework acknowledges the existence of several kinds of livelihood assets, transforming institutions and processes, factors influencing vulnerability context and forms of livelihood outcomes. However, the study only addressed the ones depicted in Fig. 5.1.

Similar to the original framework, this framework has several components, including livelihood assets, vulnerability context, transforming institutions and processes, CA as a livelihood strategy, and livelihood outcomes as discussed below.

Livelihood assets

In this framework, it is understood that people require livelihood assets as the basis for building their livelihoods. These assets,

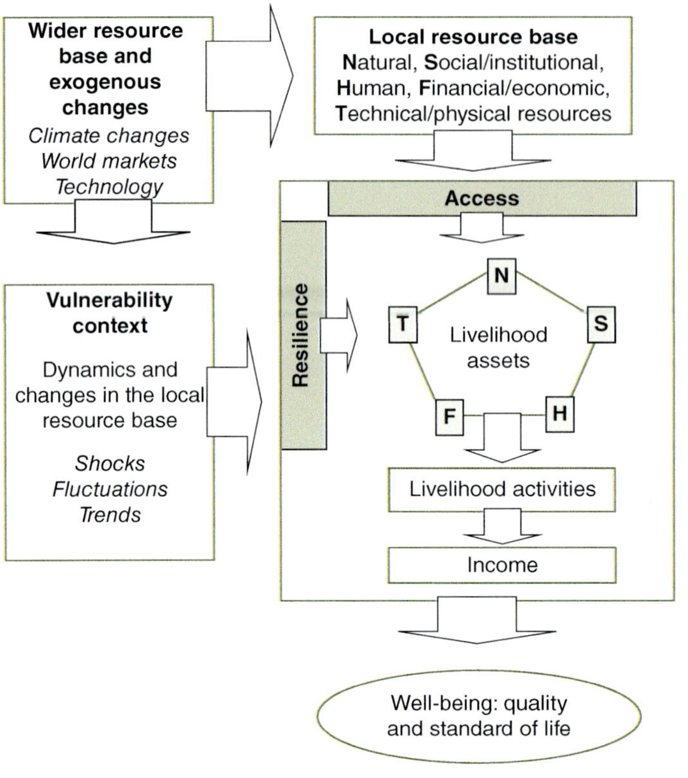

Fig. 5.1. Sustainable Livelihoods Framework. (Adapted and modified from DFID, 1999.)

which are both tangible and non-tangible, are grouped into:

- Human capital: this includes skills, age and education level which are needed when pursuing livelihoods.
- Physical capital: will include shelter, transport and communication that people need to support livelihoods.
- Social capital: membership of organizations helps in the attainment of objectives of particular livelihood strategies.
- Financial capital: salary, pension and savings are important, as money is required when embarking on a livelihood strategy.
- Natural capital: includes land ownership and access, which is important in the adoption of CA as an adaptation and livelihood strategy.

In this study, it was envisaged that households with more assets are better able to adopt CA and use it to improve their livelihoods. Furthermore, those with more assets have better access to transforming processes and structure.

Vulnerability context

The assets described above are constantly affected by external shocks and trends such as environmental degradation, pests and disease occurrence, and droughts, just to mention a few. These shocks and trends affect the quality and quantity of the assets available for exploitation by the households.

Transforming institutions and processes

These include laws and policies, institutions and culture. The significance of this component cannot be overemphasized as they influence the use of assets through a particular livelihood strategy in order to realize improved livelihoods. However, the relationship between them and the assets is two-way as the assets also influence the transformational processes and structures.

CA as a livelihood strategy

In this study CA was considered to be one of the livelihood strategies that can be used to achieve positive sustainable livelihood outcomes for the people. The adoption of the three principles of CA (namely minimum soil disturbance, maximum soil cover and crop rotations and associations) was vital for this study, without much emphasis on the extent or intensity of the actual practices. However, the viability and effectiveness of livelihood strategies depend on the availability and accessibility of assets, services and opportunities which can be positively enhanced or adversely undermined by ecological factors, social structures or institutional processes (Majale, 2002).

Livelihood outcomes

This study envisaged that several livelihood outcomes would be realized as a result of adoption of CA. However, it focused on changes in the income levels, vulnerability, soil quality, food security, social networking and access to agricultural technical information. Further, the study proposed that such changes in the livelihoods of the people not only alter the vulnerability contexts of the people but also influences the livelihood assets. This observation is vital though less emphasized by the widely used DFID Sustainable Livelihoods Framework. This in turn would give a better starting point for more improved livelihoods and therefore put rural communities in a better position to adapt to climate change.

Material and Methods

Description of the study area

The study was conducted in Bazale Extension Planning Area (EPA) located in Balaka District. This EPA was selected as it is considered to be one of the most vulnerable EPAs within the district. Balaka District is one of the 13 districts in the Southern Region of Malawi. It is predominantly in the rain shadow part of Malawi. It has a surface area of about 2193 km^2. Balaka was purposively selected as one of the districts, which experiences frequent droughts and dry spells (Mangisoni *et al.*, 2011). As such, it is one of the districts where CA activities were pioneered and have been heavily promoted since 2004. Figure 5.2 shows the location of Balaka District and the EPAs within the district.

Specifically, the study was conducted in Hanjahanja and Sawali sections found in Bazale EPA. The two sections were selected because of their differences in hydrological features. Hanjahanja has several rivers flowing through it with some places where large rivers join each other while Sawali is generally dry, with very few streams running through it. This makes Hanjahanja more prone to river flooding while Sawali section is generally dry.

Data collection methods

Data was collected through key informant interviews, focus group discussions (FGDs) and a household survey. The key informant interviews were guided by a checklist, which was administered to personnel who are supposed to be anexpert source of the information needed (Marshall, 1996). These included agriculture extension officers, the Land Resources and Conservation Officer, officers from non-governmental organizations (NGOs) promoting CA practices such as Project Concern International (PCI), National Smallholder Farmers' Association of Malawi (NASFAM), Self-Help Africa and the lead farmers. Further, two FGDs, of men and women separately, were conducted in each section. Each group constituted six to eight participants, including both adopters and non-adopters, as this has been observed to be an optimum size for a focus group (Gill *et al.*, 2008). These discussions were guided by a checklist to capture information on how CA contributes to improved livelihoods of the communities and the challenges and opportunities of CA as an adaptation strategy to climate change.

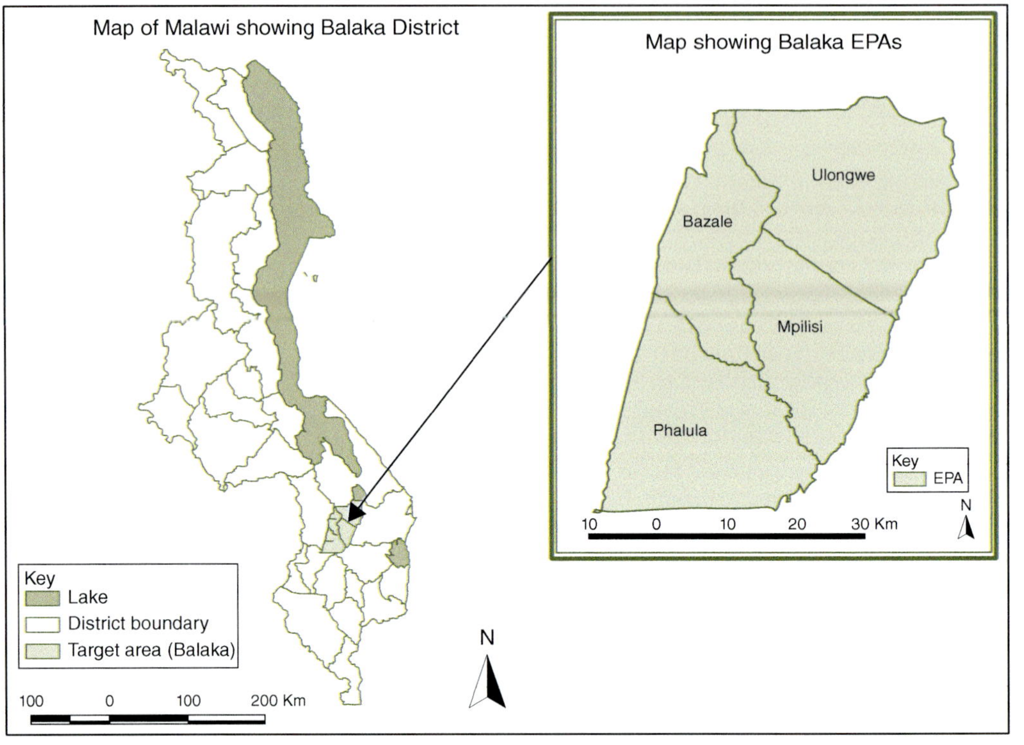

Fig. 5.2. Map of Malawi showing Balaka District. EPA, Extension Planning Area. (Produced by author, 2016.)

Household surveys were also conducted to collect primary data such as demographic information, vulnerability context information and livelihood conditions. This was done using a mixed questionnaire, which was administered to household heads. The mixed questionnaire combined both open-ended and closed-ended questions. The study used probability sampling methods in order to obtain a representative sample. A sample intensity of 5% of households in the sections was used. This is because Kajembe and Luoga as cited in Lazaro *et al.* (2013) asserted that 5% is considered adequate for statistical analysis. The total number of farming households of Hanjahanja and Sawali sections was 2877. Using a 5% sampling intensity for the respective sections, the sample size of the study was 145 households. However, since usually not all sampled households are available for the survey, an additional eight households were sampled and interviewed. As such, the total number of households included in the study was 153, of which 44 were adopters and 109 were non-adopters. The number of these adopters and non-adopters were proportional to their size in the population.

Field observations were also used to triangulate information collected through key informant interviews, FGDs and household survey. Observations on quality of shelter of adopters and non-adopters, quality of crops in the field, social and economic activities were documented to validate the information and further help to understand the livelihood patterns of the people. Further, information collected through review of published and unpublished documents was used to improve the research design, understand issues around CA implementation and farmers' livelihoods and discuss the findings of this study.

Data analysis and presentation

The Statistical Package for Social Sciences (SPSS) was used to analyse data collected using the questionnaire. It was used to come up with percentages and means. This package was chosen because it is the most widely used software for the statistical analysis of quantitative data (Greasley, 2008). Microsoft Excel was also used to develop tables, pie charts and graphs, where fitting and necessary. As for the qualitative data collected through FGDs, key informant interviews and field observations, it was transcribed, described and categorized. Then patterns and connections within and between categories were identified and then interpreted was done (Taylor-Powell and Renner, 2003). Professional judgement together with lessons of best practices from similar studies conducted in relatively similar conditions to the study area were also used to make sense of the data collected.

Results and Discussion

Vulnerability context

Vulnerability context is a vital component in the sustainable livelihoods framework. Several indicators were used to assess the vulnerability context of the farmers. Overall, the majority of the farmers, in both Hanjahanja and Sawali sections, reported an increase in climatic hazards, and pest and disease outbreaks. However, more farmers, about 80.8%, in Hanjahanja section reported an increase in forest cover. This differs from Sawali section where the majority of farmers, about 80%, reported a decline in forest cover. The difference is attributed to the vibrant tree planting programmes, which were more successful in Hanjahanja than Sawali section.

This increase in the farmers' exposure to climatic hazards highly increases their vulnerability. This is exacerbated by the low adaptive capacity associated with poverty in developing countries since adaptive capacity is known to be a function of people's access to resources (Adger *et al.*, 2004). However, for farmers in Sawali section, CA reduces their vulnerability to climate variability due to improved soil moisture retention and increased fertility as a result of mulching and incorporation of organic matter. The situation is different for farmers in Hanjahanja section where CA reduces their vulnerability only when there are dry spells. In times of heavy rains and flooding, like the 2014/15 growing season, CA increases the vulnerability of these farmers. This is so because mulching together with increased organic matter lead to waterlogged conditions. As a result, those farmers who do not mulch their fields actually realize better yields than those CA adopters, in such events.

Transforming institutions and processes

There are several institutions promoting CA in the study area. In addition to the Government of Malawi (GoM) through its agricultural extension workers, other institutions include Concern Universal, PCI, World Vision International, World Food Programme (WFP), Self-Help Africa and NASFAM, just to mention a few. The presence of these organizations at the local level is important as they are known to moderate the impact of climate hazards on livelihoods through the various roles they play, ranging from research to capacity building in different ways (Agrawal *et al.*, 2009). Most of these institutions reported that their CA activities are guided by national policies developed by the GoM. However, since these policies are broad in nature, the actual information disseminated to the farmers on the ground sometimes differs slightly from one institution to another.

Further, 96% of the farmers indicated that culture does not influence crop production in any way. This finding is in contrast to findings of a study in South Africa. Manyevere *et al.* (2014) reported that a shift in the culture of the people led to the abandonment of cultivation of some crops such as maize, beans and wheat. Therefore, the findings by this study suggest that there could be no culture-related hindrances for the farmers to adopt any new technology, including CA, introduced with the aim of improving their crop yield.

Relationship between CA and livelihood assets

Generally, most CA adopters seem to be endowed more with livelihood assets unlike non-adopters. From the FGDs, the CA adopters have more skills, in terms of agricultural production, than non-adopters. This is true for both Hanjahanja and Sawali sections. This is attributed to an improvement in information flow as a result of introduction and adoption of CA. Therefore, as Imoloame and Olanrewaju (2014) affirm, agricultural extension, using any approach, should lead to changes in farmers' skills. As regards natural capital, adopters in both sections, Hanjahanja and Sawali, own larger farm plots than non-adopters. This allows them, therefore, to use a portion of the land to test a new technology. They eventually adopt the technology if they realize the expected benefits.

In terms of social capital, the membership of social groups is different between adopters and non-adopters in both sections. For instance, in Hanjahanja section, 77% of adopters or their spouses are members of a social group within their community as compared to 32% of non-adopters or their spouses who are members of a social group in the community. The trend is similar in Sawali section where 77% of the adopters or their spouses are members of a social group as compared to 49% of the non-adopters or their spouses are members of a social group. This is important as participation in organizations is taken to be a key factor influencing adoption of technologies, such as irrigation (Ramirez, 2013). However, CA also influences participation in social groups. Though the percentages were low, 5.6% and 3.9% of adopters from Hanjahanja and Sawali sections, respectively, reported an increase in membership and participation in groups. This is compared to the finding that no one in both sections observed a decrease in participation in groups as a result of CA. This concurs with the International Centre for Agricultural Research in the Dry Areas (ICARDA) (2012) who found that CA adoption saves time, which allows farmers to engage in other diversification options.

The adoption of CA has significantly affected the financial capital of farmers in the area. This is shown by the finding that 27.8% and 26.9% of the CA adopters in Hanjahanja and Sawali sections, respectively, reported an increase in their savings while none of the adopters, in both Hanjahanja and Sawali sections, reported a decrease in savings. Money is saved from reduced tillage and also selling of extra produce due to increased crop yields. Further, adopters have been found to be well off in terms of ownership of physical assets as compared to non-adopters. These assets, such as radios and mobile phones, are not only as a result of CA, but also help them to better access vital agriculture-related information about weather forecasts and CA practices therefore being able to better adapt to climate change. Table 5.1 shows the differences in proportion of CA adopters and non-adopters owning a number of assets in the study area.

Generally, except for the television set in Sawali section, more adopters own valuable physical assets than non-adopters in the study area. This ownership of assets has an influence on whether a farmer adopts a technology or not. This concurs with Muzari *et al.* (2012) who assert that a lack of assets such as land, equipment or education limits adoption of a technology. Tables 5.2 and 5.3 summarize farmers' perceptions on the effect of CA on their livelihood assets.

This finding is consistent with that of a study conducted among smallholder farmers in Mozambique. Even though the McNair *et al.* (2015) study did not intend to develop a causal relationship, it found that farmers practising CA had more farm assets and better building materials for their homes than those of non-adopters. The findings suggest, therefore, that CA has a positive influence on the assets owned by the household.

Table 5.1. Proportion of farmers (%) owning physical assets. (From field survey, 2016.)

	Hanjahanja		Sawali	
Physical asset	Adopters (n = 18)	Non-adopters (n = 60)	Adopters (n = 26)	Non-adopters (n = 49)
Radio	50.0	33.0	50.0	24.5
Mobile phone	44.7	36.7	61.5	46.9
Motorcycle	0.0	0.0	3.8	0.0
Bicycle	61.0	40.0	53.8	29.2
Television set	5.6	3.3	3.8	4.0

Table 5.2. Farmers' perceptions (%) of the effects of conservation agriculture (CA) on livelihood assets in Hanjahanja section. (From field survey, 2016.)

Livelihood asset	Increase	Decrease	No change	Don't know	Total
Education	38.9	0.0	61.9	0.0	100
Skills	33.3	5.6	55.6	5.5	100
Quality of shelter	22.2	5.6	66.7	5.5	100
Transport	11.1	0.0	77.8	11.1	100
Communication	5.6	0.0	72.2	22.2	100
Savings	27.8	0.0	66.7	5.5	100
Soil quality	55.6	0.0	38.9	5.5	100
Group membership	5.6	0.0	88.9	5.50	100

Table 5.3. Farmers' perceptions (%) on the effects of CA on livelihood assets in Sawali section. (From field survey, 2016.)

Livelihood asset	Increase	Decrease	No change	Don't know	Total
Education	19.2	0.0	65.4	15.4	100
Skills	26.9	11.5	50.0	11.6	100
Quality of shelter	23.1	0.0	69.2	7.7	100
Transport	3.9	0.0	84.6	11.5	100
Communication	3.9	0.0	80.8	15.3	100
Savings	26.9	0.0	61.5	11.6	100
Soil quality	30.8	11.5	50.0	7.7	100
Group membership	3.9	0.0	61.5	34.6	100

Contribution of CA to farmers' livelihoods

In a nutshell, the findings on ownership of assets are evidence of better or positive livelihood outcomes of CA adopters as compared to non-adopters. Figure 5.3 shows the livelihood outcomes realized by the adopters in the study area as a result of CA. About 66.7% of the farmers have experienced an improvement in food security in their households as a result of adoption of CA in Hanjahanja section. Similarly, the livelihoods of adopters in Sawali section have improved through improved food security, as observed by 50% of the adopters. This is mainly through increased crop yields associated with the adoption of CA. Further, 22.2% and 26.9% of the farmers reported an increase in their income as a result of CA, in Hanjahanja and Sawali sections, respectively. This outcome too is not only associated with increased crop yields but also saved time, which allows the farmers to engage in other income-generating activities.

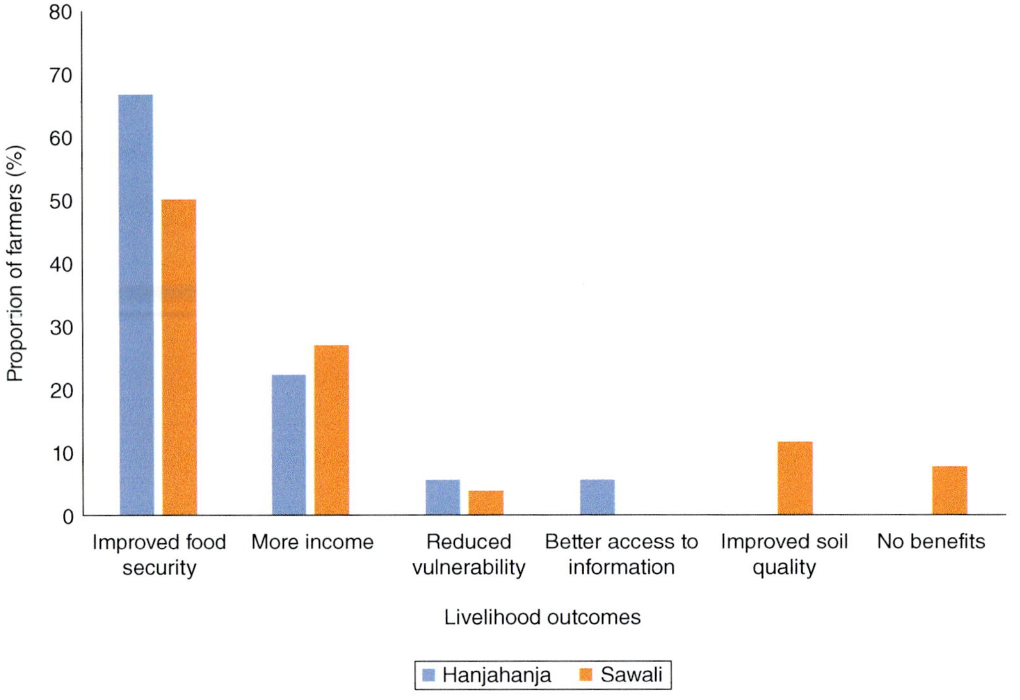

Fig. 5.3. Livelihood outcomes for farmers as a result of adoption of conservation agriculture (CA). (From field survey, 2016.)

In addition, 11.5% of the farmers in Sawali section have experienced an increase in the quality of the soil. This is not only associated with the incorporation of residues that increase the organic matter content in the soil but also the intercropping with legumes which fix nitrogen in the soil. Despite that, some farmers have not realized benefits mainly because they had just started practising CA in the 2015/16 growing season.

These findings are quite consistent with those of Jumbe and Nyambose (2016) in which they found CA to have a positive effect on maize production, an important indicator of food security in Malawi. The study concluded that CA adoption significantly improves food security among adopters. Further, Corbeels *et al.* (2013) also observed increase in income of farmers as a result of CA adoption among smallholder farmers in sub-Saharan Africa. This increase in income was attributed to the increase in yields of rice and maize under CA. Despite the benefits observed by the majority of farmers, 7.6% of the adopters have not realized any improvements in their livelihoods as a result of CA. This could be because they are new adopters and as Islam and Reeder (2014) allude, benefits of no tillage and CA are realized after years of continuous practice.

Challenges of CA in adaptation to climate change

Most farmers face challenges in implementation of a new technology. Surprisingly, in both Hanjahanja and Sawali sections, most of the farmers reported to have met no challenge as a result of CA implementation. However, some farmers still met some challenges. Figure 5.4 depicts the challenges faced by farmers in adapting to climate change. The challenges include high labour demands, rainfall variability and lack of inputs, as detailed below.

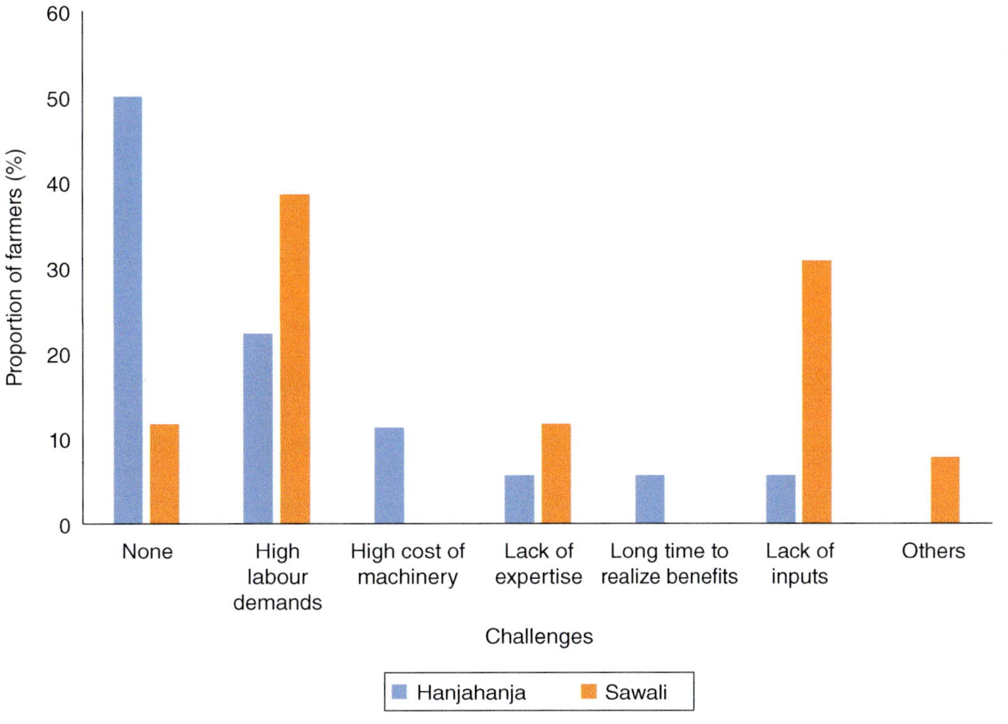

Fig. 5.4. Challenges of CA faced by farmers in adapting to climate change. (From field survey, 2016.)

High labour demands

Data from Fig. 5.4 show that 22.2% of the adopters in Hanjahanja complained about CA being labour demanding whereas more adopters in Sawali (38.5%) complain of the same. Both government extension officers and the officials from the NGOs raised the concern that this was the greatest factor hindering the adoption of CA, and in some cases leading to lack of adoption. This is mainly due to the high labour demand needed for looking for vegetative material for mulching. The condition was worsened by a problem of termite attacks, which meant more vegetative material needing to be collected to still have some left even after other materials have been broken down by termites. The high proportion of farmers complaining about the labour burden of CA in Sawali section was due to the dryness of the area, which makes it difficult to get vegetative material for mulching.

Mulching has proved to be difficult for most farmers in sub-Saharan Africa. In Zambia, crop residue retention in the field is challenging due to burning of the residues by local people looking for mice, which is a local delicacy in the Eastern Province (Umar *et al.*, 2011). This means at the start of the growing season, farmers have to look for vegetative material, which is not in abundance at that time. In order to deal with this, farmers should be further encouraged to fetch mulch materials during the rainy season, when they are in abundance, and store them in places where termites cannot destroy them. This will reduce the burden of looking for mulch materials during the dry season, as is currently the case.

Rainfall variability

This is especially a problem in Hanjahanja section where river flooding is common. Too much rain has been leading to waterlogged conditions in fields under CA, thereby leading to lower yields in such seasons. This observation is in line with a study conducted in Rwanda where CA was associated with decreased yield in high rainfall areas (Kabirigi *et al.*, 2015). Equally challenging is when there is too little rainfall in the previous season, which makes it difficult to find vegetative material for mulching in the following growing season. Such findings underline the fact that CA is not a 'one-size-fits-all' technology and caution should be taken when promoting particular practices at local level. This, therefore, calls for institutions promoting CA to consider such local environmental differences in order to promote best practices according to the conditions in the different areas.

Lack of inputs

CA, especially the principle of minimum tillage, promotes the use of herbicides for weed management. Unfortunately, due to the low-income levels of the farmers, it is difficult for them to purchase herbicides, which are mostly costly. Thierfelder *et al.* (2013) in their study of CA in Southern Africa also acknowledged that lack of access to inputs worsened the challenges already faced by adopters of CA. Considering the low income levels of most smallholder farmers, deliberate policies by the government could help deal with this challenge. This could either be through removing taxes on such inputs or inclusion of such inputs in the on-going Farm Input Subsidy Programme.

Opportunities provided by CA in adapting to climate change

Despite the challenges, CA presents several opportunities in adaptation to climate change. These include improved soil moisture retention, control of soil erosion, improvement in food security, presence of several institutions promoting CA, and provision of an enabling environment for adoption of CA, as discussed below.

Improved soil moisture retention

Almost all farmers, whether adopters or non-adopters, government extension workers and officials from the NGOs, agreed on the effect of CA on improving soil moisture retention. This was the most important thing for the farmers in Balaka, which is generally a dry area, as it is in a rain-shadow region. Mulching prevented direct heat from reaching the soil thereby reducing evaporation from the soil. Similarly, intercropping, which improves total soil cover, also prevented direct heat from reaching the soil thereby further reducing evaporation from the soil. As observed in a study in Zambia by Thierfelder and Wall (2009), improved soil moisture was associated with CA. They observed that soil moisture was higher throughout the season in most CA plots than in the conventionally tilled ones.

As such, promoters of CA should still encourage the farmers to mulch their fields. Further, application of organic manure as a way to improve soil moisture retention can also be explored. This is in line with a study in China by Wang *et al.* (2016) which found that manure application increased soil water storage. As such, farmers could explore this opportunity and realize more benefits from CA.

Control of soil erosion

CA has also been observed to control soil erosion. This is a result of the incorporation of organic matter in the soil. Soil erosion is also controlled due to mulching, minimum tillage and intercropping. Mulching reduced the raindrop impact thereby reducing splash erosion; an observation that coincides with that of Thierfelder and Wall (2009) who observed in Zambia and Zimbabwe that treatments that included reduced tillage and surface residue retention had less water runoff and erosion. Furthermore, according to Seran and Brintha (2010), intercropping controls soil erosion by preventing raindrops from directly hitting the soil, thereby reducing surface runoff, which erodes the soil. As such mulching, minimum tillage and intercropping should be promoted in order to take advantage of the opportunity these CA practices present as regards soil erosion control.

Improvement in food security

Generally, farmers practising CA were observed to have positive changes in their yields as a result of CA adoption. The increase in yields was not only attributed to improved soil moisture but also improved soil fertility as a result of crop residue retention and the consequent high organic matter content. However, it is important to acknowledge that the effects of CA on crop yield varied depending on the type of crop and the climate. According to Zheng *et al.* (2014), for example, CA was observed to increase maize yields more than rice and wheat. The Chinese study also reported an increase in yields associated with mulching in low precipitation areas. Both of these situations put Balaka farmers in a better position to realize improved food security as a result of CA, since they mainly grew maize as a staple food and they received lower levels of precipitation than their colleagues in in other districts.

Therefore, institutions promoting CA in Balaka should continue promoting mulching as an important practice to ensure improved food security. Further, the advocates of CA should also promote intercropping since those that practise it are not only better able to survive should one crop fail, but they also have more meal diversification from the crops grown. This is important as it helps to meet the dietary needs of the farmers, an important aspect of food security.

Presence of several institutions promoting CA

CA has brought many players into the study area, most of them NGOs, in addition to the existing government agricultural extension workers. NGOs are known to assist in providing critical information through information collection, analysis and dissemination (Gemmill and Bamidele-Izu, 2002). This information-based role ensures that implementation is guided by research, leading to successful implementation of activities such as CA. In view of this, new institutions intending to promote CA in the area should consider using the institutions already working there. This will ensure continuity and reduce the effect of lag in implementation activities during the period of familiarization when a new actor is introduced in a community. Exploring this opportunity could ensure increased adoption and successful implementation of CA in the study area.

Presence of an enabling environment

Generally, there was a conducive environment for the implementation of CA in the area. This was due to the inclusion of CA in national policies. Application of CA is now government policy in Malawi actively promoted by regional organizations such as the International Maize and Wheat Improvement Centre (CIMMYT) and the International Centre for Research in Agroforestry (ICRAF) (Giller *et al.*, 2015). Further, CA promotion is reflected in the strategy of the Machinga Agricultural Development Division, a government institution, under which Balaka District falls. One of the strategy's goals is to promote CA in the area. This is an opportunity since the integration of CA into relevant government institutions is crucial for the sustainability of the technology (Bhan and Behera, 2014). As such, institutions promoting CA in the area should align their activities with these strategies since strategies ensure easy implementation, monitoring and evaluation of the activities to ensure their success and realize intended outcomes.

Conclusion

The aim of this study has been to understand the benefits of CA in improving the livelihoods of farmers in Balaka District. This study was conducted in Hanjahanja and Sawali sections, in Bazale EPA. The study found that increase in droughts and floods has increased the vulnerability of the farmers in the study area. However, this vulnerability is partly reduced by the increase in income due to the adoption of CA, which increases the adaptive capacity of the people. However, CA adoption was found to increase the vulnerability of farmers in Hanjahanja section, in times of heavy rains and floods. This is due to waterlogged conditions which lead to lower crop yields by CA adopters than those obtained by non-adopters. Further, the presence of several institutions promoting CA in the area is an enabling environment for its adoption and success. As such, CA has led to improvement in the livelihoods of the farmers. This is not only through the increase in the livelihood assets of the adopters, but it has also led to the realization of positive livelihood outcomes. Most adopters have observed an improvement in food security, increase in income, improved soil quality, better access to agricultural information and reduced vulnerability.

Even though many adopters seem not to face serious challenges with CA adoption, lack of inputs, high labour requirements and rainfall variability are the challenges that other adopters are struggling with in both Hanjahanja and Sawali sections. However, opportunities are there for CA to be adopted as an adaptation measure to climate change. These include the presence of several institutions promoting CA in the area, and the benefits associated with CA such as improved soil moisture retention, soil erosion control and ultimately improved food security. As such, the GoM should consider including herbicides and pesticides among the inputs under the Farm Input Subsidy Programme in order to assist farmers to be able to buy and use herbicides that are vital for practising CA. Further, actors promoting CA should consider promoting practices based on the conditions prevalent in the area so as to minimize the issues of low yield associated as a result of mulching in flood-prone areas. In addition, further research should be done on the best CA practices for areas with different environmental conditions in order to maximize the benefits of CA in improving farmers' livelihoods and avoid maladaptation.

References

Adger, W., Brooks, N., Bentham, G., Agnew, M. and Eriksen, S. (2004) New indicators of vulnerability and adaptive capacity. *Technical Report* 7. Tyndall Centre for Climate Change Research. Available at: https://www.tyndall.ac.uk/ (accessed 4 July 2016).

Agrawal, A., Kononen, M. and Perrin, N. (2009) The role of local institutions in adaptation to climate change. *Social Development Working Papers*. Paper no. 118. Available at: http://www.researchgate.net/publication/237430572_The_Role_of_Local_Institution_inadaptation_to_Climate_Change (accessed 4 July 2016).

Bhan, S. and Behera, U. (2014) Conservation agriculture in India – problems, prospects and policy issues. *International Soil and Water Conservation Research* 2(4), 1–12.

Chambers, R. and Conway, G.R. (1991) Sustainable rural livelihoods. Practical concepts for the 21st century. *IDS [Institute of Development Studies] Discussion Paper* 296.

Corbeels, M., Graaff, J., Ndah, T., Penot, E., Baudron, F., *et al.* (2013) Understanding the impact and adoption of conservation agriculture in Africa: a multi-scale analysis. *Agriculture, Ecosystems and Environment* 187, 155–170. http://dx.doi.org/10.1016/j.agee.2013.10.011

Department for International Development (DFID) (1999) Sustainable Livelihoods Guidance Sheets Framework. Available at: http://www.livelihoodscentre.org/documents/20720/100145/Sustainable+livelihoods+guidance+sheets/8f35b59f-8207-43fc-8b99-df75d3000e86 (accessed 8 January 2016).

Food and Agriculture Organization of the United Nations (FAO) (2015) What is Conservation Agriculture? Available at: http://www.fao.org/ag/ca/1a.html (accessed 21 January 2016).

Gemmill, B. and Bamidele-Izu, A. (2002) The role of NGOs and civil society in global environmental governance. In: Esty, D.C. and Ivanova, M.H. (eds) *Global Environmental Governance: Options and Opportunities*. Yale School of Forestry and Environmental Studies, Yale University, New Haven, Connecticut, pp. 77–100.

Gill, P., Stewart, K., Treasure, E. and Chadwick, B. (2008) Methods of data collection in qualitative research: interviews and focus groups. *British Dental Journal* 204, 291–295. doi:10.1038/bdj.2008.192

Giller, K., Andersson, J., Corbeels, M., Kirkegaard, J., Mortensen, D., *et al.* (2015) Beyond conservation agriculture. *Frontiers in Plant Science* 6(870). doi.10.3389/fpls.201500870

Grabowski, P., Walker, F., Haggblade, S., Maria, R. and Eash, N. (2013) Conservation agriculture in Mozambique – literature review and research gaps. Working Paper No. 4E, (March). Available at: https://vtechworks.lib.vt.edu/bitstream/handle/10919/70134/6729_CA_Literature_Review_using_IIAM_WP_Mar20.pdf?sequence=1&isAllowed=y (accessed 5 December 2019).

Greasley, P. (2008) *Quantitative Data Analysis Using SPSS. An Introduction for Health and Social Science*. Open University Press, New York.

Imoloame, E. and Olanrewaju, A. (2014) Improving agricultural extension services in Moro Local Government Area of Kwara State, Nigeria. *Journal of Agricultural Extension and Rural Development* 6(3), 108–114.

Intergovernmental Panel on Climate Change (IPCC) (2007) Summary for policymakers. In: *Climate Change 2007: Impacts, Adaptation and Vulnerability. Contribution of Working Group II to the Fourth Assessment Report of the Intergovernmental Panel on Climate Change*. [Parry, M.L., Canziani, O.F., Palutikof, J.P., van der Linden, P.J. and Hanson, C.E. (eds)]. Cambridge University Press, Cambridge, pp. 7–22.

International Centre for Agricultural Research in the Dry Areas (ICARDA) (2012) Conservation agriculture: opportunities for intensified farming and environmental conservation in dry areas. *Research to Action* 2. Available at: https://reliefweb.int/report/world/conservation-agriculture-opportunities-intensified-farming-and-environmental (accessed 22 November 2019).

Islam, R. and Reeder, R. (2014) No-till and conservation agriculture in the United States: an example from the David Brandt farm, Carroll, Ohio. *International Soil and Water Conservation Research* 2(1), 97–107.

Julie, D., Manu, I., Bime, M., Tabi, O. and Defang, H. (2014) Impacts of climate change on crop production and development of Muyuka Subdivision – Cameroon. *International Journal of Agriculture, Forestry and Fisheries* 2(2), 40–45.

Jumbe, C. and Nyambose, W. (2016) Does conservation agriculture enhance household food security? Evidence from smallholder farmers in Nkhotakota in Malawi. *Sustainable Agriculture Research* 5(1), 118–128.

Kabirigi, M., Musana, B., Ngetich, F., Mugwe, J., Mukuralinda, A. and Nabahungu, N.L. (2015) Applicability of conservation agriculture for climate change adaptation in Rwanda's situation. *Journal of Soil Science and Environmental Management* 6(9), 241–248. doi.10.5897/JSSEM15.0508

Kamtimaleka, S. (2009) Assessment of socio-economic factors influencing farmers' adoption and intensity of conservation agriculture in dry land areas of Malawi: a case study of Chinguluwe EPA in Salima District and Bazale EPA in Balaka District. Agricultural Innovation in Dryland Africa (AIDA). Specific Support Action (SSA) of the International Co-operation 6th European Union (EU) Research Framework Programme (EU-FP6). (SSA Africa 2006). CIRAD, Montpellier, France/Technical Centre for Agricultural and Rural Cooperation ACP-EU (CTA), Wageningen, the Netherlands.

Lazaro, M., Silas, S. and Liwenga, E. (2013) Community perceptions and willingness to accept and execute REDD+ initiative: the case of Pugu and Kazimzumbwi Forest Reserves, Tanzania. *Cross-Cultural Communication* 9(3), 48–54. doi:10.3968/j.ccc.1923670020130903.2441

Majale, M. (2002) *Towards Pro-poor Regulatory Guidelines for Urban Upgrading*. A review of papers presented at the International Workshop on Regulatory Guidelines for Urban Upgrading, held at Bourton-On-Dunsmore, Warwickshire, UK, 17–18 May 2001. Available at: https://assets.publishing.service.gov.uk/media/57a08d3c40f0b64974001734/R7850_Majale_RGUU1_Review.pdf (accessed 5 December 2019).

Mangisoni, J., Katengeza, S., Langyintuo, A., La Rovere, R. and Mwangi, W. (2011) Characterization of maize producing households in Balaka and Mangochi Districts in Malawi: Country Report – Malawi. International Maize and Wheat Improvement Center (CIMMYT), Nairobi.

Manyevere, A., Muchaonyerwa, P., Laker, M. and Manken, P. (2014) Farmers' perspectives with regard to crop production: an analysis of Nkonkobe Municipality, South Africa. *Journal of Agriculture and Rural Development in the Tropics and Subtropics* 15(1), 41–53.

Marshall, M.N. (1996) The key informant technique. *Family Practice* 13(1), 92–97.

McNair, W., Lambert, D. and Eash, N. (2015) Conservation agriculture and household wellbeing: a non-causal comparison among smallholder farmers in Mozambique. *Journal of Agricultural Science* 7(1), 1–18.

Milder, J., Majanen, T. and Scherr, S.J. (2011) Performance and potential of conservation agriculture for climate change adaptation and mitigation in sub-Saharan Africa. *Eco-agriculture Discussion Paper* Number 6. Ecoagriculture Partners, Washington, DC.

Muzari, W., Gatsi, W. and Muvhunzi, S. (2012) The impacts of technology adoption on smallholder agricultural productivity in sub-Saharan Africa: a review. *Journal of Sustainable Development* 5(8), 69–77.

Nyanga, P.H. (2011) Factors influencing adoption and area under conservation agriculture: a mixed methods approach. *Sustainable Agriculture Research* 1(2), 27–40.

Ramirez, A. (2013) The influence of social networks on agricultural technology adoption. *Procedia – Social and Behavioural Sciences* 79, 101–116.

Seran, T. and Brintha, I. (2010) Review on maize-based intercropping. *Journal of Agronomy* 9, 135–145. doi:10.3923/ja.2010.135.145

Taylor-Powell, E. and Renner, M. (2003) Analysing Qualitative Data. Learning Store, Madison, Wisconsin. Available at: https://cdn.shopify.com/s/files/1/0145/8808/4272/files/G3658-12.pdf (accessed 28 May 2015).

Thierfelder, C. and Wall, P. (2009) Effects of conservation agriculture techniques on infiltration and soil water content in Zambia and Zimbabwe. *Soil and Tillage Research* 105, 217–227.

Thierfelder, C., Rusinamhodzi, L., Ngwira, A., Mupangwa, W., Nyagumbo, I., *et al.* (2013) Conservation agriculture in Southern Africa: advances in knowledge. *Renewable Agriculture and Food Systems* 30(4), 328–348. doi:10.1017/S1742170513000550

Umar, B., Aune, J., Johnsen, F. and Lungu, O. (2011) Options for improving smallholder conservation agriculture in Zambia. *Journal of Agricultural Science* 3(3), 50–62.

Wang, X., Jia, Z., Liang, L., Yang, B., Ding, R., *et al.* (2016) Impacts of manure application on soil environment, rainfall-use efficiency and crop biomass under dryland farming. *Scientific Reports* 6, 20994. doi:10.1038/srep20994. Available at: http://www.nature.com/articles/srep20994 (accessed 22 July 2016).

Zheng, C., Jiang, Y., Chen, C., Sun, Y., Feng, J., *et al.* (2014) The impacts of conservation agriculture on crop yield in China depend on specific practices, crops and cropping regions. *The Crop Journal* 2(5), 289–296. http://dx.doi.org/10.1016/j.cj.2014.06.006

6 A Comparative Cost–Benefit Analysis of Mobile and Sedentary Pastoral Production Systems in Selected Villages in Northern Tanzania

Lucas E. Yamat* **and Claude G. Mung'ong'o**

Institute of Resource Assessment, University of Dar es Salaam, Dar es Salaam, Tanzania

Abstract

Despite a growing body of evidence that highlights the economic, social and environmental benefits of mobile pastoralism, few governments are ready to tolerate mobility and many policy makers promote knowingly or inadvertently the policies of sedentarization. This production system seems not to be clearly understood by many and has been characterized as backward, environmentally destructive and economically unsustainable; and the view is that it should be replaced with more sedentary forms of livestock production or other beneficial land uses. The overriding question is whether sedentary livestock keeping is more productive and utilizes fewer resources and less space than the mobile pastoral system. This study carried out a comparative cost–benefit analysis of the two production systems in selected villages of Kiteto and Karatu districts. The aim was to come up with credible data to test this hypothesis. Two alternatives were compared in terms of their net present value (NPV) to test a null hypothesis. The alternative with an NPV greater than zero or higher than its alternative was accepted to be more viable compared with the one with an NPV less than zero or less than its alternative. Whenever the NPV of the sedentary production system in the analysis was shown to be greater than zero and/or greater than the NPV of the mobile pastoral production system the null hypothesis was accepted and vice versa. The study was conducted in Makame village of Kiteto District and Dofa village of Karatu District. Makame village represents a mobile pastoral production system while Dofa village represents a sedentary production system. The study employed a quantitative approach using a household survey in the two villages. The comparative cost–benefit analysis was carried out using monetary values derived from the livestock unit statistical approach. The findings have revealed that the average cost of maintaining a mobile pastoral and sedentary production systems are TSh90,096,333 and TSh112,295,200, respectively. The cost–benefit ratios are 1:0.5 for a mobile pastoral production system and 1:0.25 for the sedentary one.

Introduction

The mobile pastoral production system is a system in which people earn their livelihood by tending domesticated animals such as cattle, sheep, goats and donkeys kept on a strategically mobile lifestyle mainly in arid and semi-arid lands (ASALs) (Leshan and Standslause, 2013). The system is complex and dynamic as pastoralists seek to adapt

*lucasyamat@gmail.com

to evolving climatic, social, political and economic conditions at local, national and regional levels (Hesse and MacGregor, 2006). The African Union's Policy Framework for Pastoralism (AU, 2010) estimates that 268 million pastoralists live and move on 43% of Africa's land mass. Other sources show that these pastoralists are spread across 40 countries in East Africa, the Horn of Africa, the Sahel, the Middle East, South and Central Asia, Mongolia, China and elsewhere (WISP, 2007; Oxfam, 2008; UNPF, 2010).

Several ecological, sociological and anthropological studies (e.g. Behnke and Scoones, 1993; Tiffen *et al*., 1994; Fratkin and Sher-Mei Wu, 1997; Fratkin, 1998; Niamir-Fuller, 1999; Bolling and Shulter, 1999; Raynaut, 2001; Moritz, 2008; Oxfam, 2008; Tenaw, 2016) have acknowledged mobile pastoralists to be custodians of dryland environments and that they have been providing services through good rangeland management, including biodiversity and wildlife conservation. The key factors which have enabled pastoralists to manage their rangelands in adapting to the disequilibrium of their environment have been tied to the strong social organization and strategic herd mobility in rangeland management (Niamir-Fuller, 1999; Davies, 2008). The customary land management and governance structures are integral to the social safety nets and shared claims over rangelands that characterize mobile pastoral systems and which are a cornerstone of pastoral resilience and risk management (Nicolaisen, 1963; Dyson-Hudson, 1966; Gallais, 1967; Johnson, 1969; Niamir-Fuller, 1999; WISP, 2007, etc.).

The mobile pastoral production system has made significant contributions to national economies and provided most of the meat consumed in the countries where the system is practised. The system has been a livelihood support to millions of households globally (King, 2000). It is estimated that pastoralism contributes between 10% and 44% of the gross domestic product (GDP) in countries where it is practised (AU, 2010). These contributions are expressed in terms of revenues collected from marketplaces and employment for most of the rural population which may include direct values derived from livestock and indirect benefits which mobile pastoralism contributes to the ecosystem health of drylands (Hatfield and Davies, 2006; Oxfam, 2008).

There are many products that come from pastoralist lands, which are increasingly sought after. Many of these products have a high value on global markets and, for the most part, pastoralists (Hatfield and Davies, 2006; Boto and Edeme, 2012; Isolina Boto (Technical Centre for Agricultural and Rural Cooperation (CTA)) and Janet Edeme (African Union Commission), 2012 unpublished policy briefing) manage them not for business purposes. Apart from the economic and environmental contribution that pastoralism confers, it also contributes to social, cultural and institutional stability to the pastoral communities (Galvin, 2008). The social, cultural and institutional benefits forms three assets, social cohesion, social memories and collective actions, which are important for pastoralists to withstand and adapt to changing environmental conditions, including climate change and variability (Galvin, 2008).

The pastoral production system plays an important role in the economy of Tanzania. According to the Tanzania Livestock Modernization Initiative (URT, 2016), at the national level 50% of households keep livestock that directly supports 27 million people and most livestock are kept in the traditional pastoral system. This production system produces about 93% of the milk and 99% of the meat consumed in the country and is a major engine for rural livelihoods and development (URT, 2012). Apart from the supply of meat and other animal products, pastoral production systems make productive use of a large percentage of the available drylands where scarcity and variability of its natural resources have few alternative uses (Mung'ong'o and Mwamfupe, 2003).

Despite a growing body of evidence that highlights the economic, social and environmental benefits of mobile pastoralism, few governments are ready to tolerate mobility and many policy makers knowingly or inadvertently promote the policies of sedentarization (Davies and Hatfield, 2007). In addition, this production system seems not to be clearly understood by many and has been characterized as backward, environmentally destructive and economically unsustainable and the view is that it should be replaced with more sedentary forms of livestock production or other beneficial land uses (Hesse and MacGregor, 2006; WISP, 2011; Davies *et al.*, 2015). The overriding question is whether sedentary livestock raising is more productive and utilize fewer resources and space than the mobile pastoral system. This study carried out a comparative cost–benefit analysis of the two production systems in selected villages of Kiteto and Karatu districts. The aim was to come up with credible evidence-based data to test this hypothesis.

One old study that has been conducted to compare mobile pastoral production system and sedentary production appreciated that mobile pastoralism had a high value, particularly in the drylands while supported by a strategic herd mobility (Hesse and MacGregor, 2006). On the other hand, Niamir-Fuller (1999) found that sedentarization had increased environmental degradation, reduced economic potential and eroded the socio-cultural systems among the settled communities in East Africa. However, very few studies have as yet been carried out at the community level to update the situation and appreciate the social and economic benefits of mobile pastoralism against that of the sedentary system.

The theoretical framework

During the 1980s African mobile pastoralism, particularly in Eastern Africa, was said to be worse off than any other historical records due to investments influenced by romantic economists who advocated for the formation of ranching associations and group ranches which threatened the mobile pastoralists such as the Maasai (Simpson and Sullivan, 1984; Hodgson, 1999; Fratkin, 2001). Since that time to date, there have been debates about mobile pastoralism and its sustainability in African countries (Oba, 2011). These debates have evolved around four paradigms, namely: (i) the carrying capacity paradigm (CCP); (ii) the pastoral mobility paradigm; (iii) the pastoral modernization paradigm; and (iv) the resource risk, variability and governance (RRG) paradigm. These paradigms were all based on the strong feelings that mobile pastoral development had failed and largely contributed to the demise of pastoral societies because of misunderstanding of the pastoral societies and rangeland ecology (Galaty and Aronson, 1981; Goldschmidt, 1981 as cited by Moritz, 2008).

The carrying capacity paradigm (CCP)

The carrying capacity of a biological species in an environment is the maximum population size of species that the environment can sustain indefinitely, given the food, habitat, water and other necessities available in the environment (Wikipedia, 2016). In the context of pastoralism, it refers to the maximum population size of livestock and families that the environment can sustain by providing the necessities without causing degradation.

The concept of carrying capacity is related to homeostatic regulation as part of the new ecology paradigm (Brush, 1975). Hardin's concept 'tragedy of the commons' of 1968, which used a parable of individual ownership of livestock and communally used grazing land to illustrate the problems of human population growth on the environment, has played a key role in shaping the mainstream view towards mobile pastoral systems (Lamprey, 1983).

The CCP has two main diverging hypotheses: (i) equilibrium; and (ii) disequilibrium. The equilibrium hypothesis asserts that animal densities in excess of the appropriate stocking rates cause degradation of the land by altering the soil fertility, land cover change and biodiversity loss. It therefore promotes interventions of regulating livestock stocking densities to protect the rangelands. The disequilibrium view of the CCP opposes the above by recognizing livestock mobility in addressing those rangeland challenges concerning livestock densities (Oba, 2011).

Based on the carrying capacity assumption, mobile pastoralism has been renounced with the view that rangelands protected from livestock grazing will benefit from reversals of these conditions (Hardin, 1968). Consequently, many pastoral development initiatives around 1980s and 1990s (and even today) were not aimed at developing pastoral systems but at minimizing their impacts on the environment (Moritz, 2008).

The CCP neglects the role of indigenous knowledge used by pastoralists in managing their rangelands. The livestock mobility under the mobile system of land use is based on periods of abundance and periods of scarcity but in reference to resource availability and capacities of livestock (Oba, 2011). Pastoralists portion their rangelands into different parcels for different types of livestock (calves, sick animals and old animals), using their long historical knowledge of managing environmental risks which the CCP does not consider.

The CCP does not consider arid environments with stochastic climatic events in terms of space and time as well as average productivity values which have a greater impact on rangeland production than grazing alone (Coppock, 1994). It is assumed that pre-drought livestock densities can be used as indicators of density-dependent livestock die-off (Scoones, 1995). According to the paradigm, it implies that if the livestock density is high before the drought, the number of stock lost would be correspondingly high during or after the shocks which is not realistic in the presence of livestock mobility (Scoones, 1995; Hary *et al.*, 1996).

Also, the paradigm does not consider mobility as a solution to regulate herd stocking density. Accordingly, the paradigm fails to acknowledge that land degradation is the product of settling formerly mobile pastoralists through development initiatives of sedentarization and commercialization of grazing lands (Fratkin, 1998).

The carrying capacity is based on average values and assumes homogeneity of rangelands and livestock managed by herders, which are restricted to individual grazing landscapes, contrary to the counter evidence (Scoones, 1995; Hary *et al.*, 1996). For livestock management systems such as those of African pastoralists, there is no fixed carrying capacity. Multiple capacities exist, reflecting the different nutritional requirements of different livestock species during different seasons of the year (Scoones, 1995; Hary *et al.*, 1996; Moritz, 2008; Oba, 2011).

The pastoral mobility paradigm

According to Niamir-Fuller (1999), pastoral mobility refers to the time and space behavioural movement of herds and their handlers in response to variations in pasture and the distribution of other important livestock resources. This movement has never been a one-way process; rather, it involves the rotational use of spatially distributed land resources to feed the diverse species of livestock (Andriansen and Nielsen, 2002). Mobility can take different forms such as mobility relative to the gradient or an indigenous management system whereby mobility is flexible and opportunistic, guided by times of abundance and of scarcity of livestock resources in different seasons of the year and in different landscapes (Andriansen and Nielsen, 2002; Butt, 2010).

The mobility paradigm was first developed by Maryam Niamir-Fuller in 1999 and discussed in a book titled *Managing Mobility in African Rangelands*. According to Niamir-Fuller, the objective of the paradigm was to ensure appropriate policies, legal mechanisms and development programmes to allow self-evolution of pastoralism towards an economically, socially and environmentally sustainable system (Niamiri-Fuller, 1999, p. 31).

The central point of the pastoral mobility paradigm is to support mobile pastoral systems rather than interfering in their management of natural resources under the assumption that traditional extensive systems are well adaptive to the drylands of the African continent (Moritz, 2008). The proponents of the mobility paradigm are optimistic that the development of African mobile pastoral systems can be successful if the lessons from the new ecology are incorporated in project development and implementation (Scoones, 1995).

The relevant views advocated by this paradigm towards the future of pastoralism include: (i) protecting the remaining rangelands from encroachment from outsiders; (ii) supporting the management of rangelands through traditional pastoral institutions; and (iii) supporting pastoral mobility and marketing to allow pastoralists to balance between restocking and destocking to enable them ultimately to cope more effectively with disequilibrium environments. The proponents of the mobility paradigm believe that migration is a key to reduce herd losses and manage the risks through developed mobility pathways (Homan *et al.*, 2008); thus, disruptions of these natural migratory pathways might expose pastoralists to more severe stress (Behnke *et al.*, 2011).

The pastoral mobility paradigm tries to explain optimal utilization of rangeland and optimal livestock production through strategic livestock mobility at different seasons of a year. This paradigm takes into consideration the families managing stocks, which requires a certain pattern of mobility and their nutritional requirements that are reflected by the environment (Oba and Kaitira, 2006; Roba and Oba, 2009).

While mobility is associated with environmental variability, traditional herd mobility might also be a result of different socio-economic, political and cultural push and pulling factors among the mobile pastoralists (Bassett and Turner, 2007). The pushing factors for migration of the pastoralist include early mortality of calves, a shortage of forage, wildlife breeding and increased distances to reach watering points, which are important early warning signs that pastoralists should heed. The decision that pastoralists take involves a risk factor that will influence the decision to migrate (Warner *et al.*, 2010).

Despite the fact that pastoral mobility takes into consideration issues of indigenous knowledge its limitations include disruptions of herd mobility pathways by internal and external pressures. The disruptions of these mobility pathways will constrain traditional adaptation and expose mobile pastoralists to much more vulnerability to changing environmental conditions (Behnke *et al.*, 2011).

Another weakness of the pastoral mobility paradigm is that it rationalizes the sustainability of the indigenous pastoral system, but it does not explain the functional decline in mobility in recent years (Moritz, 2008). The paradigm also assumes that the mobile pastoralists act rationally, which is sometimes not realistic as some may not act rationally due to the ineffectiveness of customary institutions in safeguarding rangeland utilization or because of external interference from other livelihood mechanisms.

The paradigm also depends on the government's will to ensure appropriate policies; legal mechanisms and development programmes are developed to allow self-evolution of pastoralism towards an economically, socially and environmentally sustainable system (Niamir-Fuller, 1999). Sometimes the government may not be interested in developing the policies that recognize livestock mobility due to the worry that they may cause chaos and conflicts with settled communities.

The pastoral modernization paradigm

The modernization paradigm was developed by Mortimore (2000) and later used by Tiffen (2004 as cited by Moritz, 2008) to challenge the pastoral mobility paradigm. Their argument is not that mobile pastoral systems are not well adapted to rangelands' disequilibrium and they do not reject the theoretical framework of the mobility paradigm (Mortimore, 1998). Rather, they based their paradigm on recent pressures facing pastoral mobility including population growth, shrinkage of grazing land and agricultural expansion, which will have detrimental impacts on extensive pastoral systems in the near future. Mortimore and Tiffen proposed that the only option to sustain mobile pastoralists' livelihoods is by intensification and integrated livestock production in the mixed farming systems (Mortimore and Adams, 1998; Mortimore, 2000; Tiffen, 2004 as cited by Moritz, 2008).

The argument of the modernization paradigm is that pure forms of pastoralism (where families rely entirely on livestock produce for their livelihoods) are diminishing. Some research suggests that the transformation from pastoralism to livelihood diversification is triggered by declining livestock per capita (Desta and Coppock, 2004; Homan *et al.*, 2008). The modernization paradigm is commonly used in the mixed-farming system literature. The paradigm considers the integration of crop and livestock production in farming systems which involves diversification at the household level rather than specialization at the household level and diversification and integration at the regional level (McIntire *et al.*, 1992; Bourn and Wint, 1994).

The proponents of the modernization paradigm focus not only on the development of pastoral systems, but development of the drylands in general. This focus has much to say about pastoral system development (Moritz, 2008). The mixed-farming model of Michael Mortimore and Mary Tiffen's Drylands Research group has been most explicit about the future of pastoral systems in African drylands (Mortimore, 2000; Tiffen, 2004 as cited by Moritz, 2008). Their research challenges conventional Malthusian doom scenarios for African drylands by showing that population growth does not have to lead to environmental degradation (Tiffen *et al.*, 1994; Moritz, 2008). On the contrary, they argue that two factors, population growth and market incorporation, are necessary conditions for sustainable development of African drylands (Moritz, 2008).

The relevance of the modernization paradigm is due to the fact that mobility of pastoral systems requires a large area to be effective and any internal and external pressures on rangeland may cause a serious obstacle to mobile pastoralists. The pressures such as population growth, expansion of agricultural production, expansion of conservation and shrinkage of mobility pathways are already problematic and causing conflicts (Fratkin, 2001).

Another relevance that the modernization paradigm has, is where the fodder used to feed the livestock during times of crisis is made from agricultural by-products and this is only possible when mixed farming and livestock is in place. Agropastoralists may also opt to use their agricultural residue to feed their livestock when pastures are scarce in the era of climate change and variability.

While pastoralism is the only means of sustaining people's livelihoods in the drylands, in some areas where agriculture is a viable alternative there can be intense competition with pastoralism and consequently pure pastoral communities will be pushed to more marginal lands. This will expose mobile pastoralists to high risks of losses due to environmental changes including climate change and variability. However, commercial pastoralists will not risk long-distance migration because of a greater risk of huge financial losses when the herds are lost as a result of environmental stress (Adriansen, 2006).

The paradigm also has limitations on the grounds of variability of rangelands in both space and time with scarcity of resources and scattered rainfall patterns. In most arid areas agriculture is not a viable alternative to be integrated with livestock production without bringing water from long distances (Mung'ong'o and Mwamfupe, 2003). The small wet areas will attract human settlements and in this zone population growth, agricultural expansion and urbanization will take place leading to unbalanced rangeland utilization (Moritz, 2008).

The paradigm also fails to recognize that in integrated mixed farming, livestock will increasingly become dependent on farm-produced fodder which is a dry matter and it does not take into account issues of palatability and digestibility of this dry matter for livestock. In addition, livestock producers without title to farmland will be denied the possibility of an intensification integration trajectory and thus access to farm-produced fodder will become a big challenge to them (Mortimore and Adams, 1998, pp. 269–272).

The implication of the above scenarios is clear that, for pastoral improvement, the development policies should run with the direction of change if it is to stand any chance of success in the long run (Tiffen *et al.*, 1994; Moritz, 2008).

The resource risk, variability and governance (RRG) paradigm

As the three paradigms (carrying capacity, pastoral mobility and modernization) are weak on social perspectives of mobility and none of them demonstrates clearly the role of governance on policy decisions in resource management, the RRG paradigm was proposed (Oba, 2011). The mobility and modernization paradigms both involve some aspects of decision making but both are less clear on how the system can be holistically used to guide management. The RRG paradigm provides a contextual framework for examining the socio-ecological, environmental and governance drivers that help to explain changes in pastoral mobility patterns in changing socio-political circumstances. It does this by disaggregating the drivers (the pull factors) and the counter actions (the push factors) that determine the trajectory of change in mobility (Oba, 2011, p. 12).

This new paradigm provides an integrative holistic way of understanding how the decision can influence the drivers of pastoral mobility on the rangelands. The natural resource base for human livelihoods, the risks associated with access rights, other external drivers imposed on the environment, and the governance instruments can help one to understand the functions of pastoral mobility in reducing risks and exploiting opportunities in disequilibrium environments.

The governance part in this paradigm focuses on the management of survival mechanisms and optimal use of the environment to reduce risks and exploit opportunities brought by changes on the rangelands. Governance can be assessed in terms of performance and effectiveness. Governance includes both resources and risks as factors, which have a powerful impact on decision making. Governance also has a moral responsibility and accountability (Morton, 2010). The RRG framework makes an important contribution to understanding how pastoralists adapt through decision making that may be examined from two perspectives: (i) resource pressure points; and (ii) loss of key resources.

RESOURCE PRESSURE POINTS. Resource pressure points are of various types and may include those that are environmental and politically triggered. In the Horn of Africa, ethnic conflicts have increased the risk of political instability, and this has had a dramatic impact on herd mobility. Armed conflicts have displaced populations and separated them from their resources. Conflicts contribute to economic collapse and the appropriation of property. The immediate effect is the disruption of herd mobility pathways patterns, which in turn will affect the future of pastoralism (Eriksen and Lind, 2009; Oba, 2011).

In the escape areas, populations of internally displaced peoples together with those of residents may exceed the capacity of the available resources. This makes political stressors serve as a trigger, whose effect is felt through the pastoral production system (Bell, 2004). An obvious outcome of politically induced migration is changing the ownership of resources, including the land and its attached resources (Unruh and Bailey, 2009). There is, however, insufficient literature on the long-term consequences of conflict and the environmental changes associated with pastoralists' forced migrations and their impact on the spatial distribution of resources as the causal drivers have not been sufficiently documented (Randall, 2005; Oba, 2011).

LOSS OF KEY RESOURCES. To sustain the pastoral production system there are key resources to be considered including water points, grazing land, mobility pathways, livestock and labour force. However, different users compete for these resources, for example pastoralists, farmers and fishermen all depend on water to sustain their production and water is also needed for conservation. Pastoral herd survival is dependent on the availability of pastures during the dry season. The floodplains enable the annual transhumance of herds moving from their wet season pastures to their dry season pastures; where the fresh grass growth of the flooded pastures provides them with nutritious grazing (Scholte *et al.*, 2006; Oba, 2011).

In most cases, alterations in land use systems modify patterns of land use, as there is no attempt to integrate seasonal grazing patterns with commercial agriculture or wildlife conservation (Unruh, 1990). When the pastoral pathways are blocked livestock die-off increases because mobility to access dry biomass is restricted during the season of scarcity. The loss of these key resources including water points, grazing land, mobility pathways, livestock and labour force have detrimental impacts on the sustainability of pastoralism systems.

The relevance of the RRG paradigm is its visibility to many policy makers. The paradigm will make an important contribution to understanding how pastoralists adapt through decision making that may be examined from the two perspectives of resource pressure points and loss of key resources. In particular, the paradigm has huge benefits where the indigenous system of governance is used to participate in decision making (Oba, 2011).

While the RRG paradigm is important in maintaining the pastoral systems, consideration of the RRG paradigm requires good will and purpose from all the stakeholders. If the government ignores the roles of customary institutions in rangeland management, pastoral systems will continue being marginalized in comparison to other land users. The knowledge of decision makers about pastoral production systems is key for effective policies and enactment of regulatory frameworks that work for pastoralists to sustain their livelihood. However, this depends on the moral responsibility and accountability of the existing governance style (Morton, 2010).

Summary of mobile pastoralism and sustainable development paradigms

To summarize, the earlier paradigms such as carrying capacity and modernization have never been successful in understanding the pastoral production systems but instead have increased marginalization of pastoral people through policy influence (Moritz, 2008). However, around the 1990s and 2000s a paradigmatic shift took place in the study of pastoral systems that led to a much greater understanding of pastoral systems and rangeland ecology, which is integrated in the theoretical framework of the new rangeland ecology (Behnke and Scoones, 1993; Scoones, 1995; Niamir-Fuller, 1999; Moritz, 2008). The new rangeland ecology is currently the dominant theoretical paradigm in the study of pastoral systems and it guides the thinking of policy makers and practitioners in pastoral development. The four paradigms, therefore, have been a turning point for policy makers to design and implement policies for the livestock sector in many African countries as well as in setting priorities for development to rescue a diminishing pastoral production system (Moritz, 2008).

The studies reviewed above have focused on economic contributions of pastoralism, livelihood diversification, the environmental importance of pastoralism, the future of pastoralism and policy and legal frameworks as drivers for changing the mobile pastoral production system. In addition, some studies focus on a global, regional or national scale to understand the dynamics of pastoral production systems. However, there are few studies that have been undertaken at the community level to explore the dynamics within pastoral production systems, the drivers for those dynamics and thereby the implications on pastoral communities' welfare as well as the community members' understanding of those changes.

Frameworks for estimating the values/benefits of pastoralism

There are various developed or adopted frameworks for measuring the environmental, economic and social benefits of pastoralism around the world. These may measure the pastoral production system either quantitatively or qualitatively. Among the frameworks that are popularly used are total economic value (TEV) framework (TEV), cost–benefit analysis (CBA), social cost–benefit analysis (SCBA) and participatory economic valuation (PEV) analysis.

Total economic value (TEV)

TEV is the approach used to capture all the economic values for not only man-made capital assets, but also the natural resources. It includes non-marketed goods and services in economic analysis. The concept of TEV is a useful framework for identifying the various values associated with ecosystems which include use values and non-use values. The approach enables a holistic assessment of all the critical values of pastoralism to generate information for pro-pastoralist policy formulations or dialogue (Mdoe and Mnenwa, 2007). According to the World Initiative for Sustainable Pastoralism (WISP) (2006), the TEV framework is useful in providing a strong tool for understanding the true contribution that pastoralists make to their domestic economies. The framework also proved as a useful tool to explore the full range of costs and benefits emanating from pastoralism activity (MacGregor and Hesse, 2006).

The social cost–benefit analysis (SCBA)

SCBA is an extension of economic CBA that have been adjusted to take into account the full spectrum of costs and benefits by including social and environmental effects borne by a society as a whole resulting from an intervention (NEF, 2013).

Participatory economic valuation (PEV)

PEV is an emerging methodology used for assessing the value of intangible costs and benefits, whereby community members representing a specified population (e.g. a district or region) could assess the relative impact value of certain outcomes, dynamics and policies which they have themselves identified as being important. These relative impacts can also be estimated in absolute terms if financially measurable costs and/or benefits, such as milk production, are included in the set of parameters to be valued (WISP, 2006). The usefulness of PEV is that all direct, indirect and non-use values can be used for all value types. In estimating the PEV a thick data set is not necessary and a small data set is sufficient.

Cost–benefit analysis (CBA)

This study employed a CBA framework to compare cost and benefits of the mobile pastoral production system and a sedentary production system to test the contention that 'sedentary livestock raising is more productive than mobile systems'. Levin and MacEwan (2001) defined CBA as referring to the evaluation of alternatives according to their costs and benefits when each is measured in monetary terms. Since each alternative is measured in terms of monetary values of its benefits, each alternative can be examined on its own merit to see if it is worthwhile. In selecting from among alternatives, one would choose that particular one that had the highest benefit–cost ratio (Levin and McEwan, 2001, p. 14). Ali (2012) defined CBA as a model widely used in economics by various organizations to select among alternatives and to justify making certain investments. In this study, CBA is the evaluation of mobile pastoral production system and sedentary production system according to their cost and benefits and each is measured in monetary terms by adopting and modified Levin and McEwan (2001) definition.

Theoretically, the CBA is commonly used by government and other organizations such as the private sector or business firms to appraise the desirability of a given policy or investment. It is an analysis of the expected balance of benefits and costs including an account of foregone alternatives and status quo (NEF, 2013). The CBA assists in predicting whether the benefits of a policy outweigh its costs and by how much relative to other alternatives. Accurate CBA identifies choices that increase welfare from a utilitarian perspective.

The assumption underlying the CBA is that the accurate CBA changes the status quo by implementing the alternative with the lowest cost–benefit ratio which can improve Pareto efficiency. An analysis using CBA should recognize that perfect appraisal of all present and future costs and benefits is difficult, and while CBA can offer a useful estimate of the best alternative, performance in terms of economic efficiency and social welfare are not guaranteed (NEF, 2013; European Union, 2015).

In undertaking a CBA, some processes are followed in order to meet the objective of the study which requires a CBA. The processes relevant to this study are as follows (in this order): (i) to list the alternatives for the decisions that are to be made; (ii) to select measurements and measure all cost/benefit elements; (iii) to predict outcomes of costs and benefits over the relevant period; (iv) to convert all costs and benefits into a common currency; (v) to apply the discount rate to each alternative; (vi) to calculate net present values (NPVs) for each of the alternatives; (vii) to perform a sensitivity test; and (viii) to adopt the recommendation of the comparative analysis (European Union, 2015).

The two alternatives were compared in terms of their NPV to test the null hypothesis. The alternative with an NPV greater than zero or higher than its alternative was accepted to be more viable compared to the one with NPV less than zero or less than its alternative. In this study whenever the NPV of the sedentary production system in the analysis was shown to be greater than zero and/or greater than the NPV of the mobile pastoral production system, then the null hypothesis was accepted (and vice versa). The NPV is calculated from the standard formula which is expressed as:

$$\text{NPV} = \sum_{t=1}^{T} \frac{CF_t}{(1+r)^t} \qquad \text{(Eqn 6.1)}$$

where:
NPV = net present value
T = total time of the project or investment operation given in a number of years
t = time of cash flow
r = discounting rate
CF_t = cash flow (benefits minus costs) at a given time period (t).

For the purpose of this study, a 10% discounting rate was used. The choice for the 10% rate was due to the fact that it is more easily determined and acceptable for undertaking a real value of money (European Union, 2015).

The conceptual framework

Contributions of the pastoral production system

Pastoralism contributes to the national economies of their origin through revenues collected from market places and employment to a majority of the rural population. These contributions include the direct values derived from livestock and indirect benefits which pastoralism has to the ecosystem health of drylands (Hesse and MacGregor, 2006).

Pastoralists are the custodians of dryland environments, providing services through good rangeland management, including biodiversity conservation and wildlife tourism (Oxfam, 2008). This production system makes a significant contribution to national economies and provides the majority of meat consumed in the countries where pastoralists live. It has been a livelihood support to million of households globally (King, 2000).

Ecological, sociological and anthropological studies such as Behnke and Scoones (1993), Tiffen *et al.* (1994), Fratkin and Sher-Mei Wu (1997), Fratkin (1998), Niamir-Fuller (1999), Bolling and Schulte (1999), Raynaut (2001), Moritz (2003), Oxfam (2008) and Tenaw (2016) have acknowledged that mobile pastoral production systems are the custodians of dryland environments and that they have been providing services through good rangeland management practices, including biodiversity and wildlife conservation.

The study by J. Letara, J. MacGregor and C. Hesse, (2006, unpublished work) covering the Arusha municipality concluded that pastoralism has made a significant contribution to the economy of Tanzania. They specifically estimated that 18,949 employees were directly employed in the nyama choma supply chain and about 6.6% of the population of Arusha obtained a proportion of their livelihoods from the nyama choma business supply chain.

The study conducted by the WISP in 2006 focusing on a global review of the economics of pastoralism contended that the predominant discourse on pastoralism and the environment was more concerned with the degradation caused by pastoralists rather than the services provided by pastoralism, which usually goes unvalued (WISP, 2006, p. 23). However, there are many environmental services that are provided by pastoralists and pastoralism that are poorly understood and that national accounts do not capture. Among the contributions pastoralism has made are climate change adaptation and mitigation as in the case of grasslands, where livestock grazing contributes to maintaining healthy vegetation that captures carbon, reduces erosion, maintains soils and facilitates soils' water holding capacity (WISP, 2006, p. 25).

Mdoe and Mnenwa (2007) also conducted a study on assessing the total contribution of pastoralism to the national economy of Tanzania. The findings from the study revealed that pastoralism was the hub of the meat supply chain in the country. Annual red meat production in the country then was estimated at approximately 259,800 t out of which 98% were produced from livestock bought from pastoralist areas in the country, while only 2% came from the national ranches and smallholder producers. Trading and value addition along the supply chain contributed to a large number of livelihoods, covering costs and even providing profits for each participant (Mdoe and Mnenwa, 2007). This study made a very surprising contention, that:

> If other contributions of pastoralism were included in the calculation of GDP, such as value of subsistence production, value of inputs to agriculture, value of complementary products, value of tourism services and value of market chain linkages, the contribution of the livestock subsector would possibly be higher than that of agriculture.
>
> (Mdoe and Mnenwa, 2007, p. 45)

The above contention needs to be justified using more recent data and an appropriate framework to see if it still holds water today. The findings of a recent study by the Pastoralists Indigenous Non-Governmental Organizations (PINGOS) (2016) on the socio-economic contribution of pastoralism in Tanzania revealed that pastoral systems offered dual benefits to the country, including the cultural heritage and continuity, and to the socio-economics of the micro and macro levels (URT, 2016; PINGOS, 2016). In the latter case pastoralism made huge contributions to the micro and macro-economic development of individual pastoralists, non-pastoralists and to governmental and non-governmental institutions but has clearly not been treasured (PINGOS, 2016, p. 53).

Hesse and MacGregor (2006) in their paper developed a broad framework to assess the benefits of pastoralism. They emphasized looking beyond the immediate values of livestock and their products. The framework was further reinforced by Hatfield and Davies (2006) by identifying various values that pastoralism had but that were not often attached to pastoralism's contributions.

Hesse and MacGregor (2006) framed the economic argument for pastoralism by identifying the common preconceptions and misconceptions held by many decision makers in Africa about pastoralists and their way of life. They also addressed the invisibility of the pastoral contribution to the economy by attempting to segregate the economic statistics and employing economic valuations. They partly compared the productivity of the three pastoral production systems of nomadic, sedentary and transhumance in terms of annual reproduction rates, calf mortality rates, calf weights, average lactation days and quantity of milk production for human consumption in one lactation period (Hesse and MacGregor, 2006). The findings showed that nomadic and transhumant systems were more productive than sedentary livestock raising in all the parameters.

Origins of pastoralism and pastoralists

The origin of pastoralism has been much discussed especially in older literature, influenced, unconsciously perhaps, by Marxist historical schemas (Blench, 2001). Pastoralism was seen as an evolutionary stage in human history, a phase following hunting-gathering and finally leading to sedentarization and agriculture (Blench, 2001). The uncertainties from the literature are said to be due to lack of archaeological evidence or it may also be an unconscious reflection of the contempt in which settled communities have held nomads throughout history. The ambiguities are also said to have been influenced by earlier biblical myths, such as those of Abel and Cain, which place the burden of the original sin on pastoralists (Blench, 2001).

However, the expansion of the archaeological data and historical sources in Asia has established a more complex story, especially on the review of modern theoretical developments such as that of Cribb (1991). It would appear, for example, that in most parts of the world (except in Africa) agriculture seems to have preceded pastoralism. Pastoralism is said to have evolved from surplus as individuals accumulated too many animals to be grazed around human settlements throughout the year. Additionally, when livestock keepers learnt about the relationship between ecological and environmental types and vector spread, they gradually developed the practice of seasonal movement of their herds towards disease-free zones (Blench, 2001).

On the other hand, Blench's theory is contradictory as empirical evidence indicates that pastoralism has been practised on marginal land with little rainfall in which crop cultivation could not have been possible without bringing water from long distances away (Mung'ong'o and Mwamfupe, 2003; Yanda and Mung'ong'o, 2016). Other studies such as those by Ryan *et al.* (2000) and Dudd and Evershed (1998) proposed that pastoralism originated in early Neolithic times when hunter-gatherers took up livestock keeping.

Ryan *et al.* (2000) contended that in Laikipia, Kenya, which has relatively developed pastoral Neolithic sites, Maasai pastoralists originated from a transitional shift from hunting and gathering to pastoral farming 5000 years ago. According to these authors, the pastoral evolution had been the first activity to be adopted in Laikipia like many other areas around the world where domesticated animals appear to have preceded domesticated plants in the area.

According to Cribb (1991) and also Blench (2001), the first people to have practised pastoralism appear to have been the Amorites who herded cattle, sheep, goats and donkeys in the Near East in the first half of the second millennium BC. In China, on the other hand, the chronicles of the Hsia dynasty (2205–1766 BC) recorded that Ch'iang nomads were the ancestors of modern Tibetans who were weavers of fine wools (Scarnecchia *et al.*, 1997; see also Blench, 2001).

In sub-Saharan Africa, the history and origin of pastoralism and pastoralists are described by Pliny's 'Blood and Milk in the Horn of Africa' which was written around 23–79 AD. However, pastoralism is likely to have started earlier in the African continent than these records show. Blench (2001), for example, has this to say:

> The exact origins of pastoralism can only be gauged from archaeology and in particular from careful osteometric work demonstrating the gradual divergence between the wild form of livestock and domesticated relatives. Some claims have been made for domestic cattle in Northeast Africa as early as 9000 BP although not all scholars accept these dates and more solid dates are available for 6000 BP onwards ... However, the interpretation of osteometric evidence already depend on the assumption that early herders were controlling breeding; but it seems likely that the earlier stages of pastoralism involved the management of wild animals as reindeer pastoralism does still today in some parts of the subarctic.
>
> (Blench, 2001, p. 17)

The studies also show that pastoral culture spread from the Nile Valley and North Africa, probably through the agency of the ancestors of present-day Berber populations (Blench, 2001). The pastoral production system appears clearly in the archaeological records both in the east and in South Africa between 4500 and 4000 years ago (Marshal, 2000). The exact time, routes and dates whereby pastoralists reached Southern Africa are conflicting (Bousman, 1998). However, there seems to have been pre-Iron Age pastoralism in about 2000 BP, probably initially with sheep and shortly after with cattle. In East Africa, pastoralists have a long history. Originating from Sudan, Ethiopia and Egypt they are said to have come along the Nile River and entered Tanzania through Kenya.

The above studies have attempted to present the history of pastoralism and pastoralists in different parts of the world, including Central Asia (Russia), China and Mongolia but some descriptions of these origins remain doubtful. For instance, the llama and Alpaca are conventionally described as having started domestication of animals' way back from 6000 years ago.[1]

Forms of pastoral production system

Pastoral systems take many forms and are adapted to evolving social, cultural, political and economic environments. The key factor that defines the form of pastoralism is mobility and its patterns. Therefore, pastoral production systems are referred to as **nomadic** where a high degree of mobility is practised with irregular patterns of movement, as **transhumance** where the degree of mobility is moderate and movement is back and forth between fixed locations, and **sedentary** for settled systems (Rota and Sperandini, 2009).

However, based on their dynamic nature, pastoral production systems can also be classified as pure pastoral production systems and mixed systems when people are taking part in other livelihood activities such as farming. Agropastoralists and ex-pastoralists who hang up and switch to other alternative livelihood systems are in this category (Hesse and Odhiambo, 2006; Leshan and Standslause, 2013). The Food and Agriculture Organization of the United Nations (FAO) place pastoral production systems in four broad categories: (i) nomadic; (ii) transhumance; (iii) agropastoralism; and (iv) enclosed pastoralism/ranching (FAO, 2003).

In fact, pastoral production systems are classified according to the degree of mobility, species kept, and management system used, geography and ecology. In addition, the distinction is made between developed and developing countries as livestock are kept under different conditions, which are different among countries and continents (Blench, 2001). For instance, in both Australia and North America, extensive livestock production is practised under conditions that are very different from those found elsewhere in the world, using fenced ranges and unambiguous tenure. This creates a level of investment in land and animals that is very dissimilar from that of traditional systems (Blench, 2001).

Unique characteristics of the pastoral production system

Pastoral production systems across regions and continents have certain features which are common to all pastoral communities. The main four features are almost universal throughout European, South American and African pastoral systems. These features are: (i) the use of a common pool of resources; (ii) organized and strategic herd mobility; (iii) the use of locally adapted livestock breeds; and (iv) a system built on three pillars (i.e. livestock, people and natural resources/environment).

Pastoralists are living in fragile environments with erratic rainfall patterns, which are unpredictable in space, time and magnitude. The ASALs is the home of pastoralists at all times and across all regions of the world. These zones are very hot and dry where the availability of water, pasture and salt licks are unevenly distributed over the land (Oxfam, 2008). Given this nature of their home environments, it is very risky for the pastoralists to settle in one place for a longer period and to overcome these risks so they have developed a strategic system of herd mobility.

Mobility has always been regarded as a common and unique characteristic among the pastoral communities and their production system, particularly in Africa (Niamir-Fuller, 1999). Mobility may confer many advantages to the pastoralists, including: (i) escaping livestock diseases; (ii) the ability to find pasture, water and salt licks; (iii) minimizing environmental degradation; (iv) access to diverse markets; (v) symbiotic relationships with farming communities; (vi) cultural gatherings; and (vii) avoiding the socio-political stresses of the state (Niamir-Fuller, 1999).

Mobility is also an adaptive tool that serves several aspects of livestock production simultaneously. One benefit is the provision of fodder to livestock using minimal labour and at low economic cost. Extensive livestock production, taking livestock to pasture and water, is less costly than bringing feed and water to livestock, because there are lower labour demands and lower inputs.

However, the understanding of the relevance of pastoral mobility among policy makers, ordinary citizens and researchers from different backgrounds has been changing along with the various discourses. While the perception of new drylands researchers on pastoral mobility is relatively positive, this is not new among social scientists studying pastoralists, especially anthropologists, as they have understood the viable strategy employed by pastoralists to manage risks (Nicolaisen, 1963; Dyson-Hudson, 1966; Gallais, 1967; Johnson, 1969; Niamir-Fuller, 1999). Among the policy makers, pastoral mobility is still a serious riddle as it is not well understood and, therefore, the policies are designed to force the mobile pastoralists to settle. This attempt worsens the already fragile pastoral systems and the prospects for future sustainability are seriously compromised.

Pastoral production system challenges and drivers for change

Policies and regulatory frameworks

Herders keep most of Africa's livestock in extensive systems in the arid and semi-arid lands while smallholders keep theirs in subsistence-oriented mixed crop-livestock systems. The majority of livestock owners are poor and not commercially oriented (AU-IBAR, 2015). While mobile pastoral production systems make significant contributions to the economies and environment, as indicated above, few governments are ready to tolerate mobility and many promote, knowingly or unknowingly, inadvertent policies of sedentarization (Davies and Hatfield, 2007). The mobile livestock production system does not seem to be clearly understood and has been characterized as a backward, environmentally destructive and economically unsustainable system that should be replaced with more sedentary forms of production and by other productive land use activities (WISP, 2011; Davies *et al*., 2015).

The regional and national policies and legal mechanisms over the last decade have been promoting the commercial livestock sector and transformation of mobile pastoralism. In Tanzania, government policies and regulatory frameworks are in support of the modern livestock sectors. Mobile livestock keeping is officially seen as a backward and economically inefficient system that needs to be modernized if it is to contribute to Tanzania's vision of graduating from a least developed country to a middle-income country with a semi-industrialized economy based on a modern and highly productive agricultural sector (J. Letara, J. MacGregor and C. Hesse, 2006, unpublished work). The National Livestock Policy (NLP) (URT, 2006) and National Adaptation Programme of Action (NAPA) (URT, 2007) as well as the Tanzania Livestock Modernization Initiative (TLMI) 2015–2022 are all in favour of livestock commercialization and view traditional pastoralism as making very little contribution to the national economy.

The NLP of 2006, for example, is an instrument aimed towards achieving the Tanzania Development Vision (TDV) 2025 and the National Strategy for Growth and Poverty Reduction (NSGPR) 2005–2015 in the country, particularly by improving the well-being of communities which have livelihoods based on livestock. The general objective/goal of the policy envisions that:

> By the year 2025, there should be a participatory livestock sector which to a large extent shall be commercially run, modern and sustainable, using improved and highly productive livestock to ensure food security, improved income for the household and the nation, while conserving the environment.
>
> (URT, 2006, p. 6)

The policy is commodity oriented while being aimed at modernizing the mobile pastoral production system through sedentary modes of livestock production and settling of pastoralists. The instruments for the implementation of this policy come from the Livestock Sector Development Strategy of 2010, the Livestock Sector Development Programme of 2011 and various regulatory frameworks including: (i) the Veterinary Act, 2003; (ii) the Animal Diseases Act; (iii) the Dairy Industry Act; (iv) the Meat Industry

Act; (v) the Hides and Skins and Leather Trade Act; (vi) the Animal Welfare Act; (vii) the Livestock Identification, Registration and Traceability Act; and (viii) the Grazing-lands and Animal Feed Resources Act. To a large extent, all of these instruments place less recognition on livestock mobility and livestock herders and, therefore constrain the mobile pastoral production systems in the country.

The NAPA (URT, 2007), for its part, identifies climate-change-related vulnerabilities of key economic sectors, which form the basis of the livelihood of the rural community and backbone for national development and prosperity. The overall vision of Tanzania's NAPA is to identify immediate and urgent climate change adaptation actions that are robust enough to lead to long-term sustainable development in a changing climate. It has also identified climate change adaptation activities that most effectively reduce the risks that a changing climate poses to sustainable development. One of its specific objectives is to mainstream adaptation activities into national and sectoral development policies and strategies, development goals, visions and objectives.

In identifying the immediate and urgent climate change adaptation actions, it has identified vulnerability and potential adaptation activities, which include advocating zero grazing and restricting mobility by controlling livestock movement (URT, 2007, p. 22). This means that the mobile livestock system, which is raised under traditional classification, will be forced to settle.

The TLMI (URT, 2016) is intended to support the progressive and adaptive development of a vibrant livestock sector which is responsive to growing demands and emerging commercial opportunities, and which is economical, socially and environmentally sustainable. The TLMI aims to harness the potential of the meat, dairy and poultry sectors for poverty alleviation through improvements across the value chains including improvements in security and access to grazing, supplementary feed and water, genetic potential of traditional livestock, animal husbandry and disease control, access to markets, quality and safety of livestock products, processing and value addition. It is aimed at identifying the coordinated interventions needed across sectors to support these improvements and the necessary regulatory environments to ensure safe and healthy livestock products.

The overall goal of the TLMI is to increase food and nutrition security and food safety, create employment opportunities and contribute to the national economy, social stability and sustainable environment. The purpose of the TLMI is to improve the livelihoods of traditional and smallholder livestock farmers and the contribution of the sector to the national economy. The TLMI aims to implement the NLP of 2006 in the context of The National Five Year Development Plan, the Agricultural Sector Development Programme, Kilimo Kwanza, the NSGPR, the Rural Development Strategy (RDS), the Comprehensive Africa Agriculture Development Programme (CAADP) and TDV 2025. The interactions between these broad national development instruments may have a detrimental implication on mobility and thereby to mobile pastoral production systems, as the mobile pastoralists will be forced to settle.

Climate change

Climate change is now a global issue posing challenges to the very survival of humankind and sustainable development. The adverse impacts of climate change are now evident almost everywhere. Climate change poses a serious risk to poverty reduction efforts and threatens to undo decades of development efforts. It is widely accepted that the impacts of climate change are, and will continue to be, more pronounced in poor countries. The developing countries have contributed insignificantly to climate change, but are the ones most affected and they cannot cope with the impacts (URT, 2007).

Climate change adversity is a reality and its effects are far reaching within the pastoral area. As a result, animal feeds for livestock have deteriorated, and hence productivity is compromised, and herd mortality has increased. Apart from climate change, empirical and scholarly evidence shows that other stressors including land grabbing and weak livestock market access that weaken their adaptive capacity have pressured pastoralists and their resilience to withstand climate change stresses. Drought is a normal feature of pastoral systems in Africa. However, in the second half of the 20th century, droughts have occurred with greater frequency, for instance, those in the 1960s, 1970s, 1980s, 1990s and more recently 2000s (Ekaya, 2001).

Historically, mobile pastoralists adapted to conditions of famine and drought or low and erratic rainfall by physical mobility, diversification and dispersion of their herds. Despite these challenges, traditional livestock keepers have been surprisingly resilient as pastoralists have shown a wide variety of adaptation strategies to climate change, including periodic oscillation between pastoralism and farming and more recently wage labour (Little, 1992; Fratkin and Sher-Mei Wu, 1997; Fratkin, 2001).

Currently, there are ongoing debates as to whether pastoralists will survive climate change impacts. Two opinions prevail about the impacts of climate change on the pastoral production system. Some experts argue that pastoralists will be among the first groups to lose their livelihoods as rangelands and water points dry up. Others insist that pastoralists are the best equipped to adapt to climate change since their livelihood strategies are honed to respond to scarce and variable natural resources and to cope with uncertain ecological conditions (WISP, 2007; Oxfam, 2008).

Population growth

The countries with more mobile pastoralists such as Kenya, Tanzania and Ethiopia continue to have among the world's highest population growth rates (2.7%, 3.2% and 4.6% annual increase, respectively, in 2015).

However, ironically in the above countries, pastoralists' families have the lowest population growth rates. The rapid population growth has affected rural and urban areas, causing crop and livestock farmers in high rainfall areas to move to less productive areas in the drylands (UNCCD, 2015). Pastoralists in the drylands have also increasingly adopted farm cultivation, leading to loss of pasture and water resources for pastoral production.

In arid lands of North Kenya, for example, crop agriculture is possible only in isolated pockets in the highlands. Human and livestock population growth has led to competition with pastoral neighbours for pasture and water (Fratkin, 2001). It is also the case in Tanzania where mobile pastoralists are increasingly adopting farming activities as a means to diversify their livelihoods. Among the Maasai of Simanjiro, subsistence agriculture and other income-generating activities have been adopted to give the communities another option to survive. For instance, they have been engaging themselves in agriculture, food vending kiosks (*magenge*), shops, charcoal burning, guest-houses and small mining brokering (Yanda and William, 2010).

The ever-increasing human and livestock populations, and the expanding farming activities in the country with finite land and other natural resources have placed increasing pressure on land at different levels. When scarcity of land combines with the noticeable effects of climate change, it means that the mobile pastoralists' precious livelihoods will be at risk of being lost forever (PINGOS, 2016).

Land use conflicts

In particular, pastoral areas have been more affected by conflicts emanating from land grabbing by the government through expansion of areas under conservation, including national parks, forest reserves, game-controlled areas and hunting blocks, on the one hand, and by farmers through expansion of farming activities, on the other. It is also claimed that mobile pastoralists are causing conflicts due to their behaviour of moving from one place to another driven by push-and-pull factors, which are connected to access of resources in some restricted areas or areas that are already occupied by other land users. The ongoing land use conflicts create a hard situation for mobile pastoralists in accessing important livestock resources such as pasture, water, salt licks and disease-free grazing zones.

Outmigration of poor pastoralists

The former nomadic pastoralists are currently migrating to urban and peri-urban centres seeking livelihood diversity or alternative livelihood options. Seeking alternative livelihoods by former mobile pastoralists is not a new phenomenon in East Africa, as many nomadic livestock-keeping people have historically utilized ties with foraging, farming and recently urban communities in times of drought and conflicts (Fratkin *et al.*, 2006).

However, the settling of former pastoralists has increased dramatically in the past half century, driven mainly by impoverishment and stock loss due to constrained mobility, severe droughts, cattle raiding and political instability while simultaneously attracted to the benefits of a settled life, including human security. Former pastoralists are increasingly moving to urban or peri-urban settings to seek new livelihoods to survive the challenges of the monetarized economy. Formal education has been another driver for young people to move to urban areas to acquire the primary benefits of communities whose parents or relatives have pursued employment in government, business and non-government organizations located in towns (McPeak and Little, 2005).

Today, mobile pastoralists are under intense pressure to adopt new options, including migration to famine relief centres, urban migration for wage labour, and widespread adoption of cultivation (Fratkin, 2001). Will these options ultimately sedentarize former mobile pastoralists altogether?

Commoditization and loss of common pool resources

Since independence, some pastoralist countries and regions moved away from recognizing communal land tenure in favour of individual tenure rights. In pastoral Maasai regions of Kenya, for example, the government established 'group ranches' in the 1960s and subsequently promoted private and individual land titles since the 1980s, leading to a scramble for land similar to the American West in the 19th century. Kenyan and Tanzanian pastoralists have also lost former grazing lands to national game parks, including the Amboseli, Masai Mara, Tsavo and Samburu Parks in Kenya and the Serengeti, Ngorongoro Conservation Area (NCA), Tarangire and Mkomazi in Tanzania. While pastoralists live in more arid and less populated conditions, they are experiencing land crowding due to the privatization of rangelands into individual land parcels.

The common property regime is a key characteristic which enables mobile pastoralists to escape the risks of environmentally changing conditions due to climate variability and change as well as disease outbreaks. The loss of communal grazing land will expose mobile pastoralists to a high risk of losses during times of hard conditions or scarcity and thereby they may be forced to settle.

Future of the Mobile Pastoral Production System

Studies conducted around the 1990s and more recently have raised a debate about the future of mobile pastoralism, especially in Africa. Some researchers are pessimistic while others are optimistic as to the future of mobile pastoral production systems. This is due to the pressures on the African pastoral systems as a result of population growth and expansion of agriculture into grazing lands. The pressures on this production system are also linked to the expansion of conservation areas into pastoralists' grazing land (Fratkin and Sher-Mei Wu, 1997; Homewood, 2008 as cited by Galvin, 2009).

The study conducted by Spencer in 1998 contended that for East Africa where pastoralists lived and in which economic development exacerbated the problem of population growth, pastoralists would be confined to the most marginal lands and this would lead to increasing pressure on natural resources (Spencer, 1998). 'Pastoralism is doomed to extinction as a way of life by increasing pressure of population growth' (Spencer, 1998).

Sandford (2006) argued that there was a fundamental imbalance in the Horn of Africa where there were too many people with too few livestock. Sandford (2006) also pointed to a growing imbalance between human, livestock and natural resources which he said would lead to greater economic inequalities and impoverishment of the East African pastoral societies. 'The pastoral populations are growing steadily while there is no increase in livestock numbers due to decreasing grazing lands as a result of the expansion of cultivation and conservation' (Sandford, 2006).

On the other hand, the possibilities of increasing livestock productivity were limited, and so were the possibilities of improving household incomes through livestock marketing or livelihood diversification.

Agriculture systems scholars are pessimistic about the future of pastoral production systems. Scholars such as Mortimore and Adams (1998), who conducted a comparative study in four villages in northern Nigeria, based on their findings contended that agricultural populations would continue to increase and so would livestock populations. These increases would simultaneously encourage agricultural expansion into grazing lands, which means that livestock would no longer subsist on natural forages alone. The authors suggested the solution to be an intensification of the production of livestock and to integrate it into agricultural systems by feeding cattle with farm-produced fodder (Mortimore and Adams, 1998). However, the intensification and integration route is available only to agropastoralists who already have title to land; mobile pastoralists without land titles will have to move elsewhere to increasingly scarce pastoral areas (Mortimore and Adams, 1998).

Other studies are optimistic about the future of mobile pastoral production systems as they argue that the system has proved to be adaptive to harsh environments for many years. The comparative analysis of seven case studies by Moritz (2008) revealed that West African pastoral systems were not in crisis, despite the high population growth and densities. As evidence of successful adaptation, Moritz (2008) found out that pastoral production systems continue to exist and adapt to changing conditions and provide increases of agropastoral production and market values of their by-products. The results from this comparative study suggest an alternative to the trajectories described by Sandford (2006) and others such as Moritz (2012).

Other debates around the future of pastoralism are whether they will survive climate change impacts or not. The two opinions prevail about the impacts of climate change on the pastoral production system. Some experts argue that pastoralists will be among the first groups to lose their livelihoods as rangelands and water points dry up. Others insist that pastoralists are the best equipped to adapt to climate change since their livelihood strategies are honed to respond to scarce and variable natural resources and to cope with uncertainty in the ecologically changing conditions (WISP, 2007; Oxfam, 2008).

Materials and Methods

Description of the study areas

This study was conducted in Makame village of Kiteto District and Dofa village of Karatu District. Makame village represents a mobile pastoral production system while Dofa village represents a sedentary production system. The village profiles are as follows.

Makame village

The specific study area for the mobile pastoral production system in Kiteto District was Makame village located approximately between 80 km and 100 km from Kibaya town. Economically, pastoralism forms a major part of the economic activity of Makame village whereby 97% are mobile pastoralists and only 3% of the villagers engage in growing crops of maize, beans, sesame (*simsim* in Kiswahili) and sorghum (WWF, 2012).

Makame village is located in Kiteto District (Fig. 6.1) which is one of the six districts of Manyara Region of Tanzania with its headquarters based in Kibaya township. The district borders Dodoma Region to the

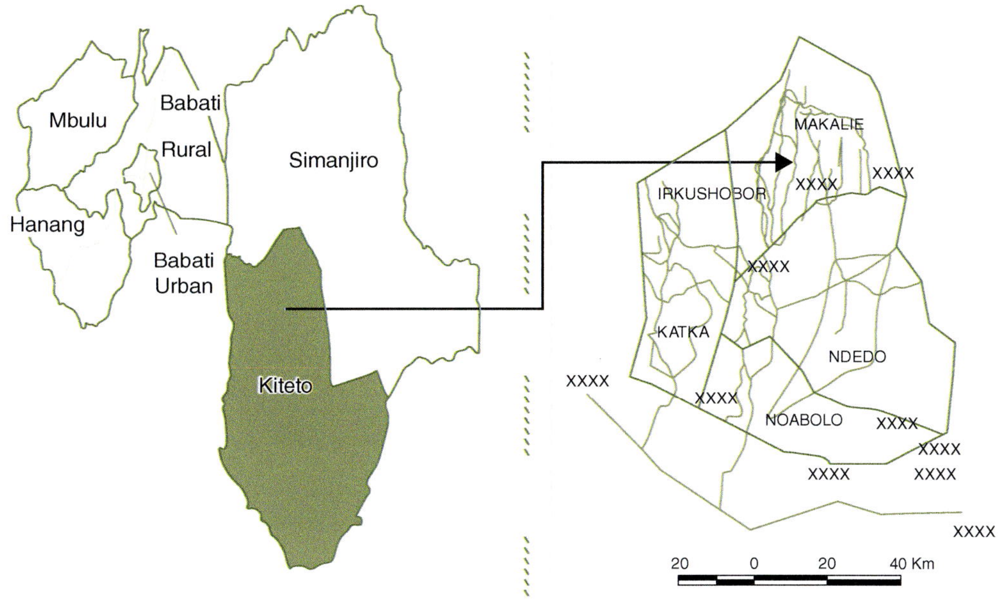

Fig. 6.1. Map showing Kiteto District and Makame village. (From WWF, 2012.)

south and west, Tanga Region to the east, and Simanjiro District to the north. The national statistics from the 2012 census show that the district had a population of 244,669 inhabitants of whom 120,233 were males and 124,436 were females (URT/NBS, 2013). The majority of dwellers (90.9%) were rurally based and the remaining 9.1% were living in urban areas. The majority of the district's settlers are aged between 15 years and 64 years and the second largest age group is 0–14 years old (URT/NBS, 2013).

Administratively, the district had 19 wards during the 2012 census but it has since been expanding and in 2015 it had 23 wards. Currently, the wards are Makame, Matui, Dosidosi, Engusero, Kibaya, Lengatei, Bwagamoyo, Dongo, Ndedo, Chapakazi, Songambele, Sunya, Loolera, Magungu Njoro and Partimbo. The major economic activities of the district are livestock keeping, crop production followed by mining activities. According to the district statistics, Kiteto has a population of 141,472 chickens, 161,612 sheep, 448,106 goats and 484,833 cattle (KDC, 2017). These figures present an exceptional investment opportunity for processing of livestock products and by-products such as hides/skins, meat, milk and horns.

Dofa village

Data for the second village was collected from Dofa village of Qurus ward located 10 km from Karatu town and 4 km from the Lodoare (Ngorongoro Conservation Area Authority) entrance gate. It is a newly established village which broke away from Njiapanda ward and village in 2015. The majority of the population here are sedentary agropastoralists who keep livestock and farm crops.

Dofa village is located in Karatu District (Fig. 6.2) one of seven districts of Arusha Region which was established in 2000. It is located 150 km from Arusha city. Geographically, it borders Iramba and Meatu districts to the north-west, Mbulu District to the north and Monduli and Babati districts to the south-east. Administratively, the district is divided into four divisions, which are Eyasi, Karatu, Mbulumbulu and Endabash that comprise 14 wards and 58 registered villages.

The official data from the National Bureau of Statistics (NBS) (URT/NBS, 2013) shows that the district has a total population

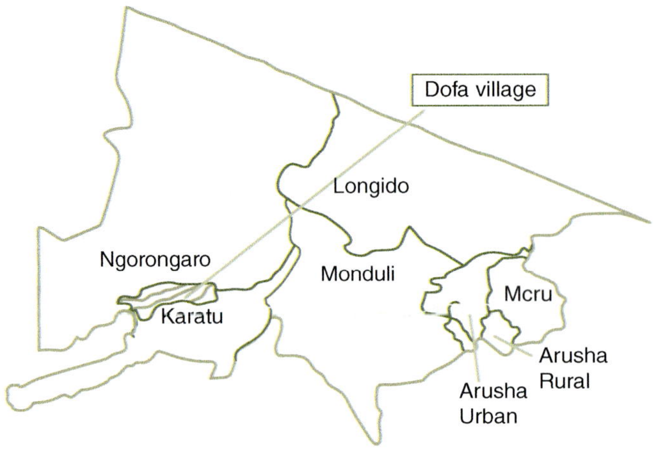

Fig. 6.2. A map of the Arusha Region showing the study areas. (Modified from Macabe5387 (CC BY-SA 4.0) (https://commons.wikimedia.org/w/index.php?curid=48416628Wikimedia Commons).)

of 230,166 residents of whom 117,769 are males and 112,397 are females and the total number of households is 45,130 with an average household size of 5.1 persons/household. The major economic activities in Karatu District are crop cultivation and keeping livestock (more than 85%) followed by petty business and cultural tourism activities. The district has arable land for crop cultivation covering 102,573 ha where both cash and food crops are cultivated (KDC, 2017). Crops grown include coffee, onions, wheat, barley, pigeon peas and sunflower as cash crops. Food crops comprise maize, beans, finger millet and sorghum.

The second largest economic activity is livestock keeping, with a total number of 271,885 cattle, 373,588 goats and sheep, 12,816 pigs, 165,887 chickens and 46,766 donkeys being kept in the district. The main livestock product sold is milk, and records show that 97,879,190 l were produced and sold in 2015 alone from both dairy and indigenous cattle. The average yield of milk produced varies from species to species and from time to time. But the records show that 8–10 l of milk/cow/day are produced from the dairy cattle and 1.5–3.0 l/cow/day from the indigenous herd (URT, 2017).

Study design

The study design is a road map on how to conduct a study with maximum control over factors that may interfere with the validity of the findings. Kothari (2004) defined a research design to mean arrangements of conditions for collection and analysis of data in a manner that aims to combine relevance to the research purpose with economy in procedure. Kothari further explained that the research design is a conceptual structure within which research is conducted and therefore constitutes the blue print for collection, measurements and analysing the data (Kothari, 2004). Polit *et al.* (2001) defined study design as the researcher's overall plan for answering the research questions or testing the research hypothesis. The study design is very important to facilitate smooth conducting of the study and therefore makes research more efficient in terms of money, effort and time while yielding a maximum desire of conducting such a study (Kothari, 2004).

For the purpose of meeting the objectives of this study, the study employed a quantitative approach. Kothari (2004) defined quantitative research design to mean the one which is based on the measurement of quantity or amount. The comparative CBA carried out for this study used monetary values to rationalize the argument that sedentary livestock keeping was more viable compared to mobile livestock keeping and vice versa.

To arrive at true costs and benefits of each production system, the livestock unit (LSU) statistical approach was used to make estimates. Furthermore, the study was conducted through a household survey in the two villages to obtain quantitative data to test the predetermined hypotheses introduced in the earlier sections of this chapter. Moreover, a cross-sectional approach was used to conduct this study with the data being collected only once in time.

Study population, sample size and sampling procedures

The study population is the total number that represents the universe of units from which a study sample is selected. The study aim was to compare mobile and sedentary pastoral production systems in terms of their costs and benefits. It was necessary to choose one mobile community village and one sedentary community village in which Makame and Dofa suited the requirements of the study. Therefore, for the purpose of those prerequisites, the study population was the total number of households in the two villages of Makame and Dofa where this study was conducted.

From the total number of households in each village, the study samples were drawn randomly from a sampling frame of 438 mobile pastoralists' and 701 sedentary heads of households. As the purpose of this study was to compare a mobile pastoral production system and sedentary production system in terms of costs and benefits, the selection of samples was an important factor in order to achieve representative results. Babbie and Mouton (2007), Field (2009) and Welman *et al.* (2005) contended that a sample should be selected in such a way as to be representative of the population from which it is drawn. To make a balance, equal samples were collected from both production systems and 60 households were interviewed in each village. The combined sample size for the two villages (120 households in total, 60 from each village) adhered to the recommended size to enable statistical analysis to be carried out and this was considered to be representative while being economical in terms of time and financial resources. Kothari (2004) contended that a sample size above 100 is statically significant for drawing conclusions as it is sufficient to make inferences and has statistical power.

Babbie (1990) defined sampling as the selection of a number of study units from a definite study population. Kothari (2004) contended that 10–15% sample size from a rural setting where the population is sparsely distributed is a representative sample. For the objectives of this study, non-probability sampling techniques were used to obtain study areas at the district level where this study was conducted. This is due to the fact that this study required a purely mobile pastoral community village on the one hand and a sedentary community village on the other. Based on the above criteria, Makame village was selected to represent the mobile pastoral community and Dofa to represent the sedentary community. After obtaining the study villages from the districts, probability sampling techniques were used to obtain a representative sample size to be studied at the household level. In Makame village, out of 438 households a total number of 60 heads of households (13.7%) were selected for the study. In Dofa village the study took 60 heads of households (8.5%) from the total number of 701 households in the village to be studied (Table 6.1).

This rather low percentage was deemed commendable according to Babbie (1990) as Dofa is a peri-urban area. The combination

Table 6.1. Distribution of sample households by village. (From survey data, 2017.)

	Population				
Village name	Number of households	Number selected	Percentage	Male	Female
Makame (Kiteto District)	438	60	13.7%	39	21
Dofa (Karatu District)	701	60	8.5%	35	25
Total	1139	120	10.5%	74	46

of the samples studied from both villages makes a total of 120 heads of households which is 10.5% of the total number of households, as recommended by both Babbie (1990) and Kothari (2004). A combination of probability and non-probability sampling techniques at district, village and household levels was used to make sure that the required data were collected without bias. The whole process was supported by village executive officers and village chairpersons of the villages concerned.

Data sources and data collection methods

For the purpose of this study, both secondary and primary data were collected. However, the study used mainly primary data to draw conclusions whether the argument that sedentary livestock keeping was more viable compared to mobile livestock keeping and vice versa. The data collection process was as follows.

Secondary data

Secondary data was collected through the review of the available international and national documents and case studies. Different published and unpublished research findings, reports and articles which highlight the contribution of mobile pastoralism to national economies compared to sedentary livestock keeping were reviewed. Moreover, national level livestock statistics, agricultural census data, trade and export of live animals and livestock products were collated from Ministry of Agriculture, Livestock and Fisheries (MALF) and NBS websites. Secondary data was used to complement the field survey data.

Primary data

Primary data was collected from the field using a questionnaire survey and field observation. The survey data was collected using structured questionnaires that were administered to the 120 respondents from Makame and Dofa villages. The questionnaire was used to collect the data in a timely and cost-effective manner, while enabling clarification of some issues that were relevant to the study. The data collected could be quantified.

On the other hand, field observation was used to clarify some costs and benefits of the target production systems according to the views of respondents in the study area. More specifically, observation in this study was employed to validate responses obtained from the respondents during the survey sessions. Field observations were used, for example, to visualize water sources (as shown in Fig. 6.3), use of feed supplements, the purposes of selling livestock and the size of cattle sheds to prove the number of animals kept in a household's herd.

Data processing, analysis and presentation

The data collected during the household survey using questionnaires was processed, analysed and presented using computer-based software packages. The data collected in the field was mainly quantitative and as such the analysis was done using the Statistical Package for Social Sciences (SPSS) version 20 and 2007 Microsoft Excel spreadsheets. The small set of qualitative data from open-ended questions or clarified issues of interest for this study were described through contents analysis.

The analysed data was presented in a suitable form such as charts, tables, pictorials, graphs and contents descriptions. The presented data was interpreted in order to provide meaning and implications while making the comparative cost–benefits analysis of the two pastoral production systems. The interpretations were provided in the form of percentages, financial numbers, explanations and ratios.

Results and Discussion

Demographic and socio-economic characteristics of respondents

This study collected data from 60 respondents from mobile pastoralists' households in Makame village (Kiteto District) and 60 households from a sedentary community of

Fig. 6.3. Maasai pastoralist women walking towards their homes carrying buckets of water on their backs. (From survey data, 2017.)

Dofa village (Karatu District). The demographic and socio-economic characteristics of the respondents covered by the study were gender, age, marital status, educational status, occupation and years that a respondent had lived in the particular village, which are discussed as follows.

Gender and age of respondents

A significant percentage of the respondents interviewed were aged above 20 years. Out of the 120 respondents, for example, 15% from the mobile pastoralists' village were aged below 30 years while in the sedentary community the respondents below the age of 30 were only 8%. In both villages, more than one-fifth of the respondents were from the age group 51–60 years (20% from the mobile pastoralists and 28% from the sedentary community, respectively) while a small percentage (3%) came from the age group >70 years in both communities (Table 6.2). Analysis of respondents in the two communities by gender further shows that, on average, 38.3% were female heads of households or their representatives (data not shown). The proportion of female heads of households disaggregated by production system was 35.0% for mobile pastoralists and 58.0% for the sedentary community, respectively (Table 6.2).

Marital status of the respondents

The marital status of respondents in this study showed some interesting patterns. Findings revealed that there was no history of divorce or widowhood among the mobile pastoralists group. Out of 60 heads of household respondents from the mobile pastoralists only 2% were single while 98% were married (Table 6.2). The cultural and historical reasons for this phenomenon are discussed in detail below. Whereas among the sedentary community heads of households, 8% were single, 77% were married, 8% were widowed and the remaining 7% were divorced or separated from their spouses (Table 6.2, see also Fig. 6.4).

The marriage problems among sedentary communities have been reported previously by researchers such as Fratkin (2001), and Fratkin *et al.* (2006), especially separation of couples and the decision to

Table 6.2. Summary of respondents' demographic and socio-economic characteristics of the two study villages (mobile pastoralists and sedentary community). (From survey data, 2017.)[a]

Category	Number		Percentage	
	Mobile pastoralists	Sedentary community	Mobile pastoralists	Sedentary community
Age				
<30	9	5	15	8
30–40	25	16	42	27
41–50	10	14	17	23
51–60	12	17	20	28
61–70	2	6	3	10
>70	2	2	3	3
Gender				
Male	39	25	65	42
Female	21	35	35	58
Marital status				
Single	1	5	2	8
Married	59	46	98	77
Widowed	0	5	0	8
Divorced	0	4	0	7
Occupation				
Mobile pastoralist	60	NA	100	NA
Sedentary agropastoralist	NA	60	NA	100
Education level				
None	60	10	100	17
Primary education	0	42	0	70
Secondary education	0	4	0	7
Post-secondary	0	3	0	5
Other	0	1	0	2
Years of living in the village				
<5	8	0	13	0
5–10	7	5	12	8
10–30	30	9	50	15
30–50	5	46	8	77
>50	10	0	17	0
Total	60	60	100	100

[a]NA, Not applicable.

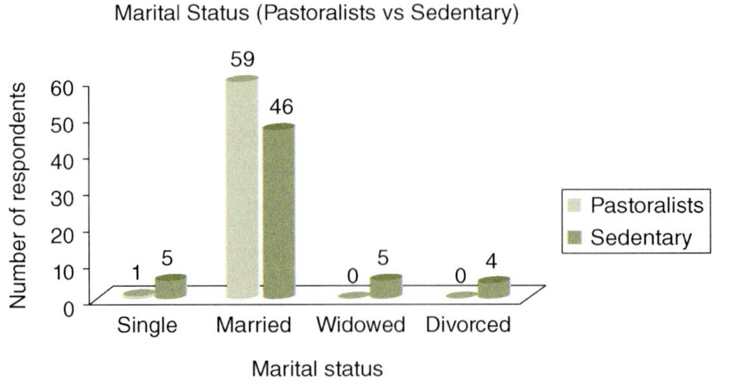

Fig. 6.4. Distribution of respondents by marital status (%). (From survey data, 2017.)

remain single. Some other studies such as the one by Coast (2005) observed the complex nature of marriages which varied between cultural settings and ethnic groups.

Marriage breakdown among sedentary communities is associated with the frequent interaction with modern society through farming and other business cooperation (Naoki, 2006). Marriage breakdown among mobile pastoral communities was not commonly observed. This was due to social, cultural and institutional stability (Galvin, 2008). The old anthropological studies show that men and women above the age of 30 among the Maasai mobile pastoralists were rarely single due to arrangements made by elders according to their customs and traditions (Thomson, 1885; Taraya, 2004). On the other hand, widows were adopted or inherited by close family or ethnic members among the Maasai mobile pastoralists to raise children for them. In non-industrialized societies, marriage is more often a relationship between groups rather than individuals and is arranged on a practical basis rather than being based on a romantic one. It is this way that polygamy is common among the Maasai mobile pastoralists (Minnesota State University, n.d.). The differences in marriage status between the two communities in the present study appear to be explained by evidence reported in these earlier studies.

Educational levels of respondents

Data from this study revealed great variation between the two study villages. In Makame village, which is dominated by mobile pastoralists, access to formal education was almost zero. For instance, out of 60 heads of households interviewed during the survey, no one had received formal education (Table 6.2). This was the case even for those who were under 30 years of age. Meanwhile, the situation was different for those from the sedentary community who had some access to formal education. The survey data in Table 6.2 shows that out of 60 respondents, only 17% had no formal education, 70% had received primary education, 7% had received secondary, and 5% had attended some tertiary institution (colleges and universities) (see also Fig. 6.5) Moreover, some 2% had received some other level of education (e.g. adult learning).

The same scenario was observed in the study by Mung'ong'o and Mwamfupe (2003) whose findings showed that the majority of mobile pastoralists in Kilosa District had no formal education at all. The data also corroborates the findings of the study by Yanda and William (2010) on livelihoods diversification and implications on food security and poverty levels in Maasai plains using a case of Simanjiro District which showed that of the respondents interviewed, 46.6% had not received any formal education and instead they relied on traditional teachings/knowledge. It should, however, be noted that Simanjiro District in recent years has been transforming to a sedentary community due to a majority of households engaging in agriculture and businesses such as brokering of Tanzanite gems (Yanda and William, 2010).

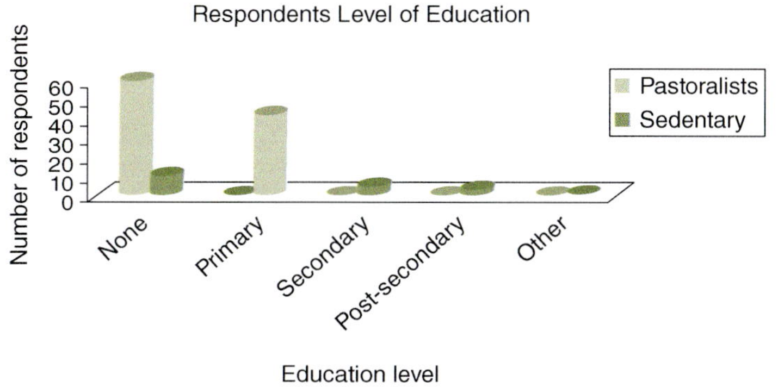

Fig. 6.5. Respondents' levels of education. (From survey data, 2017.)

This may not reflect the clear situation of mobile pastoralists' access to education. Furthermore, an earlier study by Taraya (2004) revealed that although the mobile Maasai pastoralists of Kenya and Tanzania had no formal education they were able to understand contemporary issues of evolving traditions, culture and values over time.

Occupation and experience of the respondents in the villages

The experience of the respondents in the area can be measured by the number of years that respondents stayed in a particular village. The data from this study show that respondents in both communities have a long history in their home villages but with slight differences between the two production systems. Generally, the findings indicate that the mobile pastoralists had a shorter history in their village than the sedentary ones. For example, while some 13% of the respondents reported to having been in the village for a period of less than 5 years, only 17% of heads of households had stayed in their village of domicile for more than 50 years (Table 6.2). In comparison only 8% of the respondents in the sedentary community of Dofa had stayed there for a moderate time of 5–10 years, and the majority (77%) had stayed there for between 30 and 50 years (Table 6.2).

Types of livestock kept and herd sizes

Data from the study show that people keep different types of livestock in both villages. The species targeted by this study were those common to both production systems, which included cattle, goats and sheep. From Makame village, the sample of 60 heads of households participated in this study revealed to have a total of 6506 herds of which 3462 are large stock (cattle) and 3044 small stock (sheep and goat). In Dofa village, 60 heads of households reported to have a total of 793 herds of which 216 were large stock (cattle) and 577 small stock (goats and sheep). Despite the fact that respondents kept different types and numbers of livestock species, a very small number of them in both study villages kept only goats or sheep. Furthermore, the majority of respondents (93%) from the mobile pastoralists kept all three types of livestock. This was against only 40% of those from the sedentary community. The highest proportion of agropastoralists (58%) kept cattle and goats and 47% kept cattle and sheep.

In comparison to the sedentary production system, mobile pastoralists kept more sheep, cattle and goats all together. From the respondents, there were no pastoralists keeping only sheep only under the mobile system. With regards to the number of livestock kept and owned by households generally, this ranged from 0 (zero) to over 1000. However, under the sedentary production system, the number of livestock kept was fewer than 500. A closer look at the findings reveals that although 75% of the respondents in general kept livestock ranging between 0 (zero) and 20, only 38% of the respondents under the mobile production system represented this category. Moreover, while no household under the sedentary production system owned more than 1000 head of livestock, 3% of mobile pastoralists indicated they owned livestock populations of above 1000 animals (Fig. 6.6).

LIVESTOCK GRAZING AREAS AND WATER SOURCES. In this study, both communities employed different techniques to make their livestock have access to resources, including pasture/forage and water.

The research intended to find out whether the community graze their livestock under an extensive production system (i.e. livestock are grazed on open rangelands and mobility determines access to resources), indoors (the livestock keepers collect pastures and other resources and feed the livestock in the shed), a semi-intensive production system (a combination of grazing livestock sometimes on open rangeland and at other times livestock keepers feed them in the shed) or had other means of grazing their herds. The findings revealed that 100% of the mobile pastoralists grazed their livestock under extensive systems, while 73% of respondents from the sedentary community fed their livestock indoors and only 27% used a semi-intensive grazing

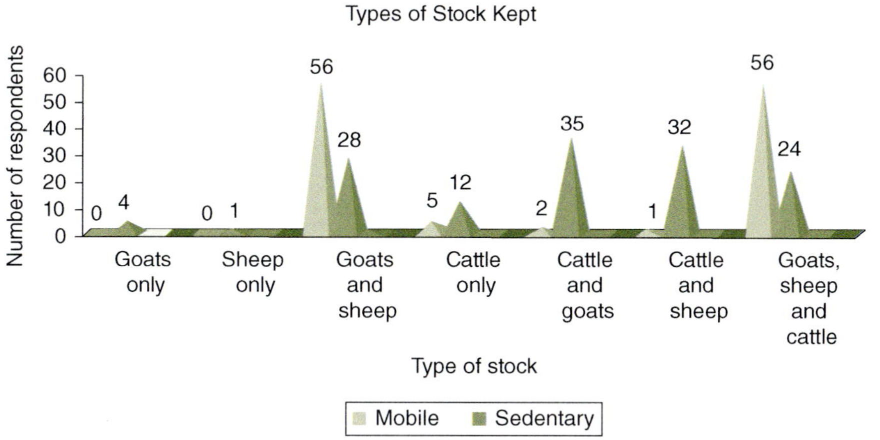

Fig. 6.6. Types and number of livestock owned by households/*bomas*. (From survey data, 2017.)

system. There were no records of using any other methods in both communities apart from those outlined above.

Water for livestock was obtained from different sources, ranging from modern to traditional watering points. Under the mobile pastoral system, all (100%) respondents interviewed revealed that water used for watering their livestock was obtained from local traditionally dug wells (Fig. 6.7).

The situation was different in the sedentary community, the majority (73%) of whom obtained water from other sources, including individual boreholes and neighbours' boreholes. A moderate number of respondents (27%) obtained water from Dofa village boreholes. Table 6.2 provides a summary of results generated during this study.

LIVESTOCK MORTALITY RATES, CAUSES AND AVERAGE LOST VALUES. The levels of livestock mortality were measured on a scale of high, medium or low in both production systems. The data from the survey shows that livestock mortality under the mobile pastoral production system was very high as more than half of respondents (57%) reported experiencing high livestock mortality. The same number of respondents reported the major cause of livestock die-off being severe droughts. The remaining 43% reported facing medium livestock mortality rates attributed to a combination of diseases and droughts. Specifically, the respondents reported common diseases to be anthrax, East Coast fever (ECF), foot and mouth diseases (FMD), contagious bovine pleuropneumonia (CBPP) and the hepatic disease called *olodwa* or *emonywa* in Maasai language.

Due to the high livestock mortality rates caused by droughts in Makame, the mobile pastoral production system seemed to suffer huge losses in terms of average economic value compared with the sedentary production system. The findings revealed, for example, that the combined heads of households (i.e. mobile and sedentary) lost up to TSh17 million/year as a result of livestock mortality attributed mainly to droughts. The major income loss per year for the mobile pastoral system, however, was reported to range between TSh10 million/year and TSh20 million/year by about 47% of the respondents. It was also evident that a significant number of respondents (about 42%) lost up to TSh2 million/year in the same village due to livestock mortality.

The sedentary community, on the other hand, reported insignificant livestock mortality rates as only 2% reported to have observed high mortality rates, about 15% reported low mortality rates and the remaining 83% said they had experienced no livestock mortality at all for the last 3 years. The reason given by the respondents for the low livestock mortality is frequent vaccination and dipping for tickborne disease control. The death of livestock in Dofa was

Fig. 6.7. Pastoralist women from Makame village fetching water from a local well and transporting it using donkeys. (From survey data, 2017.)

mainly accidental, reportedly attributable to livestock feeding on poisonous plants called *fokorai* in the local language. Other causes of livestock mortality could be diseases such as anthrax, ECF, trypanosomiasis (*ndorobo*), diarrhoea, skin cancer, eye infections, complications during delivery of young and attacks by wild animals. In some cases, livestock died due to unknown causes and the livestock were not examined post-mortem by the livestock extension officers. However, such occurrences were not frequent in Dofa.

Because of such low mortality rates in the village, there has been a low or insignificant income loss due to livestock deaths. The data show in this respect that on average 58% of the respondents reported to have lost not more than TSh2 million/year. Furthermore, the remaining 42% reported having never experienced livestock die-off for the last 3 years.

Under the mobile pastoral production system, mobility is a very important tool for coping with or adapting to changing climatic conditions. In Makame village respondents reported national policies and village land laws to be stressors for their adapting capacity to climate change. The establishment of Wildlife Management Areas (WMAs) (Makame WMA) and Mkungunero Game Reserve blocked the livestock migratory pathways used for withstanding extreme climate change events such as droughts. This created a loss of key resources that pastoralists used to cope with the changing environment and thereby resulted in high livestock mortality over the past few years.

Tanzania has a number of policies which can either help or disempower pastoralists in their efforts to deal with climatic oscillations in arid and semi-arid lands. The land policy, livestock policy, forest policy, mining policy and tourism policy, as well as their statutes, are creating detrimental impacts on livestock routes that are critical for mobility, which is a crucial coping mechanism for mobile pastoralists.

Earlier studies like those by Oba (2011) and Thornton (2009) confirm that the impacts of droughts on the mobile pastoral production system have historically been traumatic due to: (i) the increase in water stress (especially at lower latitudes); (ii) the decrease in quality and quantity of feeds; (iii) the increase in livestock diseases; and (iv) emerging poisonous vegetation. These challenges have had implications for metabolic behavior of livestock, and sometimes have resulted in diarrhoea and low productivity of the animals. This has been the case particularly when traditional adaptation and coping mechanisms have proved ineffective.

The study by Koromo (2016) also reported high livestock mortality rates in Longido District that was attributed to the severe droughts of 2009/10. The study found that about 93% of households in the district lost their livestock due to those droughts and effectively destocked 60% of the district. It was also reported through the media the loss of livestock due to severe drought in many parts of the country. The *Mtanzania* newspaper on 20 May 2017 reported that Tanzania had lost 132,329 livestock of which 102,987 were cattle, 14,881 were goats, 13,815 were sheep and 646 were donkeys (*Mtanzania*, 2017). This only reflects statistics of reported livestock mortality but the unreported losses may be higher than what was captured by the media.

Livestock markets and sales

Findings from the present study revealed that the communities under both production systems reported selling livestock for different purposes. About 85% of the heads of households interviewed in the sedentary community reported sales of some livestock while 100% of the mobile pastoralists reported on sales of livestock. The purpose of selling livestock varied between the two communities but mainly it was for buying food, paying school expenses for their children, purchasing livestock drugs, maintaining family health and meeting other family needs. The other family needs according to the respondents were for facilitating petty business, construction of modern houses, buying more livestock (bulls or heifers) as well as to control the number of livestock to reach manageable proportions. However, people who had recently joined the sedentary production system reported selling their livestock cautiously, mostly waiting for prices to improve.

Generally, 98% of the respondents from the mobile pastoral community sold their livestock to buy food, especially grains. On the other hand, about 32% of the sedentary community sold their livestock to finance educational needs for their children against 12% in the mobile pastoral community (Table 6.3). Also, 17% of the respondents in the mobile pastoral community sold their livestock to maintain their family health while there was no evidence from the mobile community of doing the same. Conversely, 77% of mobile pastoralists sold their livestock to buy animal drugs to maintain the health of their herds with only 5% of the sedentary community doing the same. The sale of livestock to meet other family needs, including facilitating petty business, construction of modern houses, buying more livestock and control of bovine populations stood at 5% and 48% for the mobile pastoralists and the sedentary one, respectively.

Furthermore, the livestock selling stations or points initially asked about were sales to neighbours, to local auctions, to cross border markets and others. The survey findings in this respect, however, showed that from the sedentary community about 8% sold their livestock to neighbours while 77% did so to local auctions. Some 15% of the respondents did not sell their livestock at all, and no one sold to cross border markets or other channels. The situation was, nevertheless, different for the mobile pastoralists. It was generally reported that 100% of households sold their livestock only through local auction markets (Table 6.3). The respondents pointed out the absence of nearby cross-border market outlets to sell their livestock left local auctions as the only option. Additionally, they exchange livestock for livestock (e.g. bulls for heifers and vice versa) with their neighbours and as such no cash transaction is involved to make this count as a sale.

Analysis of the cost of running livestock production systems

One of the objectives of the study was to analyse the costs of running the two (mobile

Table 6.3. Respondents' history of selling livestock and market places in the study villages. (From survey data, 2017.)

Information	Number		Percentage	
	Mobile pastoralists	Sedentary community	Mobile pastoralists	Sedentary community
History of selling livestock				
Yes	60	51	100	85
No	0	9	0	15
Reasons for selling livestock				
Buying food	59	0	98	0
Paying for education	7	19	12	32
Buying animal drugs	46	3	77	5
Family health	10	0	17	0
Meeting other family needs	3	29	5	48
Livestock selling stations/points				
Neighbours	0	5	0	8
Local markets/auctions	60	46	100	77
Cross border markets	0	0	0	0
Other	0	0	0	0
Total	60	60	100	100

and sedentary) production systems. To achieve this objective the costs parameters were clustered under six categories. These categories represented costs for: (i) water for livestock; (ii) forages; (iii) labour for managing livestock; (iv) animal health; (v) access to markets; and (vi) those related to additives for the livestock. It also included the livestock loss values due to mortality as a cost related to managing the production system. All costs presented in this section are based on average financial values generated from 120 samples measured in Tanzanian shillings (TSh). The estimates under each cost cluster are converted into common monetary value and time for consistency.

The mobile production system

In Makame village the sample of the 60 heads of households interviewed indicates that the total costs of managing a mobile pastoral production system is about TSh90 million/year. The amount entails the costs related to water, animal feeds, labour allowances, animal health, access to markets and losses from livestock mortality. Under the mobile pastoral production system, a larger portion of costs comes from maintaining animal health and loss of livestock due to mortality compared with the sedentary production system. The results from the analysis of average costs incurred to run a mobile pastoral production system in the village as a representative of the mobile pastoral community is shown in Table 6.4.

The sedentary production system

In Dofa village, costs of about TSh112 million/year are incurred to run a sedentary production system. Like the mobile pastoral production system, these costs are related to water, animal feeds, labour allowances, animal health, access to markets and losses from livestock mortality. A big chunk of the cost incurred to manage the sedentary production system comes from the purchase of animal feeds such as fodders to feed livestock under the indoor or semi-intensive system. The results from the analysis of average costs incurred to run a sedentary pastoral production system in Dofa village as a representative of the sedentary (agropastoralist) community are as presented in Table 6.4.

Cost–benefit ratios

The study found that the cost of maintaining a mobile pastoral production system is lower compared with that of a sedentary production

Table 6.4. Costs of managing the livestock production systems in the study villages. (From survey data, 2017.)

	Average costs (TSh/year)	
Cost measuring item	Mobile pastoral system	Sedentary system
Costs of animal feeds	9,903,000	84,885,600
Cost of labour	0	12,840,000
Cost of water	6,222,000	8,830,800
Cost of access to markets	580,000	432,000
Lost value due to mortality	17,333,333	2,535,000
Animal health management costs	56,058,000	2,771,800
Other costs per respondent	0	0
Total cost	90,096,333	112,295,200

Table 6.5. Benefits accrued from the mobile pastoral system and the sedentary production system. (From survey data, 2017.)

	Average benefits (TSh/year)	
Benefit	Mobile pastoral system	Sedentary system
Sales from live animals	285,090,000	151,300,000
Sales from milk and associated by-products	48,465,300	34,799,000
Sales from manure	0	1,391,960
Sales from other livestock by-products	729,600	2,538,000
Sales from hides and skin	1,126,667	304,200
Ox power[a]	0	12,000,000
Other revenues from livestock (savings from labour use)	36,480,000	28,200,000
Average value of stock	1,363,900,000	222,200,000
Total	1,735,791,567	452,733,160

[a]Value of having the opportunity to use ox power instead of cost of tractor.

system. The cost–benefit ratios (C/B) are 1:0.5 for the mobile pastoral production and 1:0.25 for a sedentary production system (for the benefits accrued from the two systems see Table 6.5).

Benefits accrued from the production systems

The average value of a single small stock animal in the mobile pastoral community, particularly Makame village, was TSh50,000 while a large stock animal was worth at least TSh350,000. To obtain the average value of a household flock/herd, the average value was multiplied by the number of livestock that a household owns. From the sample of 60 heads of households who participated in this study it was revealed that they owned a total of 6506 herds, of which 3462 were large stock animals (cattle) and 3044 were small stock animals (sheep and goats) worth TSh1211 million and TSh152 million, respectively. The average value of total livestock collected from a sample of 60 respondents was TSh1.36 billion.

In Dofa village, the average value of a single small stock animal (sheep and goat) was TSh80,000 while a large stock animal was worth at least TSh800,000. The total average value of a household flock/herd was calculated by multiplying the number of livestock that a household owns with the average prices. From the sample of 60 heads of households from whom the data was collected it was revealed that they owned a total of 793 herds of which 216 were large stock (cattle) and 577 were small stock animals (goats and sheep) worth TSh176 million and TSh46 million, respectively. The average value of total livestock collected from the sample of 60 respondents was worth TSh222 million.

The analysis of benefits was measured on items such as sales from live animals, of milk, manure, hides/skins, ox power (value of having the opportunity to use ox power instead of cost of tractor), and average flock/herd value and the saving from not having to hire labour. In both production systems, more benefits were derived from the stock value possessed by households than benefits from sales and savings. The results of benefits accrued from each production system are presented in Table 6.5.

One of the benefits accrued from the sedentary production system was ox draught power which represents an opportunity cost of using a tractor. The majority of households use bullocks, steers or oxen for transporting water, crops and building materials to their homes. The use of bullock power for transportation purposes was reported to give a benefit of TSh12,000,000 in the sample covered by this study.

Viability and sustainability of the two alternative systems

The viability and sustainability of each production system are measured by running discounted cost–benefits of both production systems to know which one was worth more compared with the other. The costs and benefits discussed in the previous section were discounted at a 10% rate for a period of 10 years. The decision was made based on the 10th year of existence of these production systems. The analysis revealed that, in year 0 the NPV values for mobile pastoral and sedentary production systems are TSh1,645,695,234 and TSh340,437,960, respectively (Table 6.6). In the next 10 years at a constant discount rate of 10%, the NPV values fall to TSh221,231,852 for the mobile pastoral production system and TSh45,765,290 for the sedentary production system.

The NPV at a discount rate of 10% for a period of 10 years revealed the mobile pastoral production system gives more returns to those engaged in it. The data indicate that the mobile pastoralist is 4.83 times more productive than the sedentary one. Put another way, the sedentary production system is only 0.22 times as productive as that of the mobile pastoral system. However, both production systems exhibit positive NPVs at the 10th year but that of the mobile pastoral system bears a higher value than that of sedentary production system. There is no doubt that both production systems are sustainable and continue to yield a positive return to those engaging in them due to their positive NPV signs (Table 6.7).

The findings from this study concur with other studies conducted that compare the pastoral production system and the sedentary one. For instance, Hesse and MacGregor (2006) compared the mobile pastoral production system and sedentary production and they appreciated that mobile pastoralism had a higher value, particularly in the drylands where it is supported by strategic herd mobility. These researchers partly compared the productivity of three pastoral production systems – nomadic, sedentary and transhumance – in terms of annual reproduction rates, calf mortality rates (for under 1 year), calf weights, average lactation days and quantity of milk production for human consumption in one lactation period. The study concluded that nomadic and transhumance (partial mobile pastoral production system) were more productive than the sedentary system for all three parameters. Their study's conclusion shows that mobile pastoral systems are three times more productive than the sedentary ones.

Alternatively, Niamir-Fuller (1999) found that sedentarization had increased environmental degradation, reduced economic potential and eroded the socio-cultural systems among the settled communities in East Africa. This study also revealed the difference in economic potential between the systems and highlighted family problems such as widowhood and separation of partners in Dofa village unlike those from Makame village.

Besides the above, another study found that former nomadic pastoralists are currently migrating to urban and peri-urban centres looking for livelihood diversity or alternative livelihood options. The seeking of alternative

Table 6.6. Net present value (NPV) (TSh) of the mobile pastoral and sedentary production systems over time. (From survey data, 2017.)

	Future costs (FCs)		Future benefits (FBs)		Net benefits (FB – FC)		Net present values (NPVs) at 10%	
Year	Mobile	Sedentary	Mobile	Sedentary	Mobile	Sedentary	NPV–Mobile at 10%	NPV–Sedentary at 10%
0	90,096,333	112,295,200	1,735,791,567	452,733,160	1,645,695,234	340,437,960	1,645,695,234	340,437,960
1	81,086,700	101,065,680	1,562,212,410	407,459,844	1,481,125,711	306,394,164	1,346,477,919	278,540,149
2	72,978,030	90,959,112	1,405,991,169	366,713,860	1,333,013,140	275,754,748	1,101,663,752	227,896,486
3	63,680,227	81,863,201	1,265,392,052	330,042,474	1,199,711,826	248,179,273	901,361,251	186,460,761
4	59,112,204	73,676,881	1,138,852,847	297,038,226	1,079,740,643	223,361,346	737,477,387	152,558,804
5	53,200,984	66,309,193	1,024,967,562	267,334,404	971,766,579	201,025,211	603,390,590	124,820,840
6	47,880,885	59,678,273	922,470,806	240,600,963	874,589,921	180,922,690	493,683,210	102,126,142
7	43,092,797	53,710,446	830,223,726	216,540,867	787,130,929	162,830,421	403,922,626	83,557,752
8	38,783,517	48,339,401	747,201,353	194,886,780	708,417,836	146,547,379	330,482,149	68,365,434
9	34,905,165	43,505,461	672,481,218	175,398,102	637,576,052	131,892,641	270,394,485	55,935,355
10	31,414,649	39,154,915	605,233,096	157,858,292	573,818,447	118,703,377	221,231,852	45,765,290
102[a]	1,938	2,416	37,345	9,740	35,407	7,324	2	0
110[a]	834	1,040	16,076	4,193	15,241	3,153	0	0

[a]Indicates that this is out of the scope of the study time period. The time beyond 10 years has only been used to show how long greater returns are exhibit by each production system.

Table 6.7. Evaluating the two alternative livestock production systems in terms of costs, benefits and net present value (NPV). (From survey data, 2017.)

	Production system	
Item	Mobile pastoral system	Sedentary system
Costs	90,096,333	112,295,200
Benefits	1,735,791,567	452,733,160
Cost–benefit ratios ($^C/_B$)	0.52	0.25
NPVs at 10th year	221,231,852	45,765,290
NPV sign	Positive	Positive
NPV ratios at 10th year	0.22	4.83

livelihoods by former mobile pastoralists has increased impoverishment and stock loss due to constrained mobility, severe droughts, human security and political instability in East Africa, while mobile pastoralists are simultaneously attracted to the benefits of a settled life in towns (Fratkin *et al.*, 2006). The trends for future NPVs indicate an upward sloping for the first 3 years and the start of a decline thereafter (see Table 6.6). This is due to the effects of discounting to determine the future value of money which depreciates as time passes. When the analysis continues, the NPV for the sedentary production system becomes unsustainable at the 102nd year while a mobile pastoral production system can still provide two returns until it reaches zero at the 110th year (see Table 6.6). The analysis does not, however, indicate at what time both production systems result in negative NPVs, even beyond 300 years. In the long run (beyond 100 years) the two graphs will converge towards each other implying that the two alternatives have equal net worth in terms of NPVs.

Hypothesis testing

Overview

At the start of this study two hypotheses were formulated to be tested using the data collected from the two villages. The hypotheses were defined as follows:

$\mathbf{H_0}$ = Sedentary livestock raising is more productive than mobile pastoral systems.

$\mathbf{H_1}$ = Sedentary livestock raising is less productive than mobile pastoral systems.

The study findings revealed that the average cost of maintaining a mobile pastoral and sedentary production systems are TSh90,096,333 and TSh112,295,200, respectively. The cost–benefit ratios are 1:0.5 for the mobile pastoral production system and 1:0.25 for the sedentary system. The average benefits accrued from each production system are TSh1,735,791,567 for the mobile system and TSh452,733,160 for the sedentary one. Furthermore, the NPVs for both production systems at a 10% discounting rate in the next 10 years are TSh221,231,853 for mobile and TSh45,765,290 for the sedentary. Both production systems exhibit a positive NPV in the next 10 years at a constant discounting annual rate of 10%. However, the NPV for the mobile pastoral production system after 10 years is higher than that of the sedentary one by more than four times. Nevertheless, in terms of costs of managing the production systems, these are lower in the mobile system and at the same time the benefits are greater compared with the sedentary system. Although, the cost–benefit ratio is lower in the sedentary (1:0.25) production system than that of mobile one, which is 1:0.5/year.

The analysis also indicates that the mobile pastoral production system has a longer period of giving positive returns to the community of Makame village. The differences are that while the sedentary production system gives no more benefits after 102 years, a mobile pastoral production system continues producing positive benefits up to 110 years. Summing up, there is no doubt that unlike the claims that it is an economically unsustainable system and that it needs to be taken over by other production systems, the mobile pastoral production system remains more viable and more sustainable than its rival, especially so in the dryland ecosystems of Eastern Africa. Conversely, the conclusion drawn here is strictly based on the methodology used and parameters studied and analysed in this study.

Decision criteria

The bottom line for accepting the investment alternatives is by comparing them in terms of their NPV and internal rate of returns (IRR) or sensitivity. The alternative with a positive NPV will be accepted to be more viable than its alternative (European Union, 2015). However, in a case where the two alternatives have NPV greater than zero (positive), the one with the higher net benefit will be accepted to be more viable than the other (European Union, 2015).

This analysis is, however, limited to comparing the NPVs of the sedentary and mobile pastoral production system such that it will not evaluate based on IRR and draw any plausible conclusions. The findings of this study show that both production systems exhibit positive NPVs but the mobile production system has higher benefits at all times. It rather gives a net benefit which is more than four times that of the sedentary production system. Based on these findings, there is not enough evidence to support the null hypothesis (H0) that 'sedentary livestock raising is more productive than the mobile pastoral system'.

The claims made earlier that the mobile pastoral production system is economically unsustainable, that it should be replaced with more sedentary forms of livestock production or any other beneficial land uses as reported by WISP (2011), Davies *et al*., (2015) and Hesse and MacGregor (2006), does not at present hold any water. From this study it is sufficiently safe to conclude that 'sedentary livestock raising is less productive than the mobile pastoral system' in the two study areas.

Conclusion and Recommendations

Conclusion and key messages of the study

This study undertook a CBA specifically by: (i) analysing the costs of running each production system in the study areas; (ii) estimating the benefits accrued from each production system; (iii) comparing costs and benefits of each production system; and (iv) producing a discounted income stream to determine the NPVs of each production system in the next 10 years. The performance of the two systems was then compared to test the viability and sustainability of the two alternatives.

Cost of running the production systems

The study findings have further revealed that the average cost of maintaining a mobile pastoral and sedentary production systems are TSh90,096,333 and TSh112,295,200, respectively. The cost–benefits ratios are 1:0.5 for a mobile pastoral production system and 1:0.25 for the sedentary one. The total costs for the mobile pastoral production system constitute TSh9,903,000 for animal feeds, TSh6,222,000 for water, TSh580,000 for access to markets, TSh17,333,333 for lost value due to mortality, and TSh56,058,000 as a cost for maintaining livestock health.

On the other hand, the total costs for the sedentary production system constitute TSh84,885,600 for animal feeds, TSh12, 840,000 for hiring labour, TSh8,830,800 for water, TSh432,000 for access to markets, TSh2,535,000 for lost value due to mortality, and TSh2,771,800 as a cost for maintaining livestock health.

Benefits accrued from the production systems in the study areas

The average benefits accrued from each production system in the study areas were TSh1,735,791,567 for the mobile pastoral system and TSh452,733,160 for the sedentary one. In the mobile pastoral system,

TSh285,090,000 were benefits from the sale of live animals, TSh48,465,300 from sales of milk, TSh729,600 from other livestock by-products, TSh1,126,667 from sales of hides/skins, and TSh36,480,000 as a saving from not hiring labour. The total benefits accrued from sedentary production system, on the other hand, were TSh151,300,000 from sale of live animals, TSh34,799,000 from sales of milk, TSh1,391,960 from sales of manure, TSh2,538,000 from other livestock by-products, TSh304,200 from sales of hides/skins, TSh28,200,000 as a saving from not hiring labour and TSh12,000,000 from ox power.

Viability and sustainability of the two alternative systems

The viability and sustainability of each production system are measured by running discounted cost–benefits of both production systems to know which one is worth more compared to the other. The costs and benefits discussed in the previous sections were discounted at 10% discounting rate for a period of 10 years. The decision was made based on the 10th year of existence of these production systems. The analysis revealed that in year 0 the NPV values for the mobile and sedentary production systems were TSh1,645,695,234 and TSh340,437,960, respectively. In the next 10 years at a constant discounting rate of 10%, the NPV values fall to TSh221,231,852 for the mobile production system and only TSh45,765,290 for the sedentary one.

The NPV at a discount rate of 10% for a period of 10 years revealed that the mobile pastoral production system gave more returns to those engaged in it, with the data indicating that the mobile pastoralist was 4.83 times more productive than the sedentary producer. On the contrary, the sedentary production system was only 0.22 times of that of the mobile pastoral system. Nevertheless, both production systems exhibited positive NPVs at the 10th year with the mobile system bearing a higher value than that of the sedentary production system.

When the analysis continues beyond the 10th year, the NPV for the sedentary production system becomes unsustainable at the 102nd year while a mobile pastoral production system can still provide returns until it reaches zero at the 110th year. The analysis does not, however, provide a tipping point in time when both production systems give negative NPVs, even beyond 300 years.

Recommendations for policy and further research

This study undertook a comparative CBA of mobile and sedentary production systems in Makame and Dofa villages in Kiteto and Karatu districts, respectively. The key findings from this study revealed some areas which have policy implications. The general recommendation would, therefore, be that policies should be enacted to sustain both pastoral production systems in order to benefit from the greater economic potentials of both systems.

Second, this study was limited to comparing the financial costs and benefits of mobile pastoral production system vis-à-vis the sedentary one. It was also conducted at village level through a cross-sectional approach using a household questionnaire. It is clear that the study has left several important areas which need to be explored or analysed further. The areas for further learning that are recommended by this study are to expand these findings to include social benefits and costs as well as how much space is needed to sustain each production system. The two villages covered by this study may not give a comprehensive conclusion for the comparative CBA of the two production systems, as such the studies which cover wider geographical areas are recommended.

Third, the study only employed a CBA framework and employed a cross-sectional approach to collect data to achieve its objectives. Thus other frameworks such as TEV, PEV analysis and SCBA could not be tested due to constraints of time and resources. Hence, further studies that will also use these other frameworks of analysis while employing designs such as longitudinal surveys are recommended.

Note

[1] Although differentiating the bones of domesticated animals from wild ancestors pose doubts (Blench, 2001).

References

Adriansen, H.K (2006) Continuity and change in pastoral livelihoods of Senegalese Fulani. *Agriculture and Human Values* 23: 215–229.

Ali, A. (2012). A Framework for Using Cost-Benefit Analysis in Making the Case for Software Upgrade. *Issues in Informing Science and Information Technology* 9.

Andriansen, H.K. and Nielsen, T.T. (2002) Going where the grass is greener: on the study of pastoral mobility in Ferlo, Senegal. *Human Ecology* 30, 215–225.

African Union (AU) (2010) *Policy Framework for Pastoralism in Africa: Secure Protecting and Improving Lives, Livelihoods and Rights of Pastoralist Communities*. Department of Rural Economy and Agriculture, African Union, Addis Ababa with support from Technical Centre for Agricultural and Rural Cooperation (CTA), Addis Ababa. Available at: https://au.int/sites/default/files/documents/30240-doc-policy_framework_for_pastoralism.pdf (accessed 6 January 2020).

AU-IBAR (2015) The Livestock Development Strategy for Africa (LiDeSA) 2015-2035 Executive Summary. Nairobi, Kenya.

Babbie, E. (1990) *The Practice of Social Research*. Wadsworth, New York.

Bassett, T.J. and Turner, M.D. (2007) Sudden shift or migratory drift? FulBe herd movements to the Sudano-guinea region of West Africa. *Human Ecology* 35, 33–49.

Behnke, R.H. and Scoones, I. (1993) Rethinking range ecology: Implications for rangeland management in Africa. In: *Range Ecology at disequilibrium* (eds) R. Behnke, I. Scoones and C. Kerven, 1-30. Overseas Development Institute, London.

Behnke, R.H., Scoones, I. and Kerven, C. (eds) (1993) *Range ecology at disequilibrium: new models of natural variability and pastoral adaptation in African savannas*. Overseas Development Institute, Nottingham.

Behnke, R.H., Fernandez-Gimenez, M.E., Turner, M.D. and Stammler, F. (2011) Pastoral migration: mobile systems of livestock husbandry. In: Milner-Gulland, E.J., Fryxell, J.M. and Sinclair, A.R.E. (eds) *Animal Migration: a Synthesis*. Oxford University Press, Oxford, pp. 145–170.

Bell, D.R. (2004) Environmental Refugees: What Rights? Which Duties? *Res Publica* 10(2): 135. https://doi.org/10.1023/B:RESP.0000034638.18936.aa

Blench, R. (2001) 'You can't go home again': Pastoralism in the new millennium. Overseas Development Institute, London. Available at: https://www.odi.org/sites/odi.org.uk/files/odi-assets/publications-opinion-files/6329.pdf (accessed 7 January 2020).

Bollig, M., and Schulte, A. (1999) Environmental Change and Pastoral Perceptions: Degradation and Indigenous Knowledge in Two African Pastoral Communities. *Human Ecology* 27, 493–514 doi:10.1023/A:1018783725398

Bourn D.M and Wint G.R.W. (1992) 'Nigerian National Livestock Resource Survey: The IT Implication in Implementation and in Future Use'. Paper presented to the 1992 Summer Colloquium of the BCS Developing Countries Specialist Group, London.

Bousman, C.B. (1998) The chronological evidence for the introduction of domestic stock into southern Africa. *African Archaeological Review* 15: 133–150.

Brush, S.B. (1975) The Concept of Carrying Capacity for Systems of Shifting Cultivation. American Anthropological Association. 77: 4. https://doi.org/10.1525/aa.1975.77.4.02a00040

Butt, B. (2010) Seasonal space–time dynamics of cattle behaviour and mobility among Maasai pastoralists in semi-arid Kenya. *Journal of Arid Environments* 74, 403–413.

Coast, E. (2005) *Maasai Marriage: a Comparative Study of Kenya and Tanzania*. Department of Social Policy, London School of Economics, London.

Coppock, D.L (1993) The Borana Plateau of southern Ethiopia: Synthesis of pastoral research, development and change, 1980-91. ILCA Systems Study. International Livestock Centre for Africa, Addis Ababa, Ethiopia.

Cribb, R. (1991) *Nomads in Archaeology*. Cambridge University Press, Cambridge.

Davies, J and Hatfield, R (2007). The Economics of Mobile Pastoralism: A Global Summary. *Nomadic Peoples* 11(1): 91–116. https://doi.org/10.3167/np.2007.110106

Davies, J., Ouedraogo, R., Hagelberg, N. and M. Niamir-Fuller, M (2015) Sustainable Pastoralism for the Post 2015 Agenda. Brief for GSDR 2015. Available at: https://sustainabledevelopment.un.org/content/documents/626970-Davies-Sustainable%20Pastoralism%20for%20the%20Post%202015%20Agenda.pdf (accessed 6 January 2020).

Davies, K.W. (2008) Medusa head dispersal and establishment in sagebrush steppe plant communities. *Rangeland Ecology and Management* 61: 110–115.

Desta, S. and Coppock, D.L. (2004) Pastoralism under pressure: tracking system change in southern Ethiopia. *Human Ecology* 32, 465–486.

Dudd, S.N. and Evershed, R.P. (1998) Direct demonstration of milk as an element of archaeological economies. *Science* 282(5393), 1478–1481.

Dyson-Hudson, N (1966) *Karimojong Politics*. Clarendon Press, Oxford.

Ekaya, W. (2001. Nutritional Characteristics of Selected Grass and Browse Species From Kenya's Pastoral Ecosystems. *Journal of Human Ecology*. 12. 171–175.

Eriksen, S. and Lind, J. (2009) Adaptation as a Political Process: Adjusting to Drought and Conflict in Kenya's Drylands. *Environmental Management* 43, 817–835

European Union (2015) *Guide to Cost–Benefit Analysis of Investment Projects: Economic Appraisal Tool for Cohesion Policy 2014–2024*. European Commission, Directorate-General for Regional and Urban Policy, Brussels.

Food and Agriculture Organization of the United Nations (FAO) (2003) The State of Food Insecurity in the World. Available at: http://www.fao.org/docrep/006/j0083e/j0083e03.htm (accessed 23 November 2019).

Fratkin, E. (1998) Pastoralism: governance and development issues. *Annual Review of Anthropology* 26, 235–261. http://dx.doi.org/10.1146/annurev.anthro.26.1.235

Fratkin, E. (2001). East African pastoralism in transition: Maasai, Boran, and Rendille cases. *African Studies Review* 44: 1–25.

Fratkin, E. and Sher-Mei Wu, T. (1997) Maasai and Barabaig herders struggle for land rights in Kenya and Tanzania. *Cultural Survival Quarterly* 21(3), 55–61.

Fratkin, E., Nathan, M.A. and Roth, E.A. (2006) Is settling good for pastoralists? The effects of pastoral sedentarization on children's nutrition, growth, and health among Rendille and Ariaal of Marsabit District, northern Kenya. Presentation for 'Pastoralism and Poverty Reduction in East Africa: a Policy Research Conference', International Livestock Research Institute, Nairobi, 27–28 June 2006.

Galaty, J.G., Aronson, D., Salzman, P.C., and Chouinard, A. (1981) The Future of Pastoral Peoples. Institute of Development Studies and Ottawa, Commission of Nomadic Peoples, Nairobi.

Galvin, K.A. (2008) Transitions: Pastoralists Living with Change. Annual Review of Anthropology. 38: 185–98

Hary, I., Schwartz, H.J., Pielert, V.H.C. and Mosler, C. (1996) Land degradation in African pastoral systems and the destocking controversy. *Ecological Modelling* 86, 227–233.

Hardin, G. (1968) The Tragedy of the Commons. *Science* (162): 3859, 1243-1248.

Hatfield, R. and Davies, J. (2006) *Global Review of the Economics of Pastoralism*. International Union for Conservation of Nature (IUCN), Nairobi.

Hesse, C. and MacGregor, J. (2006) *Valuing Pastoralism in East Africa*. International Institute for Environment and Development (IIED), London.

Hesse, C. and Odhiambo, M.O. (2006) Strengthening pastoralists' voice in shaping policies for sustainable poverty reduction in ASAL regions of East Africa. Paper presented at the regional conference on Pastoralism and Poverty Reduction in East Africa, June 27-28th, Nairobi, Kenya. ILRI.

Homan, S., Rischkowsky, B., Steinback, J., Kirk, M. and Mathias, E. (2008) Towards endogenous livestock development: Borana pastoralists' responses to environmental and institutional changes. *Human Ecology* 36, 503–520.

King, A (2000) Joint Donor Agencies Study on Performance of and Growth Prospects for Strategic Exports in Uganda: Annex to Case Study on Livestock and Livestock Products. Kampala: Delegation of the European Commission.

Kiteto District Council (KDC) (2017) *Investment Opportunities available in Kiteto District Council*. Kiteto District Council, Tanzania. Available at: http://www.kitetodc.go.tz/storage/app/uploads/public/58f/3b4/a31/58f3b4a3115c1676769003.pdf (accessed 16 March 2017).

Koromo, P.J. (2016) Impact of climate change and variability on pastoral community livelihoods in Tanzania: the case of the 2009/2010 drought in Longido District. MSc dissertation, University of Dar es Salaam, Dar es Salaam, Tanzania.

Kothari, C.R. (2004) *Research Methodology Methods and Techniques*, 2nd edn. New Age International, New Delhi, pp. 1–2.

Lamprey, H.F. (1983) Pastoralism and Today: The Overgrazing Problem. In: Bouriliere, F (ed). *Tropical Savannas*. Elsevier, Amsterdam, pp 643–666.

Leshan, M.T. and Standslause, O.E.O. (2013) Pastoralism: a livelihood system in conflict. *Asian Journal of Management Sciences and Education* 2(4), 249–258.

Levin, H.M. and McEwan, P.J. (2001) Cost-effectiveness analysis: methods and applications. Sage, Thousand Oaks, CA.

Little, P.D. (1992) *The Elusive Granary: Herder, Farmer, and State in Northern Kenya*. Cambridge University Press, Cambridge.

MacGregor, J. and Hesse, C. (2006) Pastoralism: dry lands' invisible asset? *IIED*. Issue Paper No. 142

Marshall, F. 2000. The Origins and Spread of Domestic Animals in East Africa. In: *The origin and development of African livestock*. (eds) R.M. Blench and K.C. MacDonald, University College Press, London, pp. 191–221.

McIntire, J., Bourzat D. and Pingali, P. (1992) Crop-Livestock interaction in Sub-Saharan Africa. The World Bank Washington, D.C

Mcpeak J., Little P.D. (2005) Cursed If You Do, Cursed If You Don't. In: Fratkin E., Roth E.A. (eds) *As Pastoralists Settle. Studies in Human Ecology and Adaptation*, vol 1. Springer, Boston, MA

Mdoe, T. and Mnenwa, R. (2007) Study on options for pastoralists to secure their livelihoods: assessing the total economic value of pastoralism in Tanzania; A report Submitted to TNRF. Available at: https://www.tnrf.org/files/E-INFO-RLTF_VOL2-PART5_Mdoe-N_Mnenwa-R_2008_Assessing_the-total_economic_value_in_Tanzania.pdf (accessed 6 January 2020).

Minnesota State University (n.d.) Kinship & Family. Available at: http://web.mnstate.edu/robertsb/380/Kinship%20&%20family.pdf (accessed 9 April 2017).

Moritz, M. (2008) Competing paradigms in pastoral development from the far north of Cameroon. *World Development* 36, 2243–2254.

Mortimore, M. (1998) *Roots in the African Dust: Sustaining the Sub-Saharan drylands*. Cambridge University Press, Cambridge.

Mortimore, M (2000) Hard Questions for Pastoral Development: A Northern Nigerian Perspective. In: Tielkes E., Schlecht E and Hiernaux P (eds). Elevage et gestion de parcours au Sahel implications pour le development. Verlag Grauer, Stuttgart, Germany. pp. 101–114.

Mortimore, M. and Adams, W.M. (1999) *Working the Sahel: Environment and society in northern Nigeria*. Routledge, London.

Morton, J. (2010) *Pastoralism, Drought and Planning: Lessons from Northern Kenya and Elsewhere*. Natural Resources Institute, Department for International Development (DFID), London.

Mtanzania (2017) Bajeti Ya Kilimo Pasua Kichwa. 20 May 2017. Available at: http://mtanzania.co.tz/bajeti-ya-kilimo-pasua-kichwa/ (accessed 20 November 2019).

Mung'ong'o, C.G. and Mwamfupe, D. (2003) *Poverty and Changing Livelihoods of Migrant Maasai Pastoralists in Morogoro and Kilosa Districts, Tanzania*. Research on Poverty Alleviation (REPOA), Mkuki na Nyota Publishers, Dar es Salaam, Tanzania.

New Economic Foundation (NEF) (2013) Social Cost–Benefit Analysis (SCBA) and Social Returns on Investment (SROI). Available at: http://b.3cdn.net/nefoundation/ff182a6ba487095ac6_yrm6bx9o6.pdf (accessed 26 July 2017).

Niamir-Fuller, M. (1999) *Managing Mobility in African Rangelands: the Legitimization of Transhumance*. IT Publications, London.

Oba, G (2011) Mobility and the Sustainability of Pastoral Production Systems in Africa: Perspectives of Contrasting Paradigms. Paper prepared for the Future of pastoral peoples in Africa organized by "Pastoralism" in the Future Agriculture Consortium at ILRI in Addis Ababa, Ethiopia from 21-23 March 2011

Oba, G. and Kaitira, L.M. (2006) Herder knowledge of landscape assessments in arid rangelands in northern Tanzania. *Journal of Arid Environments* 66, 168–186.

Oxfam (2008) Survival of the fittest: pastoralism and climate change in East Africa. Oxfam Briefing Paper No. 116. Oxfam Policy and Practice. Available at: https://policy-practice.oxfam.org.uk/publications/survival-of-the-fittest-pastoralism-and-climate-change-in-east-africa-114607(accessed 23 November 2019).

PINGOS (2016) Socio-Economic Contributions of Pastoralism as Livelihood System in Tanzania: Case of Selected Pastoral Districts in Arusha, Manyara and Dar es Salaam Regions. PINGO's Forum

Randall, S (2005) The demographic consequences of conflict, exile and repatriation: a case study of Malian Kel Tamasheq. Paper presented at the IUSSP Seminar on the Demography of Conflict and Violence, held in Oslo, Norway, 8 - 11 November 2003.

Raynaut, C (2001) Societies and nature in the Sahel: ecological diversity and social dynamics. *Global Environmental Change* 11 (2001) 9,18

Roba, H.G. and Oba, G. (2009) Community participatory landscape classification and biodiversity assessment and monitoring grazing land in northern Kenya. *Journal of Environmental Management* 90, 673–682.

Rota, A. and Sperandini, S. (2009) Livestock and pastoralists, Livestock Thematic Papers Tools for project design. IFAD. p.18.

Ryan, K., Karega-Münene, S.M., Kahinju and Kunoni, P.N. (2000) Ethnographic perspectives on cattle management in semi-arid environments: a case study from Maasailand. In: Blench, R.M. and MacDonald, K.C. (eds) *The Origins and Development of African Livestock: Archaeology, Genetics, Linguistics and Ethnography*. UCL Press, London, pp. 463–477.

Sandford S. (2006) Too many people, too few livestock: The crisis affecting pastoralists in the greater Horn of Africa.

Scarnecchia, D.L., Miller, D. and Craig, S.R. (1998) Rangelands and Pastoral Development in the Hindu Kush-Himalayas. *Journal of Range Management*. 51(6):723

Scholte, P., Kari, S., Moritz, M. and Herbert, P. (2006) Pastoralist Responses to Floodplain Rehabilitation in North Cameroon. *Human Ecology*. 34, 27–51.

Scoones, I. (1995) *New Directions in Pastoral Development in Africa. Living with Uncertainty*. IT Publications, London.

Spencer, P. (1998) *The Pastoral Continuum: The Marginalization of Tradition in East Africa*. Clarendon Press, Oxford.

Thornton , P.K., van de Steeg, J., Notenbaert, A., and Herrero, M. (2009) The impacts of climate change on livestock and livestock systems in developing countries: A review of what we know and what we need to know. *Agricultural Systems* 101, 113–127.

Tiffen, M., Mortimore, M., and Francis Gichuki, F. (1994) More People, Less Erosion. Environmental Recovery in Kenya. John Wiley & Sons Ltd. Chichester, UK.

Tenaw, Z.PlnT. (2016) Indigenous institutions as an alternative conflict resolution mechanism in eastern Ethiopia. The case of the Ittu Oromo and Issa Somali clans. African Centre for Constructive Resolution of Disputes (ACCORD).

UN Convention to Combat Desertification (UNCCD) (May 2014-April 2015). UNCCD's support to the New Partnership for Africa's Development (NEPAD).

United Nations Population Fund (UNPF) (2010) State of world population 2010. From conflict and crisis to renewal: generations of change. Available at: https://www.unfpa.org/sites/default/files/pub-pdf/EN_SOWP10.pdf (acceseed 07 January 2010).

United Republic of Tanzania (URT) (2006) *Tanzania National Livestock Policy (NLP)*. Ministry of Water and Livestock Development, Government of United Republic of Tanzania, Dar es Salaam, Tanzania.

United Republic of Tanzania (URT) (2007) *National Adaptation Programme of Action (NAPA)*. Vice President's Office (VPO), Division of Environment, Government of United Republic of Tanzania, Dar es Salaam, Tanzania.

United Republic of Tanzania (URT)/National Bureau of Statistics (NBS) (2013) *The Household Survey for Tanzania Mainland*. NBS, Ministry of Finance, Dar es Salaam, Tanzania.

Unruh, J. (1990) Integration of transhumant pastoralism and irrigated agriculture in semi-arid East Africa. *Human Ecology* 18(3): 223–246. https://doi.org/10.1007/BF00889154

Unruh, J. and Bailey, J. (2009) Management of spatially extensive natural resources in post-war contexts: Working with the peace process. GeoJournal. 74, 159–173. 10.1007/s10708-008-9232-y.

URT (2014) Second National Communication To The United Nations Framework Convention On Climate Change. Vice President's Office, Dar es Salaam, Tanzania.

URT (2016) Tanzania Livestock Modernization Initiative. Ministry of Livestock and Fisheries Development, Dar es Salaam, Tanzania.

URT (2017) Karatu District Council Strategic Plan 2017/18-2021/2. Karatu District Council.

Warner, K., Hamza, M., Oliver-Smith, A. Renaud, F. and Julca, A. (2010) Climate change, environmental degradation and migration. *Natural Hazards* 55, 689–715.

Wikipedia (2016) Carrying capacity. Available at: https://en.wikipedia.org/wiki/Carrying_capacity (accessed 21 February 2019).

World Initiative for Sustainable Pastoralism (WISP) (2006) World Initiative for Sustainable Pastoralism: Review of the literature on pastoral economics and marketing, Kenya, Tanzania, Uganda and the Sudan.

World Initiative for Sustainable Pastoralism (WISP) (2007) Change of Wind or Wind of Change? Climate Change, Adaptation and Pastoralism. Available at: https://www.iucn.org/sites/dev/files/import/downloads/c__documents_and_settings_hps_local_settings_application_data_mozilla_firefox_profile.pdf (accessed 7 January 2020).

World Initiative for Sustainable Pastoralism (WISP) (2010) Building climate change resilience for African livestock in sub-Saharan Africa – World Initiative for Sustainable Pastoralism (WISP): a program of IUCN – The International Union for Conservation of Nature, Eastern and Southern Africa Regional Office, Nairobi, March 2010 Available at: https://www.iucn.org/sites/dev/files/content/documents/resilence2.pdf (accessed 7 January 2020).

World Wide Fund for Nature (WWF) (2012) *Maps of Wildlife Management Areas (WMAs) in Tanzania*. A consultancy Report submitted by Boniface P. Mbilinyi, Japhet J. Kashaigili, Reuben Mwamakimbullah and Alexander N. Songorwa to WWF-TCO.

Yanda, P.Z. and Mung'ong'o, C.G. (eds) (2016) *Climate Change and Pastoralism in East Africa*. Mkuki na Nyota Publishers, Dar es Salaam, Tanzania.

Yanda, P.Z. and William, C. (2010) Livelihoods diversification and implication on food security and poverty levels in Maasai plains. A case of Simanjiro District. *African Journal of Environmental Science and Technology* 4(3), 154–166.

Part II

Climate Change and the Socio-ecological System

7 Locally Based Responses to Impacts of Climate Change in Pastoral Landscapes of Northern Tanzania

Edmund B. Mabhuye* **and Pius Z. Yanda**

Institute of Resource Assessment, University of Dar es Salaam, Dar es Salaam, Tanzania

Abstract

The impacts of climate change and variability have manifested themselves throughout the world, but considerable temporal and spatial variations exist across various places and countries. Given the variation in vulnerability, this study was undertaken in pastoral landscapes in northern Tanzania to assess the impacts of climate change, adaptation strategies and their implications to communities' livelihoods and ecosystem integrity. It examined: (i) climate trends and associated impacts on communities' livelihood options; (ii) climate change coping and adaptation strategies adopted by selected communities to reduce the severity of climate change impacts; and (iii) the challenges associated with climate change adaptation strategies in the pastoral landscape. Primary data were collected using household surveys, interviews with key informants, focus group discussions, direct field observation using transect walks and institutional analysis. Secondary data were obtained through documentary review and theme-content analysis. Results indicate that there are slight increases in temperature and wind speed as well as decreasing trends and erratic patterns of rainfall which cause drought and extended dry spells. Fluctuation in temperature and rainfall patterns affects livestock keeping through recurrent drought that has negative implications on pasture and water availability. Communities are responding to the changes through traditional response mechanisms and have embraced a few new adaptation strategies against these climate extremes, particularly drought. Generally, strategies for adaptation are likely to be successful in the near future, subject to review and harmonization of policies, institutional and legal frameworks to harness existing opportunities for management of natural resources for sustainable development and build the long-term balance between ecosystem integrity and human needs.

Background

The livelihood systems of semi-arid Africa are rooted predominantly in rainfed agriculture.[1] Rainfed crop production and free rangeland management systems form the major livelihood systems of the rural population in most semi-arid lands especially in sub-Saharan Africa (Fischer *et al*., 2002). However, impacts of climate change and variability have been undermining the livelihood systems, resulting into low income, food insecurity, hunger and famine. In most cases, changes in rainfall and temperature patterns are associated with drought, extended dry spells and floods, which eventually threaten human activities and resilience of ecosystems especially in semi-arid areas.

The Intergovernmental Panel on Climate Change (IPCC, 2007) indicates that hotter and drier conditions and less predictable

*edmund.mabhuye@gmail.com

rainfall and prolonged droughts are likely to increase by 5–8% in most semi-arid areas of Africa. These changes have direct effects on ecosystems and human activities, especially pastoral production systems. Significant negative consequences will include loss of livestock through heat stress, loss of rangeland due to expansion of boundaries for protected areas, expansion of crop farms, and increasing livestock pests and diseases that thrive due to increasing temperature and abrupt change from a hot or long dry season to a wet season (Oxfam, 2008).

The National Adaptation Programme of Action (NAPA) of Tanzania (URT, 2007) ranks agriculture as the most vulnerable sector and requires immediate interventions to reduce the severity of the impacts given their importance to the economy, especially for the majority of people in rural areas. Projections for Tanzania show that climate change will cause a general decline in the production of most of the subsistence crops and livestock (URT, 2003, 2014). Livestock keeping, especially the pastoral system, is one of the agricultural subsectors highly affected by the impacts of the changing climate.

Apart from stresses induced by the changing climate and variability, the vulnerability of the pastoral system is largely aggravated by institutional and policy frameworks which determine access to natural resources. An unclear definition of the pastoral production system and the location of pastoral land are some of the key factors affecting access to grazing landscapes and livelihood diversification among pastoral communities (Agrawala *et al.*, 2003; Lynn, 2010; Yanda and Mubaya, 2011; Yanda and Mung'ong'o, 2016). As a result, because of their high reliance on climate for stock foraging and general livelihood strategies, the current climate extremes mark an unblessed situation to pastoralists' socio-economic integrity.

Northern Tanzania is one of the semi-arid lands, which is largely inhabited by the pastoralists who keep a combination of different livestock including cattle, goats and sheep (Mortimore *et al.*, 2009). Significantly, there have been noticeable changes in the pastoral mode of production, including increased mobility induced by unpredictable and shortage of rains. However, though acknowledged to be a viable option for healthier traditional livestock keeping, the mobility of pastoralists has been associated with various conflicts related to competing land uses (e.g. pastoralists versus farmers and pastoralists versus government over the use of pasture and water resources).

Observed and anticipated climate change and other environmental changes make it difficult for pastoral communities to cope because these changes are too rapid and intense in nature, dynamic, more frequent and unpredictable (Mortimore *et al.*, 2009). Furthermore, the magnitude and rate of current climate change, combined with additional environmental, social and political issues, are making many traditional coping strategies ineffective and/or unsustainable, amplifying environmental degradation and food insecurity. In particular, there is limited knowledge on the real effects of climate change on micro-pastoral systems and the nature of vulnerabilities generated when climate change interacts with existing vulnerabilities and the wider socio-economic and political systems. How these changes are internalized at the community level and the nature of responses to these changes are the issues that need to be studied and reflected on.

The purpose of this study was to undertake a holistic and comparative analysis of the climate-related vulnerability and impacts in selected landscapes of pastoral communities in order to contribute to the existing debate and knowledge. This entails looking at the wider socio-economic environment in which pastoralism is embedded and how this environment is enabling or hindering communities' response to climate change challenges. Specifically, the study: (i) examined climate trends and associated impacts on communities' livelihood options in selected sites in pastoral landscapes; (ii) assessed climate change coping and adaptation strategies adopted by selected communities to reduce the severity of climate change impacts;

and (iii) examined the challenges associated with climate change adaptation strategies in pastoral landscapes.

Theoretical and Conceptual Framework

This study was guided by the action theory of adaptation to climate change, which links climate change impacts and adaptation, developed by Eisenack and Stecker (2011). The purpose was to undertake systematic analysis of climate change in pastoral landscapes, examine the impacts of climate change in pastoral landscapes, analyse processes and enabling conditions for adaptation and how the adaptation actions are conditioned to be successful or unsuccessful and the associated challenges and opportunities.

The action theory of adaptation to climate change was chosen as it lays a benchmark for understanding climate change impacts and adaptation pathways. Climate change is considered to be one of the major drivers of natural and social systems change that make adaptation a necessity among communities and other actors. Changes in climate directly affect human or natural systems through variability in rainfall trends that affect community livelihoods such as livestock keeping through diminishing quality and quantity of pasture. As a result, the impacts induce affected communities to devise adaptation measures to recover from the impacts. In this regard, adaptation is considered to be actions or strategies implemented in response to climate change impacts experienced by human and natural systems (Pelling and High, 2005). Brooks and Adger (2004) also expounded that adaptation can be actual and potential depending on the capacity and level of resources endowed by the responding entity (e.g. individual or household). The level or quality of the strategies devised depends largely on the institutional framework, strategies and plans, programmes' financial capacity, and technology available. Therefore, the theory of adaptation to climate change was adopted to guide the identification of key climate parameters that have been changing in the study area, impacts of the changes in climate and how communities are responding to the changes through locally based knowledge and how other actors are supporting adaptation measures at the study sites (see Fig. 7.1).

Methodology

Study area

The study was conducted in Monduli and Ngorongoro districts in Arusha Region and Simanjiro District in Manyara Region, which hosts a wide range of pastoral communities in northern Tanzania (Fig. 7.2).

In terms of climatic conditions, some parts of Monduli District are characterized by a cool and subhumid climate as the district includes some mountains, which at high altitudes capture more rain. It has an average temperature of 20°C and the rainfall average is 1000 mm/year. Other parts of the district are characterized by arid conditions with a warm climate and temperatures ranging from 23°C to 35°C. Rainfall in most parts of the district is less than 500 mm/year (Monduli District, 2015). Ngorongoro District has a tropical climate whereby some areas such as Sale division are characterized by hot and dry conditions, while other areas receive high rainfall which varies between 400 mm/year and 600 mm/year. The district also experiences strong, dry winds that normally blow from east to west. The land is covered by open grassland with scattered *Acacia*-type shrubs especially in central, eastern and western parts. Simanjiro District is semi-arid with an average rainfall of 500 mm/year. It has a dual period of unreliable rainfall comprising of short rains in November–December and long rains in March–April. The cold months are May–July while the hot months are August–February. Temperatures range between 13°C and 30°C (Simanjiro District, 2015).

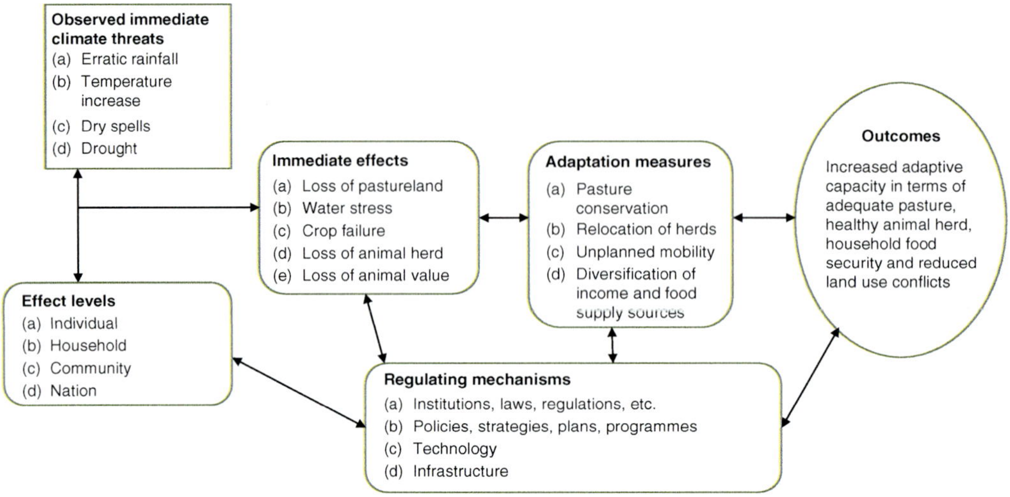

Fig. 7.1. Interlinkages between climate change impacts, adaptation and communities' livelihoods. (Modified from Eisenack and Stecker, 2011.)

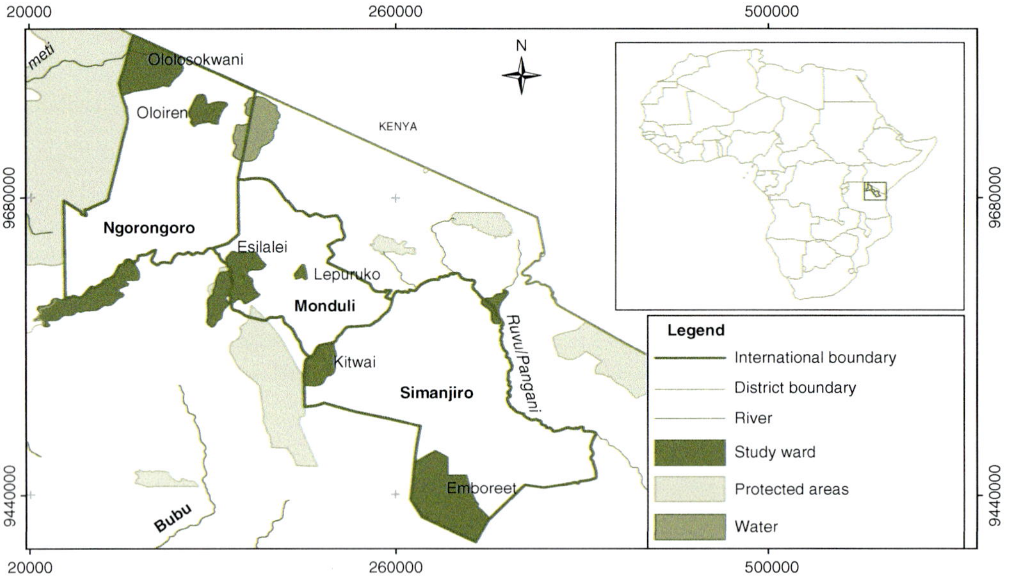

Fig. 7.2. Location of study area. (From Institute of Resource Assessment, Geographic Information System (GIS) Laboratory, 2016.)

Demography and socio-economic activities

For Monduli District, the Population and Housing Census of 2012 recorded 158,929 people, whereas the district growth rate was estimated to be 3.9%, which is higher than of the Arusha Region (2.7%) and the national average population growth rate of 2.7%. The major ethnic group in Monduli is the Maasai who constitute about 40% of the entire population. Their main activity is livestock keeping. The second ethnic group is the Waarusha, who are estimated to constitute about 20% of the entire population in the district. The main economic activities are livestock keeping and farming. The remaining 40% of the population consists of ethnic groups not indigenous to Monduli District. Their main activities are farming and trading (Monduli District, 2015). According to the Population and Housing Census of 2012, Simanjiro District had a total population of 178,693 out of which 88,975 were males and 89,718 were females (Simanjiro District, 2015). The dominant ethnic group in Simanjiro District is the Maasai, constituting 90% of the population. The other ethnic groups include the Chagga, Arusha, Meru, Pare, Fipa, Iraqw, Rangi, Sukuma and other tribes from within and outside Tanzania. On the other hand, Ngorongoro District is highly dominated by Maasai, while other tribes make up a smaller part of the inhabitants.

Sampling of respondents

This study focused on various categories of respondents. Major sample categories of respondents were district technical officers, village government officials, households and individuals. These were sampled through purposive and simple random sampling techniques. District technical officers, village government officials and individuals (key informants and focus group participants) were sampled using purposive sampling. Key informants were considered to have knowledge on the topic or they were directly involved in management of natural resources affected by climate change or involved in community livelihood improving endeavours. Simple random sampling was deployed to select 10% of the total households in each sample village in order to achieve a representative sample recommended for study (Kothari, 2004). Household respondents were selected for interview using a questionnaire. In this regard, a total of 202 households were selected as a representative sample for three districts and six villages for this study. As already stated, the districts were Simanjiro in Manyara Region, and Monduli and Ngorongoro in Arusha Region. Two wards and two villages were selected from each district, making a total of six wards and six villages for the study (Table 7.1).

Data collection methods

Both secondary and primary data were collected to obtain adequate information to address the study objectives. Secondary data and information were obtained through a review of published and unpublished literature available online and offline in order to establish the background knowledge related to the current study. Primary data were collected through direct field observation, household surveys using a questionnaire, key informant interviews and focus group discussions (FGDs). The household surveys were organized using a semi-structured questionnaire to obtain qualitative and quantitative information from selected household heads (male or

Table 7.1. Sample respondents and districts, wards and villages in the study area. (From field survey, 2016.)

		Number of sample respondents in each village						
Districts	Wards	Ololosokwani	Mageiduru	Esilalei	Nanja	Kitwai A	Emboreet	Total
Ngorongoro	Ololosokwani	34	0	0	0	0	0	34
	Oloirein	1	38	0	0	0	0	39
Monduli	Esilalei	0	0	33	0	0	0	33
	Lepuruko	0	0	0	32	0	0	32
Simanjiro	Kitwai	0	0	0	0	32	0	32
	Emboreet	0	0	0	0	0	32	32
Total		35	38	33	32	32	32	202

female) or any representative household member provided he/she was involved in family decision making and was knowledgeable about the topic. Key informant interviews were conducted with targeted experts in pastoral communities' livelihoods and resources management, and climate change. Key informant interviews also had the objective of analysing various programmes and plans relevant to pastoral production systems, climate change adaptation and natural resources management in pastoral landscapes.

The sought experts included the district environmental officials, livestock officers, community development officials, extension officers, government representative experts from the Tanzania Meteorological Agency (TMA), community leaders, civil society actors and community-based organizations with a focus on the topic under study. FGDs were used to collect information on climate change impacts, response strategies and implications on livelihoods and natural resources management. Discussions with selected participants from each of the study villages about their various experiences of pastoral production systems and climate change were held for 1–2 h sessions in each village. Discussions were guided by an open-ended checklist, which allowed free expression and flexibility of responses from the participants.

FGD participants included elders, women, youth and community leaders, civil society actors and other stakeholders. Arrangements were made to ensure that each FGD had representatives from all possible and relevant local social groups in each study village. The objective was to guarantee collection of a broad range of information to address the key research questions. FGD sessions also present a concrete spot for triangulation, validation and verification of information collected through other methods. Direct field observation was organized with the help of experienced field assistants selected from the study villages. Photos and/or videos were taken during field observations as evidence of the conditions and phenomena learned from the study sites, household surveys and interviews over the issues investigated. The observed conditions were then compared with the pre-established information and any deviations noted were discussed to better understand or explain the situation. The ongoing projects, programmes or activities related to climate change adaptation in the pastoral communities and/or management of pastoral resources existing in the study area were also carefully studied during the direct field observation.

Data analysis

Quantitative data collected through the questionnaire were analysed using Statistical Package for Social Sciences (SPSS) version 20. Before analysis, data were coded and entered into the SPSS data sheet. Cleaning of data was also conducted before generating frequency and percentage tables, which provided descriptive statistics. Qualitative data collected through interviews and FGDs were analysed using a theme-content approach. This entailed summarizing the information and assigning meanings in relation to the key research questions this study intended to answer. Results from both quantitative and qualitative data were presented in figures, tables, graphs, plates and narrative terms.

Results and Discussions

Climate trends in the study area

Changes in climate conditions are increasingly becoming evident; with many of the respondents in the study area reporting dynamic rainfall and temperature trends and patterns. Various categories of respondents from different age groups were reached in order to get their insights and determine any significant differences in terms of perceived climate trends in the study area. Inclusion of different categories of age groups presupposed that age has an influence on knowledge related to climate change impacts and determines adaptation practices, especially the decision on choices of agricultural inputs (Belay *et al.*, 2017). About 31.2% of the respondents were aged from 35 to 44 years, 24.8% were aged from 25 to 34 years, 17.3% were aged 45–54 years, 13.4% were 55–64 years, very few (7%) respondents were aged 15–24 years, and 7% were aged 65 years and above. Respondents aged between 35 years old and more than 65 years old were considered in order to gain their long-term and accumulated experience regarding timelines of major climate-related hazards and some response measures adopted by pastoral communities at the study sites.

Discussion with FGD participants indicated that different age groups had different perceptions regarding climate change impacts. Findings through FGDs and household surveys revealed that the majority of respondents aged below 30 years had mixed sources of knowledge about climate change, both traditional and modern knowledge gained through research feedback/community meetings, seminars and a few had attended workshops on climate-change-related issues. In general, members of this age group (below 30 years) could associate climate change with human activities such as industrial pollution and deforestation. On the other hand, the majority of the FGD and household respondents aged above 45 years associated climate change with natural processes that affect rainfall and temperature conditions. Their long-term 'survival' in harsh environmental conditions (e.g. drought) partly informed this perception. During discussions with household respondents across the study area, it became clear that trends and patterns of major climate elements such as temperature and rainfall have been changing over time. Pronounced changes were decreasing patterns in rainfall and increasing trends in temperature. Additionally, rainfall patterns were reported by the majority of respondents to be unpredictable in terms of onset and cessation (Table 7.2).

Similarly, the majority of respondents indicated that rainfall has been decreasing overtime, while temperature is perceived to be increasing (Fig. 7.3). As illustrated in Fig. 7.3, 96% of the respondents reported that rainfall amounts had been decreasing over the past 20 years. Additionally, as summarized in Table 7.2, the trend for rainfall was reported by 65.8% as haphazard and unpredictable with some onset of rainfall seasons becoming late or early. Interviews with key informants in Ngorongoro indicated that rainfall seasons between the 1970s and 1983 were reliable and predictable. The key

Table 7.2. Respondents' perception of rainfall patterns over the past 20 years. (From field survey, 2016.)

Perceived rainfall patterns	Frequency	Percentage
Unpredictable	133	65.8
Early rainfall onset	19	9.4
Late rainfall onset	18	8.9
Early rainfall cessation	16	7.9
Prolonged dry spells	16	7.9
Total	202	100.0

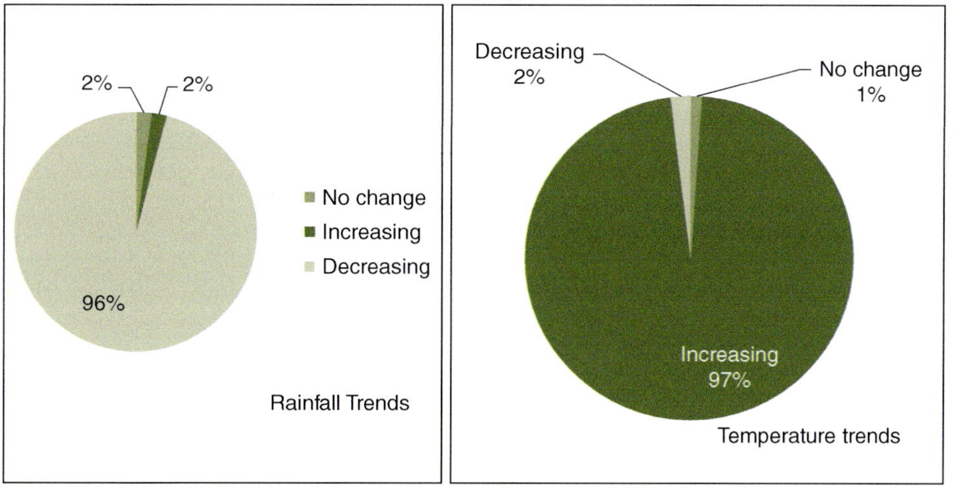

Fig. 7.3. Respondents' perception of the trends of (a) rainfall and (b) temperature. (From field survey, 2016.)

informants also indicated that there have been noticeable rapid changes in rainfall and temperature trends since the 1990s, when extreme events such as droughts and heavy rains started to emerge in Magaiduru and Ololosokwan. During the discussion with focus group participants, it was reported that extreme droughts and consequently acute hunger occurred in 1984, and it became commonly known as the year of *Yanga* or *olari lo yanga* (year of yellow maize), which reflected maize aid from the USA. It was also reported that since 1984, there have been unpredictable rain patterns. The focus group participants remembered timelines of extreme events as droughts that occurred in 1991, 1994, 2003 and 2005.

By comparison, in Monduli District, change in some weather parameters like temperature, wind and rainfall were found to be of interest in the perceptions of inhabitants during the household survey. During discussions with FGD participations in selected villages in Monduli District, they indicated that they had been experiencing a gradual increase in temperature, a decrease in rainfall amount and frequency, unpredictability in rainfall patterns, and an increase in drought intensity and reduction of the time interval between droughts. Major timelines of drought reported by FGD participants in Monduli District were 1933–1935, 1948–1950, 1953–1956, 1964–1967, 1973–1977, 1991–1994, 2003–2004 and 2009–2010. In Simanjiro District, FGDs and key informant interviews indicated that evidence for changing climate was clearly observed, and drought was one of the major indicators linked to climate change. FGD participants reported unpredictable trends in rainfall, increase in the frequency of drought and increase in temperature. These were particularly linked to the phenomenon of repetitive hunger in the district.

With regard to temperature trends and patterns, about 97% of the household respondents across the study sites claimed to be experiencing an increasing trend in temperature, while 2% reported a decreasing trend in temperature and 1% claimed that temperature has been constant in the study area. Consistent with this study, Ndesanjo (2017) indicates that there has been decreasing trends in rainfall in the study area over the past 40 years (1972–2013). The minimum temperature was estimated to have increased by 2°C while the increase in the maximum temperature was about 1.5°C for the period from 1972 to 2013[2] (Ndesanjo, 2017). Ndesanjo (2017) also indicates low average minimum and maximum temperatures in the late 1970s, mid-1980s and early 1990s, while the late 2000s had relatively higher minimum and maximum temperatures. Furthermore, some household respondents reported that wind has been increasing across the study area compared with the climatic conditions over the last 20 years. About 53.5% of the respondents were of the opinion that wind has been increasing, while 36.6% of the respondents detected no change in wind speed in the study area, and 9.9% of the respondents reported that wind speed was decreasing in the study area. FGD participants reported that winds had the highest velocity during the rainy seasons of the year and the lowest velocity during dry periods of the year. As discussed in subsequent sections, changes in rainfall and temperature trends and patterns had significant effects on communities' livelihoods, mainly crop production and livestock-keeping activities.

Major impacts related to climate change and variability

Disruption of communities' livelihoods

Communities' livelihoods, especially the livelihood of pastoral communities, are highly vulnerable to the impacts of the changing climate. In this study, the majority of the respondents in the study area believed that increasing variability of rainfall and temperature has been affecting crop production and livestock keeping. Likewise, Waiswa (2003) and Jonsson *et al.* (2003) report that impacts of climate change and variability place agriculture as one of the most vulnerable sectors across East Africa.

About 97.5% of the respondents across the study sites reported that changes in rainfall were affecting their livelihoods, signified by the declining crop yields, death of livestock and diminishing pasture availability. Discussions with FGD participants and interviews with key informants revealed that the severe drought which occurred in 2008/09 across the study districts disrupted their socio-economic systems, including loss of livestock. Accordingly, the key informants and FGD participants indicated that the drought in 2009 caused significant loss of livestock largely due to lack of pasture and drying up of rivers and other important water sources such as nemilango well and sariani stream in Ngorongoro District. The view of the key informants and FGD participants was that eventually the drought effects were perpetuated as they resulted in high food prices and hunger among pastoralists. Studies by Rama (2013) and Fyumagwa *et al.* (2011) also indicated that apart from deaths of livestock and human beings, the drought in 2008/09 also led to the restriction of export of live animals to other countries.

Increasing resource use conflicts

Increasing conflicts over resource use were also reported to be linked to the effects of the changing climate in the study area. Discussions with key informants and FGD participants indicated that population growth, which placed increasing demands on resources, and expansion of boundaries for conservation of wildlife resources coupled with shrinkage of grazing lands largely due to the recurrent drought were accelerating conflicts over resource use. Key informants indicated that droughts push some pastoralists to move their livestock to national parks searching for pasture. As a result, encroachment into wildlife areas such as Tarangire National Park and the Serengeti National Park by livestock keepers has been inevitable during the dry season, and invasion of wildlife into villagers' farms in areas such as Emboreet village has been highly pronounced.

However, this is against the conservation policies and laws. The Wildlife Conservation Act (2009) prohibits any human activity and settlement in game-controlled areas. As a result, effective adaptation options such as innovative strategies for pasture production and conservation and increasing access to water for domestic use and for supporting crop and livestock keeping can counteract the impacts and enhance the adaptive capacity of local communities. Such adaptation measures are needed because climate change impacts are already marginalizing the livelihoods of pastoral communities in many parts of the study districts and these impacts may become more severe as future climate conditions are largely uncertain.

Emergence of pests and diseases

Apart from the effects of the recurrent drought, this study revealed that heavy rains, which occur occasionally, have been resulting in unpleasant conditions for local communities' livelihood activities and in particular the emergence of pests and diseases. The impacts of climate change on livestock and human beings have also been reported by other studies (e.g. Few, 2007; Page, 2007; Blaikie *et al.* 2014).

Discussions with key informants across the study area indicated that an abrupt change in temperature is one of the conditions that favours the emergence of Rift Valley fever (RVF). FGD respondents reported heavy rains resulted in an outbreak of RVF and bone marrow diseases (*nado endolit*) in 2006 that led to the death of livestock and human beings. Increasing occurrence of diseases such as RVF, crop pests such as grasshoppers, and drying out of streams and rivers resulting in occurrence of some new weeds in grazing lands (e.g. '*cocobile*', *makutian* and *olberbag* especially in Wasso and Loliondo grazing areas) which were unsuitable for livestock were major climate-change-related impacts reported in Ngorongoro District.

Interviews with key informants in Monduli District also indicated that malaria and cholera diseases are increasingly becoming persistent due to increase in temperature that favours malaria pathogens and floods which disrupt waste management systems and influence rampant spread of wastes. With regard to effects on crops, interviews with key informants in Monduli indicated that there have been increasing frequency pests and diseases that affect crops, with beans being the most affected. Emergence of pests and diseases results in low crop yields. Vermeulen *et al.* (2012) projected that climate change will affect agricultural yields and earnings, food prices, reliability of delivery, food quality and, notably, food safety.

Generally, key informants were of the opinion that most of the programmes relating to climate change adaptation were largely short term and focused on a single village or only a few villages. Despite the observed impacts of the changing climate, there was no long-term initiative, whether village wise or district wise, to adapt to climate change impacts. These challenges might have significant negative implications on adaptive capacities at household and community level in the study area.

Communities' response measures to impacts of the changing climate

This study contributed to the scientific consensus that responding to climate change impacts is an inevitable business among the pastoral communities in northern Tanzania. This study found that communities were practising various strategies to respond to the impacts of the changing climate. One of the FGD participants remarked:

> During the previous years, keeping a large number of cattle was praised by different tribes and symbolized wealth and was relatively not threatened by environmental-related stresses. However, increase of drought incidences is affecting our production systems, compelling [the] majority of us to turn to new lifestyles.

Common measures reported by the sample respondents included seasonal to semi-permanent mobility, taking up mixed farming and construction of charcoal dams (*malambo*) to ensure water supply for domestic or human use and livestock throughout the year. Other strategies included: (i) destocking by selling during dry seasons and later or soon after drought periods restocking their herds; (ii) shifting livestock to other locations; (iii) selling livestock; (iv) buying water and food supplements for livestock; (v) seeking assistance from neighbours and relatives; (vi) carrying out casual work; and (vii) running petty businesses (Fig. 7.4).

Survey findings indicated that adoption of the response measures was influenced by several factors; about 32% of respondents reported more than one factor influenced such decisions, and these varied from access to extension services, access to inputs, and access to credit to improved knowledge and information. Key informants interviews in Ngorongoro and Monduli districts revealed that the adaptation to the impacts of climate change

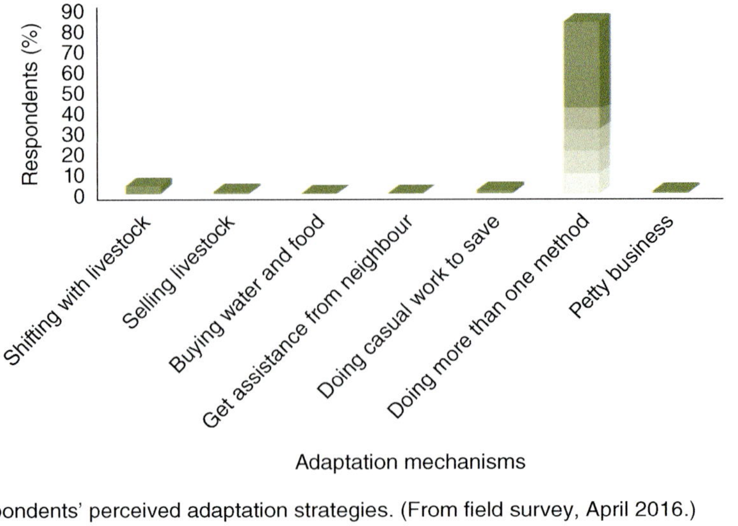

Fig. 7.4. Respondents' perceived adaptation strategies. (From field survey, April 2016.)

had been enhanced by several initiatives such as the United Nations Educational, Scientific and Cultural Organization (UNESCO) mobile clinics and RAMAT community radio, famously known as *'sauti ya wafugaji'*. The radio was principally used to sensitize livestock keepers and farmers, including providing information about new disease breakouts, their vaccine or medicine, if available, as well as where and how to find pesticides.

It was also reported that Oxfam is supporting farmers in adopting modern farming methods, seeds, diseases control and other extension services. The local government authority through District Agricultural Development Programmes (DADPs) plays a role in provision of early maturing seeds and food aid during droughts. The central government has also been helping with restocking of their cattle since 2011/12 following the death of livestock in 2009/10. Interviews with key informants indicated that Oxfam also facilitates the establishment of horticulture and provision of better bulls bred in Magaiduru village in order to improve livestock and agricultural productivity.

Mobility

Even though mobility of pastoral communities in the study area has been normal practice throughout history as part of spatial utilization of resources and the response to environmental-related stresses, field survey findings indicated that migration is now used as a coping strategy to mitigate the impacts of changing climate conditions. According to FGDs and key informant interviews in Ngorongoro District, over time, communities were able to respond to these changes through seasonal mobility as a strategy to utilize spatial variations in wet and dry seasons as well as pasture resources. Based on the inherited historical livelihood systems of pastoral societies, nomadic life was not a prominent pastoral production system. Interviews with key informants indicated that, as of 1967 net migration contributed 9.5% to the population of Arusha Region. Accordingly, household survey findings indicated that perceived reasons for in-migration into the current settlement were marriage and following their relative (11.9%), 6.5% of the respondents migrated for the purpose of searching for arable land, 4% of the respondents were searching for employment, and 2.5% of the respondents were motivated by change of settlement.

According to Monduli District Environmental Plan, the Waarusha, a sedentary agriculture group of Maasai origin from the Mount Meru area, were among the first groups to settle in the former Maasai plains. They mainly settled in the southern part of the district and around the major towns. In the 1950s and 1960s people were attracted by wage employment on commercial farms. It was reported by key informants that migrants come mainly from densely populated fertile highlands to the sparsely populated and less fertile lowlands. However, during the field surveys, it was reported that patterns of pastoral mobility have become more unpredictable as well as encountering various challenges.

It was reported by key informants that mobility has largely become haphazard, increasingly involving few or selected household members, and associated with various land use conflicts, especially between livestock keepers and small-scale farmers. In most cases, the objective for mobility has been for diversification of food and income sources, and coping with environmental stresses such as drought. FGD participants indicated that migration has increasingly become selective, given the challenges related to land tenure, whereas some household members remain home for household responsibilities such as taking care of children and elders, managing small herds and maintaining the 'identity of the land holder'.

According to the FGD participants, in most cases, women remain with household responsibilities while other household members such as youth migrate with livestock in search of pasture and water during stress. FGD participants added that other middle-aged household members can migrate to towns in search of off-farm activities to generate income for purchasing food and other household needs. Findings from the household surveys across the study indicated that in-migration to the villages has varied over time as 5.9% immigrated to the current village in the 1970s, 3% of the respondents migrated in the 1980s and 7.9% moved to the village in both the 1990s and the 2000s. So generally, the study villages experienced more immigrants during the 1990s and 2000s (Table 7.3).

Generally, this study reveals that migration as a coping strategy in pastoral communities is still evident, adoption of permanent residence is increasingly becoming evident and selective migration of household members might be a current strategy. The study also reveals that rural–rural migration as a coping strategy is still practised and climate change might increase the pressure that makes people migrate from resource-poor to resource-rich areas. However, planned mobility, especially for pastoralists, needs to be promoted in order to enhance spatial utilization of resources, such as benefiting from landscape variations in wet and dry seasons and the resulting effects these have on pastures.

Alternative pasture sources

FGD participants across the study area indicated that some livestock keepers have a number of alternative pasture sources to curb the impacts of the changing climate. FGD participants indicated that in most cases, alternative pasture sources are supported by the development of service infrastructure such as construction of charcoal dams (*malambo*) to ensure water supply for domestic or human use and livestock throughout the year and cattle dips to control tickborne diseases. According to discussions with FGD and interviews with key informants in Ngorongoro District, over time, communities were able to respond to these changes through seasonal mobility as a strategy to utilize spatial variations in wet and dry season as well as pasture resources. FGD participants reiterated that with increasing stresses related to climate change, there have been difficulties to cope given the severity of the impacts they pose.

Table 7.3. Timelines for respondents' immigration to the current village. (From field survey, 2016.)

Category	Frequency	Percentage
Born in the village	152	75.2
1970s	12	5.9
1980s	6	3.0
1990s	16	7.9
2000s	16	7.9
Total	202	100.0

Therefore, new strategies, especially those recommended by elders, have been the only course every household is trying to reach. These include adoption of feeding livestock with maize bran in periods of drought while most pastoralists mix bran with salt to make the feeds palatable for cattle. Putting aside grazing so it is reserved for use in dry seasons, rationing of livestock watering days (i.e. livestock are not watered every day) and planting of reserved pasture (e.g. the UNESCO project at Ololosokwan) have been a successful story among pastoral communities across the study area (Fig 7.5). Berhe *et al.* (2016) indicated that pastoralists have also embarked on multiple adaptation strategies including: (i) water harvesting schemes; (ii) fodder production; (iii) feed (e.g. maize bran) purchase; (iv) migration; (v) livestock diversification; and (vi) animal restocking. Income sources are also diverse coming from livestock, cropping, sales of fuelwood, agricultural wages, remittance and relief aid.

Cutting tree branches to feed livestock during drought seasons and providing grazing assistance to people with a large number of livestock have become common response mechanisms to unpromising climate extremes for these pastoral societies. It was also reported by FGD participants that local communities have learnt that keeping large numbers of livestock which require extensive pastures and a lot of water during dry seasons is not a viable livestock-keeping strategy during the current climate trends. Accordingly, distribution of livestock among several herders has been adopted during dry seasons or drought periods, in order to reduce the risk of losing all the livestock, and as a mechanism of distributing the burden of searching for pasture. The FGD participants also emphasized that they have learnt that convergence of livestock from different areas in a common place increases the spread of pests such as tsetse fly and ticks, and their associated diseases. Therefore, they have

Fig 7.5. Cattle feeding on hay during the dry season in Monduli District. (From field survey, 2016.)

learnt to reduce the chances of converging with wildlife in order to avoid the spread of pests and diseases.

One of the FGD participants emphasized that distribution of livestock herds can be done through cash and in-kind payment; in-kind payment involves agreement that there will be a certain number of livestock after the drought season. Agreement about the number of livestock varies considerably depending on the situation of the drought and efforts required to serve the number of livestock given. Findings from this study support those of Morton and Barton (2002), that pastoralists destock their herds as a strategy in times of drought. Apart from distribution, pastoralists also sell their livestock during drought as a mechanism to get cash for maintaining smaller herds throughout the crisis. This strategy has also been reported by Mwakaje (2013), that livestock culling and revolving herds by selling and purchasing before and after the crisis has been a common practice especially in response to recurrent drought induced by climate change.

Livelihood diversification

Livelihood diversification is one of the common response strategies in the study area, elsewhere in Tanzania and other parts of the world. Enyew and Hutjis (2015) and Wangu (2014) discuss that livelihood diversification can enhance resilience to the impacts of the changing climate. Through household surveys, the study found that most people across the study sites were using various strategies for responding to the impacts of climate variability and change. Discussions with FGD participants in Ngorongoro District indicated that livelihood diversification has become common given the increase of stresses that affect livestock-keeping activities. It was revealed that diversification of economic activities has become inevitable in some villages (Kitwai A and Kitwai B villages in Simanjiro District) and these villages are slowly adopting an agropastoral system, and Emboreet village has become completely agropastoral.

About 84% of the respondents reported that they were applying a mixture of strategies, which enabled them to diversify their livelihoods and cope with the impacts of weather extremes such as recurrent drought. Adoption of new food sources (including consuming goat's and sheep's milk, which was a rare diet item among the Maasai) and borrowing money from rich people in the community for business and paying back with a small amount of interest were some of the common livelihood diversification strategies reported by the respondents. Other strategies included seeking salaried jobs in town (especially as watchmen) and seeking paid activities such as the construction of livestock *bomas* (fortifications that protect livestock from predation) for wealthy households.

Discussions with respondents in Kitwai villages in Simanjiro District revealed that in many pastoral communities in northern Tanzania, possession of large numbers of livestock was used to define wealth which encourages more stock keeping. However, currently communities have resorted to other means of survival. Livestock keepers are now practising crop production, growing maize and beans as a way to diversify their living. In some wards and villages (e.g. Emboreet in Simanjiro District) more than 60% of the villagers grow maize and beans. Other community members in Simanjiro have moved most of their efforts to mining (e.g. the tanzanite mine in Mererani) in order to diversify their income. Beekeeping is also practised by a small group of communities (informally known as *N'ndorobo* among the Maasai) as an important economic activity. Small businesses (kiosk, food vendors and crop vendors) were also becoming popular among pastoral communities as a way of diversifying sources of income.

Discussions with FGD participants in Monduli indicated that livelihood diversification was clearly divided into three categories according to how well endowed people were with resources among three segments in the community. These three segments were the rich people, the middle-class segment and the poor group. The rich people (*Orkasis*) were mainly livestock keepers, farmers using tractors, business owners and lenders of money and cattle to others. The middle-class people (*Olanyuaani*) engaged in livestock keeping and crop cultivation using hand hoes and family labour. The poor people (*Ndorobo/Oloisinani*) were mainly considered to be workers on other people's farms, looking after other people's livestock to get paid and sometimes they borrowed livestock for milk only. In a study carried out in three villages in Simanjiro District, Mung'ong'o (2016) provided a more detailed study about the social stratification of these pastoral communities.

These categories were also partly reflected through disparities in land ownership that had influence on adaptive capacity, types of household activities and livelihood improvements under changing climate conditions. Field survey findings indicated that the majority (56.4%) of respondents owned 1–5 acres, the next largest group were respondents (23.2%) who owned 20–100 acres, then 17.8% of the respondents owned 6–19 acres, and finally a few respondents (2.5%) owned less than 1 acre (Fig. 7.6).

Variations in land size were largely influenced by the types of household activities and reflected the household economic base and assets to sustain the respective farm size. Discussions with FGD participants indicated that most of the rich people had large farms, largely because they were able to hire casual labour during growing seasons. It was also mentioned by a few FGD members that large farm size was a reflection of livelihood transition, whereby some pastoral communities were gradually adopting mixed farming, combining both crop production and livestock keeping. As shown in Table 7.4, most (29.6%) of the respondents occupied their land in the 1990s. Other respondents had occupied their land from the 1980s (22.6%), 2000s (24.6%) and 2010s (11.8%). Additionally, about 11.3%

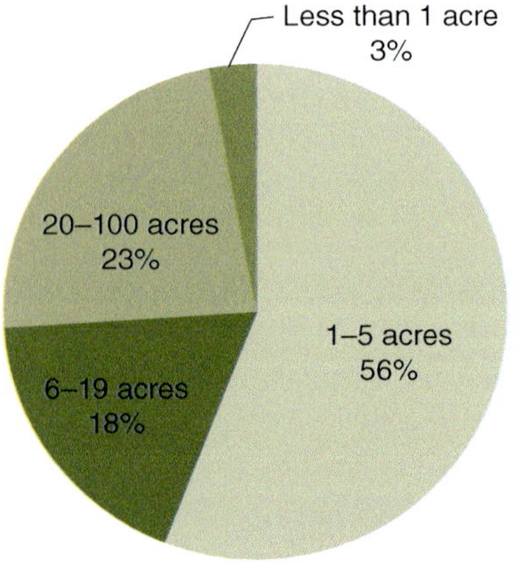

Fig. 7.6. Respondents' perceived land size ownership. (From field survey, 2016.)

Table 7.4. Respondents' views on timelines for land occupation. (From field survey, 2016.)

Years	Frequency	Percentage
1980s	46	22.6
1990s	60	29.6
2000s	50	24.6
2010s	24	11.8
Not known[a]	23	11.3
Total	203	100.0

[a]Respondents who had no idea about when they acquired their land.

Table 7.5. Respondents' major economic activities. (From field survey, 2016.)

Activities	Frequency	Percentage
Crop cultivation	9	4.5
Livestock keeping	18	8.9
Mixed farming[a]	175	86.6
Total	202	100.0

[a]A mix of crop production and livestock keeping.

of the respondents were not sure of when they occupied their land.

According to the household interviews, the majority (46.5%) of respondents acquired their land through village government, 44.6% of the respondents inherited their land, 4.5% bought their land, and 4.5% of the respondents rented their land. Discussions with FGD participants indicated purchase of land in pastoral communities was a new phenomenon, but it was increasingly becoming a common business largely due to formalization of land rights, adoption of mixed farming among pastoral communities and establishment of permanent settlement among the pastoral communities. Accordingly, interviews with key informants indicated that adoption of a mixed farming system characterized the transition of the pastoral communities towards agropastoralism. Livelihood transition from pastoralism to agropastoralism was also considered by key informants as one of the response strategies to diminishing pasture as a result of recurrent drought. FGD participants added that diversification of household livelihood options was largely influenced by the need to increase sources of food and income in the studied villages. Evidence of livelihood transitions among the pastoral communities was also revealed during the household survey. According to household survey findings, 86.6% of the respondents were practising both crop production and livestock keeping, commonly known as mixed farming (Table 7.5).

In terms of crops produced, household respondents indicated that maize and beans were produced by the majority (80.2%) of the respondents, followed by those producing more than maize and beans (14.4%), and other respondents (5.4%) reported to be producing maize only. It was also reported by the majority (91.6%) of respondents that they were keeping both cattle and goats. Other livestock kept were sheep, which was only reported by a few (8.4%) respondents. According to FGD findings, goats were kept as a buffer to risks, especially droughts, because goats were believed to be more resistant to harsh environmental conditions compared with some cattle varieties. This partly implies that, if effectively promoted, keeping goats can be used to increase the adaptive capacity of most pastoral communities in the study area. This study also found that the majority of respondents had less than 100 cattle (84.7%) and goats (75.5%) in their households (Table 7.6).

Challenges related to the response measures

Communities' efforts to respond to the impacts of the changing climate have faced several challenges. Mobility challenges, gender inequality, illiteracy and the nature of households are some of the challenges observed by this study, as discussed in the following subsections.

Unsustainable benefits of response measures

This study found that most of the response measures to changing climate had short-term benefits for pastoral communities. However, FGD participants and key informants indicated that some coping strategies

Table 7.6. Number and type of livestock owned. (From field survey, April 2016.)

Number of livestock	Livestock (%)		
	Cattle	Goats	Sheep
Less than 10	25.4	18	24
10–49	43.3	40	34.6
50–99	16	17.5	18.5
100–500	14.4	24.5	20.5
Above 500	1	0	2

were observed to be unsustainable. For example:

- Using tree branches to feed livestock when pasture was limited was suitable for small herds but could not be practised with a large number of livestock, partly because it is hard to cut enough tree branches to feed a large herd (e.g. more than a thousand animals). Key informants also added that use of tree branches to feed animals was not environmentally friendly as it could require complete clearance of small trees.
- Use of short-term crop seeds was not viable for many livestock keepers because it was reported that most of the short-term crop cultivars were very expensive. The high price of seeds meant some of the population could not afford them at all while for others it was not easy to afford to cultivate the whole farm so often only small portions of farms were cultivated. Further, managing farm operations was difficult, especially for those households with quite large farms.
- Adoption of a fattening centre and revolving livestock herds through selling during the onset of drought and reviving herds after cessation of drought was also considered as unsustainable among pastoral communities. FGD participants reported many livestock keepers could not afford to do this. Interviews with key informants indicated that in some cases when cattle were kept rather than sold at the onset of a drought, the cattle may lose value (as they loose condition due to restricted food and water intake) and may even die as the drought becomes more severe. Key informants also reported that the strategy of selling and rebuying cattle is mostly practised by the young generation who were especially conscious of economic value. Members of the older generation were still reluctant to apply this strategy so therefore it had not been fully adopted.

Mobility challenges

Mobility is increasingly becoming common among livestock keepers in response to the impacts of the changing climate. FGD participants indicated that mobility is largely accelerated by the search for new pastures and water by the livestock keepers especially during the dry season. Discussions with key informants and FGD participants indicated that mobility is no longer viable and helpful due to various challenges. Land use conflicts, especially between crop farmers and livestock keepers, and sometimes between livestock keepers and conservationists, was mentioned by the majority of respondents as the key challenge. In some instances, convergence of livestock from different areas in a common place increases spreading pests such as tsetse fly, ticks and their associated diseases.

Gender inequality

This study made efforts to include both male (54.5%) and female (45.5%) respondents. Among the interviewed respondents, 64.4% were heads of households, while 35.6% were representing their households. Inclusion of both male and female respondents is considered to be vital as gender is one of the important cross-cutting issues in many development policies and different spheres of social welfare, including climate change adaptation, natural resources access and use, planning and management (Atela *et al.*, 2018).

The study also considered that men and women have different livelihood dimensions that make them vulnerable to climate change impacts. Additionally, men and women have inherent cultural differences that affect their adaptive capacity and implementation of adaptation strategies (see, for example, Ngigi *et al.*, 2017; Wrigley-Asante *et al.*, 2017; Atela *et al.*, 2018).

This study found that gender disparity was one of the challenges affecting adaptation efforts across the study districts. Interviews with key informants revealed significant gender inequality in access to climate change information, and participation in planning and the decision-making process related to natural resources management and climate change adaptation. It was revealed that despite ongoing efforts to ensure gender balance in planning and the decision-making process, most of the leaders are yet to be sensitized on gender aspects in the planning process, including interventions related to climate change adaptation. For instance, interviews with key informants in Monduli District indicated that, due to strong cultural norms and traditions among the Maasai, male leaders dominate the district planning process, particularly at the lower administrative levels (village, ward and division). It was also found that female representation is low in technical and professional fields, except in education and community development.

Consistent with this, Nelson *et al.* (2002) writes that gender issues can be significant in climate change adaptation particularly in communities where the division of labour and gender roles are distinct, based on the resource ownership and other responsibilities related to caring for the household. As a result, male dominance and exclusion of females in the planning process, especially climate-change-related interventions can have negative implications on climate change adaptation. It was reported by key informants that women, especially those who are married, are responsible for daily household chores such as taking care of the children, the sick and the elderly, providing water and food and procuring firewood. Similar roles were also stipulated in the District Environmental Plan for Monduli District of 1996.

It was also reported by key informants and FGD participants that women's work load has been increasing due to the gradual change of diet among pastoralists from milk to grains because more firewood is needed and cooking requires more time. The Adaptation Learning Programme (2006) indicated that changes in pastoralist societies have resulted in additional social, cultural and political constraints to resource access and adaptive decision making. Findings from FGDs also indicated that, until recently, most men rarely accepted responsibilities for the domestic chores. Inclusion of females in the planning process could have a positive outcome in enhancing adaptation, especially in agriculture which is largely occupied by women.

Wrigley-Asante *et al.* (2017) indicate that climate change adaptation initiatives at the local level must take gender differences into consideration in planning and implementation of adaptation interventions in order to enhance communities' resilience. Increasing gender sensitization programmes among the Maasai community can help to empower women to adapt to the impacts of the changing climate. Inclusive policies for natural resources management are also essential for increasing women's participation in planning and management of natural resources and enhance well-being of communities and sustainability of adaptation practices. Other studies (Thaler and Priest, 2014; Holland, 2017; Schlosberg *et al.*, 2017) indicated that, if not well planned and implemented, climate change adaptation strategies can favour the 'minority' while placing negative consequences on the 'larger' group.

Key informant interviews further added that the majority of women are not involved in planning and decision making related to resource use, and this might have negative impacts on their empowerment and future adaptation to climate change impacts. Thaler *et al.* (2018) also recognize that unequal access and distribution of resources is among the key factors that increase vulnerability to climate change impacts, and affects the adaptive capacity of the 'excluded' social groups. Vulnerability of some social groups (e.g. women) can also increase relative exclusion in planning and implementation of adaptation interventions. Therefore, it is important to promote interventions that increase social inclusiveness in resource ownership and empowerment for wider community adaptation in pastoral communities.

Inadequate knowledge

This study found that about 45% of the respondents had no formal education while 40% had primary education. Other respondents (10.4%) had secondary education and 3% had attended additional training besides their secondary education, while a few (2%) had a university education (Fig. 7.7). Even though traditional knowledge (informal education) can influence good adaptation strategies, formal education has a high propensity to influence effective and sustainable adaptation strategies in a community (Raygorodetsky, 2011; Sharma *et al.*, 2013; Banda *et al.*, 2015).

As already stated and shown in Fig. 7.7, the majority of respondents (45%) had no formal education and 40% of respondents had only received primary education. A low level of education is one of the hindrances to increasing awareness of the impacts of climate change and to implementation of interventions for climate change adaptation (URT, 2015; Pandey *et al.*, 2018). Additionally, as reported in other studies, a high level of education increases the chances of accessing better employment that could contribute to higher income levels, good health conditions and the power to make decisions as well as making the community less vulnerable to the impacts of climate change (URT, 2015).

Given the importance of education in socio-economic development, discussions with FGD participants across the study area revealed that the government and other development partners have established adequate infrastructure for formal education from primary to secondary education levels in order to increase literacy levels among the communities. For instance, discussion with FGD participants in Monduli District indicated that improved education infrastructure had increased the enrolment of children from agropastoral communities into primary and secondary education compared with previous years. It was reported by the FGD participants in Monduli District that primary and secondary schools have been made accessible in most parts of the district, especially in Esilalei Ward, where four primary schools and one secondary school have been built.

During field surveys in Ngorongoro District at Ololosokwan village, it was observed that the government, in collaboration with UNESCO, has supported the building of various education infrastructures, including a kindergarten, primary and secondary schools. The schools were built using traditional architecture and there were digital facilities for teaching science subjects. The villages also received health services through a mobile clinic and a dispensary owned by Clens Company. These interventions have the potential to increase the adaptive capacity and reduce vulnerability to the impacts of climate change among the studied communities.

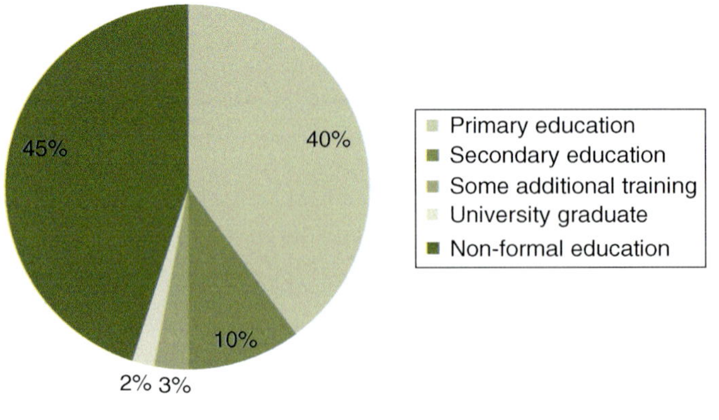

Fig. 7.7. Level of education of the respondents. (From field survey, April 2016.)

Conclusion and Recommendations

This study reveals that pastoral communities are being affected by decreasing and unreliable rainfall patterns. These affect pastoral communities through diminishing availability of pasture and water as well as inducing recurrent drought. Loss of livestock herds and disruption of socio-economic systems suffered by the majority of households were reported by different categories of respondents. As a result, communities responded to the impacts through various strategies. Livelihood diversification has been adopted by many livestock keepers in order to reduce the severity of climate change impacts. In some instances, livelihood diversification included a shift from the pastoralism mode of production to agropastoralism. However, this study observes that livelihood diversification strategies, especially crop production and livestock keeping, could likely accelerate land use conflicts and degradation of natural resources in a few years time. Conflicting land use due to livelihood diversification could also be triggered by rapid population growth which will exert pressure for expansion of settlement areas and farms to meet the demands of food supply and commercial purposes.

Expansion of crop farms is also likely to induce conversion of pasturelands and block wildlife and livestock corridors which were being used by wildlife and pastoralists for spatial utilization of resources, especially searching for water and pasture. A new land use system in pastoral landscapes may also necessitate increasing the seasonal movement of livestock keepers in search of pasture and water, especially during dry seasons. Without clear mobility paths for pastoralists with their herds of cattle, there might emerge resource use conflicts, especially conflicts between pastoralists and agriculturalists. Therefore, solutions for mobility and harmonized land uses need to be a primary priority for the government and key actors in order to ensure sustainable pastoral landscape management, effective adaptive capacity among pastoral communities and viability of livelihoods.

Acknowledgements

Sincere thanks for the financial support for research from the project for 'Building Knowledge to Support Climate Change Adaptation for Pastoralist Communities in Eastern Africa' funded by the Open Society Initiative for Eastern Africa (OSIEA). The project is implemented by the University of Dar es Salaam through the Centre for Climate Change Studies. We would like to acknowledge contributions by all who were involved in data collection and data management, including local government authorities for research clearance and logistical support.

Notes

[1] Agriculture in this context is conceptualized as the science or practice of farming, including cultivation of the soil for the growing of crops and the rearing of animals to provide food, wool and other products (URT, 2013).

[2] Findings through Standardized Precipitation Index (SPI) tests.

References

Adaptation Learning Programme (2006) *Climate Change Vulnerability and Adaptive Capacity in Garissa County, Kenya*. CARE International, Nairobi.

Agrawala, S., Moehder, A., Hemp, A., Van Aalst, M., Hitz, S., *et al.* (2003) *Development and Climate Change in Tanzania: Focus on Mount Kilimanjaro*. Organisation for Economic Co-operation and Development (OECD), Paris.

Atela, J., Gannon, K.E. and Crick, F. (2018) Climate change adaptation among female-led micro, small, and medium enterprises in semiarid areas: a case study from Kenya. Centre for Climate change Economics and Policy Working paper no. 338. Grantham Research Institute on Climate Change and the Environment Working Paper No. 304. Available at: http://www.lse.ac.uk/GranthamInstitute/wp-content/uploads/2018/10/working-paper-338-Atela-et-al.pdf (accessed 12 December 2019).

Banda, C., Namafe, C.M. and Chakanika,W.W. (2015) Traditional environmental knowledge among Lozi adults in mitigating climate change in the Barotse Plains of Western Zambia. *International Journal of Humanities Social Sciences and Education* 2(9), 222–239.

Belay, A., Recha, J.W., Woldeamanuel, T. and Morton, J.F. (2017) 3 Smallholder farmers' adaptation to climate change and determinants of their adaptation decisions in the Central Rift Valley of Ethiopia. *Agriculture and Food Security* 6(24). doi:10.1186/s40066-017-0100-1

Berhe, M., Hoag, D., Tesfay, G., Oniki, S. and Kagatsume, M. (2016) Effects of adaptation to climate change on income of cattle owners in the pastoral and agro-pastoral communities of northern Ethiopia. In: *2016 AAAE Fifth International Conference, September 23–26, 2016, Addis Ababa, Ethiopia* (No. 246974). African Association of Agricultural Economists (AAAE). Available at: https://ageconsearch.umn.edu/record/246974/files/310.%20Afar%20Pastoral%20and%20Agro-pastoral.pdf (accessed 12 December 2019).

Blaikie, P., Cannon, T., Davis, I. and Wisner, B. (2014) *At Risk: Natural Hazards, People's Vulnerability and Disasters*. Routledge, London.

Brooks, N. and Adger, W.N. (2004) Assessing and enhancing adaptive capacity. In: Lim, B. (ed.) *Adaptation Policy Frameworks for Climate Change: Developing Strategies, Policies and Measures*. UNDP and Cambridge University Press, Cambridge, pp. 165–181.

Eisenack, K. and Stecker, R. (2011) An action theory of adaptation to climate change. *Earth System Governance Working Paper* No. 13. Earth System Governance Project, Lund and Amsterdam.

Enyew, B.D. and Hutjis, R. (2015) Climate change impact and adaptation in South Omo Zone, Ethiopia. *Journal of Geology and Geophysics* 4(208), 2.

Few, R. (2007) Health and climatic hazards: framing social research on vulnerability, response and adaptation. *Global Environmental Change* 17(2), 281–295.

Fischer, G., Mahendra Shah and Harrij van Velthuizen (2002) Climate Change and Agricultural Variability. A special report, prepared by the International Institute for Applied Systems Analysis under United Nations Institutional Contract Agreement No. 1113 as a contribution to the World Summit on Sustainable Development. Johannesburg 2002 (Global, agriculture).

Fyumagwa, R.D., Ezekiel, M.J., Nyaki, A., Mdaki, M.L., Katale, Z.B., *et al*. (2011) Response to Rift Valley fever in Tanzania: challenges and opportunities. *Tanzania Journal of Health Research* 13(5), 332–339.

Holland, B. (2017) Procedural justice in local climate adaptation: political capabilities and transformational change. *Environmental Politics* 26, 391–412. doi:10.1080/09644016.2017.1287625

Intergovernmental Panel on Climate Change (IPCC) (2007) *Climate Change 2007: Impacts, Adaptation and Vulnerability. Contribution of Working Group II to the Fourth Assessment Report of the Intergovernmental Panel on Climate Change*. [Parry, M.L., Canziani, O.F., Palutikof, J.P., van der Linden, P.J. and Hanson, C.E. (eds)]. Cambridge University Press, Cambridge.

Jonsson, L.O., Mawenya, E. and Rockstrom, J. (2003) Conservation tillage I: management practices for animal drawn systems in Tanzania. *Working Paper* No. 16. Regional Land Management Unit (RELMA) Working Paper series. RELMA, Nairobi.

Kothari, C.R. (2004) *Research Methodology: Methods and Techniques*, 2nd edn. New Age International Publishers, New Delhi.

Lynn, S. (2010) The pastoral to agro-pastoral transition in Tanzania: human adaptation in an ecosystem context. Available at: http://economics-of-cc-in-tanzania.org/images/Stacy_Lynn_Pastoralism_TZ_Draft_2010_08-09_draft_2_v2.pdf (accessed 12 December 2019).

Monduli District (2015) District Socio-economic Profile. Monduli District, Arusha Region, Tanzania.

Mortimore, M., Anderson, S., Cotula, L., Davies, J., Faccer, K., *et al*. (2009) *Dryland Opportunities: a New Paradigm for People, Ecosystems and Development*. International Union for Conservation of Nature (IUCN), Gland, Switzerland, International Institute for Environment and Development (IIED), London and United Nations Development Programme (UNDP)/Drylands Development Centre (DDC), Nairobi.

Morton, J. and Barton, D. (2002) Destocking as a drought-mitigation strategy: clarifying rationales and answering critiques. *Disasters* 26(3), 213–228.

Mung'ong'o, C.G. (2016) A tale of three villages: a study of social stratification in three pastoral communities in Simanjiro District, northern Tanzania. In: Yanda, P.Z. and Mung'ong'o, C.G. (eds) *Pastoralism and Climate Change in East Africa*. Mkuki na Nyota Publishers, Dar es Salaam, Tanzania, pp. 227–248.

Mwakaje, A.G. (2013) The impact of climate change and variability on agro-pastoralists' economy in Tanzania. *Environmental Economics* 4(1), 30–38.

Ndesanjo, R. (2017) Pathways to enhance climate change resilience among pastoral households in Simanjiro District-northern Tanzania. PhD thesis, the University of Dar es Salaam, Dar es Salaam, Tanzania.

Nelson, V., Meadows, K., Cannon, T., Morton, J. and Martin, A. (2002) Uncertain predictions, invisible impacts, and the need to mainstream gender in climate change adaptations. *Gender & Development* 10(2), 51–59. doi:10.1080/13552070215911

Ngigi, M.W., Mueller, U. and Birner, R. (2017) Gender differences in climate change adaptation strategies and participation in group-based approaches: an intra-household analysis from rural Kenya. *Ecological Economics* 138, 99–108.

Oxfam (2008) Survival of the fittest: pastoralism and climate change in East Africa. *Oxfam Briefing Paper* 116. Oxfam, Nairobi.

Page, E.A. (2007) *Climate Change, Justice and Future Generations*. Edward Elgar Publishing, Northampton, Massachusetts.

Pandey, R., Kumar, P., Archie, K.M., Gupta, A.K., Josh, P.K., *et al.* (2018) Climate change adaptation in the western-Himalayas: household level perspectives on impacts and barriers. *Ecological Indicators* 84, 27–37.

Pelling, M. and High, C. (2005) Understanding adaptation: what can social capital offer assessments of adaptive capacity? *Global Environmental Change* 15, 308–319. doi:10.1016/j.gloenvcha.2005.02.001

Rama, R.E. (2013) Factors affecting multisectoral collaboration response to Rift Valley fever outbreak in Kongwa District. Doctoral dissertation, Muhimbili University of Health and Allied Sciences, Dar es Salaam, Tanzania.

Raygorodetsky, G. (2011) Why traditional knowledge holds the key to climate change. United Nations University. Available at: https://unu.edu/publications/articles/why-traditional-knowledge-holds-the-key-to-climate-change.html (accessed 26 November 19).

Schlosberg, D., Collins, L.B. and Niemeyer, S. (2017) Adaptation policy and community discourse: risk, vulnerability, and just transformation. *Environmental Politics* 26, 413–437. doi:10.1080/09644016.2017.1287628

Sharma, U., Patwardhan, A. and Patt, A.G. (2013) Education as a determinant of response to cyclone warnings: evidence from coastal zones in India. *Ecology and Society* 18(2), 18. doi:10.5751/ES-05439-180218

Simanjiro District (2015) *District Socio-economic Profile*. Simanjiro District, Manyara Region, Tanzania.

Thaler, T. and Priest, S. (2014) Partnership funding in flood risk management: new localism debate and policy in England. *Area* 46(4), 418–425. doi:10.1111/area.12135

Thaler, T., Zischg, A., Keiler, M. and Fuchs, S. (2018) Allocation of risk and benefits – distributional justices in mountain hazard management. *Regional Environmental Change* 18, 353. doi:10.1007/s10113-017-1229-y

United Republic of Tanzania (URT) (2003) *Initial National Communication under the United Nations Framework Convention on Climate Change*. Vice President's Office, Dar es Salaam, Tanzania.

United Republic of Tanzania (URT) (2007) *National Adaptation Programme of Action (NAPA)*. Vice President's Office (VPO), Division of Environment, Government of United Republic of Tanzania, Dar es Salaam, Tanzania.

United Republic of Tanzania (URT) (2009) The Wildlife Conservation Act. Tanzania Publishing House, Dar es Salaam, Tanzania.

United Republic of Tanzania (URT) (2013) *National Agriculture Policy*. Ministry of Agriculture, Food Security and Cooperatives, Dar es Salaam, Tanzania.

United Republic of Tanzania (URT) (2014) *Second National Communication under the United Nations Framework Convention on Climate Change*. Vice President's Office, Dar es Salaam, Tanzania.

United Republic of Tanzania (URT) (2015) *Pangani District Council (2014), Pangani District Socio-economic Profile (2013)*. Prime Minister's Office, Regional Administration and Local Government, Dodoma, pp. 22–55.

Vermeulen, S.J., Campbell, B.M. and Ingram, J.S. (2012) Climate change and food systems. *Annual Review of Environment and Resources* 37(1), 195.

Waiswa, M.M. (2003) *Strategic Choices for Enhancing Capacity of Rural Communities Adapt to Climate Variability: a Case for Uganda*. Department of Meteorology, Ministry of Water, Lands and Environment, Kampala.

Wangu, J.M. (2014) Pastoral land privatization and community adaptability to climate change in Maji Moto, Kenya: on opportunities and negative implications of the tenure reform. Master's thesis, Utrecht University, Utrecht, the Netherlands.

Wrigley-Asante, C., Owusu, K., Egyir, I.S. and Owiyo, T.M. (2017) Gender dimensions of climate change adaptation practices: the experiences of smallholder crop farmers in the transition zone of Ghana. *African Geographical Review* 38(2), 126–139. doi:10.1080/19376812.2017.1340168

Yanda, P.Z. and Mubaya, C.P. (2011) *Managing a Changing Climate in Africa: Local Level Vulnerabilities and Adaptation Experiences*. African Books Collective, Oxford.

Yanda, P.Z and Mung'ong'o, C.G. (2016) *Pastoralism and Climate Change in East Africa*. Mkuki na Nyota, Dar es Salaam, Tanzania.

8 Assessment of Socio-ecological Resilience of Agropastoralists to Climate Change and Variability Impacts in Bariadi District, Tanzania

Eugen Cyrilo and Claude G. Mung'ong'o*

Institute of Resource Assessment, University of Dar es Salaam, Dar es Salaam, Tanzania

Abstract

In developing countries like Tanzania, societal vulnerability to the risks of climate change and variability (CC&V) exacerbate ongoing social and economic challenges because people's livelihoods are largely dependent on resources that are sensitive to climate change such as agriculture. Although studies show that most communities in Africa have low adaptive capacity, for centuries people have developed traditional adaptation strategies to face climate inter-annual variability and extreme events based on their long-term experiences. Various studies show how CC&V have impacted the socio-economic and and environmental conditions among the pastoral and agropastoral societies. However, little emphasis has been given to studying the community's resilience status to CC&V impacts. Much of the focus has been placed on studying the community vulnerability and impacts of CC&V as well as coping and adaptation strategies to avert CC&V impacts. Little is known on how the interaction between society and nature can enhance or reduce community resilience under changing climate. The study was conducted in two villages, Ibulyu and Mahaha, in Bariadi District. The main objective of the study was to deepen our understanding of the socio-ecological resilience of agropastoral communities to CC&V impacts in a semi-arid district. The study employed both qualitative and quantitative research designs. Quantitative data were captured through a household survey whereas qualitative data was collected through focus group discussion, key informant interviews and field observation. The results show that CC&V have negatively affected the farming system in the study area. The ecological setting of the area has significantly been altered to the extent that it cannot provide the required ecosystem services and products that are important for human and livestock sustainance. Changes in the production system have negatively affected community resilience and increased their vulnerability.

Introduction

Social–ecological systems, where ecosystems and humans are inextricably linked, are facing unpredictable pressures and shocks due to global change and unsustainable human use of resources. These shocks may be internal to the system, such as overuse of a particular natural resource, or external, such as possible impacts of climate change. Such climatic shocks are problematic since they may result in the irreparable loss of ecosystem functions and services, and a subsequent collapse of dependent human livelihoods.

Rural livelihoods in many parts of the world are dramatically affected by climate change and variability (CC&V) and the corresponding impacts on provision of ecosystem services. This is particularly the case in

*claude_gasper@yahoo.com

the semi-arid tropics (SAT), which contain 22% of the world's population and high concentrations of chronic poverty and inadequate food consumption. Much of the vulnerability of smallholders within the SAT is driven by surface hydrological dynamics; both directly through rainfall variability and indirectly through additional human- or climate-induced land and water degradation. As a result of CC&V, social and ecological systems are at greater risk of adverse effects such as: (i) decreased cereal productivity and coastal flooding in low latitudes; (ii) animal and plant species extinction threats; and (iii) annihilation of marine ecosystems particularly coral bleaching (IPCC, 2007; Yanda and Mubaya, 2011).

A loss of resilience and adaptive capacity is loss of opportunity and response options. This in turn could increase the probability that the socio-ecological system will move into undesirable trajectories. In a resilient socio-economic and natural system, events such as disturbances can create opportunities for development and innovation. Carpenter *et al.* (2001) argued that in a system where resilience is high, human actors have the ability to sustain the combined system of humans and nature in a desirable state, along a desirable course, in response to changing conditions and disturbance events. In a vulnerable social–ecological system even a small event may be devastating. Therefore, this requires focus on the capacity of actors in a socio-ecological system to cope with changes without limiting future options (Berkes *et al.*, 2003). This tight coupling between social–ecological and hydrological systems in the SAT make them an ideal setting to conduct fully integrated research between social and physical sciences.

Statement of the research problem

CC&V remain a major challenge to rural livelihoods globally. In developing countries like Tanzania, societal vulnerability to the risks associated with CC&V may exacerbate ongoing social and economic challenges because people's livelihoods are largely dependent on resources that are sensitive to changes in climate such as agriculture. Although studies show that most communities in Africa have low adaptive capacity, for centuries people have developed traditional adaptation strategies to face the great climate inter-annual variability and extreme events based on their long-term experiences.

Various studies show how CC&V have impacted the socio-economic and environmental conditions among the pastoral and agropastoral societies (Mwandosya *et al.*, 1998; Mung'ong'o and Mwamfupe, 2003; Tumbo, 2007; Kazembe, 2009; Lema and Majule, 2009; Lyimo and Kangalawe, 2010). However, little emphasis has been given to studying the community's resilience status to CC&V impacts. Generally, much focus has been given to studying community vulnerability and impacts of CC&V as well as coping and adaptation strategies to avert CC&V impacts.

Also, there is no place-specific study and little is known on how the interaction between the society and nature can enhance or reduce community resilience under changing climate.

It is on this premise that the study embarked on assessing the socio-ecological resilience of agropastoralists to CC&V impacts in Bariadi District. The main objective of the study was to deepen our understanding of the socio-ecological resilience of agropastoral communities to CC&V impacts of Bariadi District, Tanzania. The study specifically focused on assessing the impacts of CC&V to livelihood activities as well as assessment of vulnerability of different livelihood options to CC&V. Lastly, the resilience status of the agropastoralists was assessed by looking at the significance of CC&V impacts, the level of vulnerability, as well as assessment of the efficiency and effectiveness of coping and adaptation strategies.

Conceptual framework

The study employed the Sustainable Livelihoods Framework (SLF) developed by the

Department for International Development (DFID, 1999) to assess the resilience of the agropastoralists to CC&V impacts in Bariadi District. The SLF provides an approach to understanding vulnerability, current adaptive capacity (in terms of assets) and how these are converted into strategies and outcomes, particularly in adapting to CC&V (Majule *et al.*, 2013). The SLF shown in Fig. 8.1 provides an entry point for assessing the current levels of vulnerability and resilience of the community. It helps to assess livelihood construction using assets and the ways these assets are affected by vulnerability to socio-ecological shocks, trends and seasonality. It also relates livelihoods to government and private structures as well as processes, which affect people's lives (in terms of laws, culture, norms, etc.) and finally to outcomes people expect to achieve from engaging in certain activities as envisioned by Lugenja *et al.* (2006). The SLF uses the concept of capital assets as a central feature and considers how these are affected by the 'vulnerability context' in which they are derived (Mung'ong'o and Mwamfupe, 2003). These capital assets are human capital, social capital, natural capital, physical capital and financial capital.

This approach assumes that people require a range of livelihood assets in order to achieve positive livelihood outcomes and that no one single asset can function independently of others in order to generate livelihood outcomes (Lugenja *et al.*, 2006). A strong asset base is necessary with sufficient conditions for an adaptive livelihood, since the strongest asset base cannot yield an adaptive and resilient livelihood unless an enabling environment (policy, world

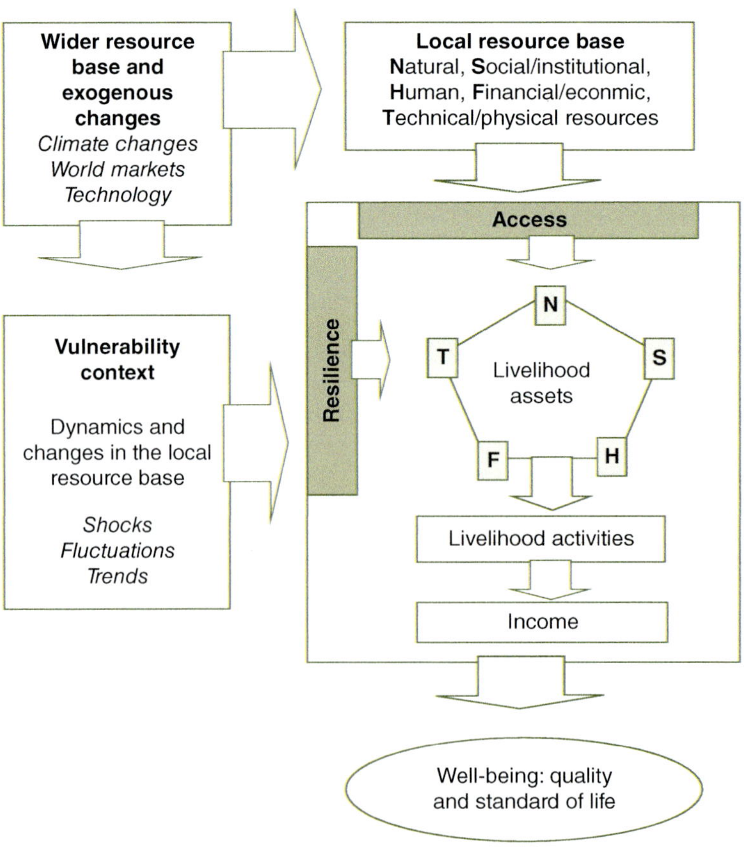

Fig. 8.1. Sustainable Livelihoods Framework. (Adapted and modified from DFID, 1999.)

markets and technologies) also exists. Measuring such assets is essential as this enables analysis of the ways people construct their livelihoods and substitute one capital to replace others in order to cope during periods of socio-ecological shocks and/or stresses.

In the context of this approach human capital (H) represents the skills, knowledge, ability to work and good health, which enable people to pursue different livelihood strategies and achieve their livelihood objectives. It also refers to the amount and quality of labour available at household level. Social capital (S) in this case entails the social resources upon which people draw in pursuit of their livelihood objectives. These are developed through: (i) networks and connectedness, membership of more formalized groups; (ii) rules, norms and sanctions; and (iii) relationships of trust, reciprocity and exchange. Natural capital (N) is a form of capital asset which represents the natural resource stock from which resource and services, which are useful for livelihoods, flow (e.g. nutrient cycling, erosion protection, etc.). These include a wide range of resources from intangible public goods (e.g. the atmosphere and biodiversity) to divisible assets used directly for production (e.g. land, trees, etc.).

The technical/physical capital (T) comprises the basic infrastructure which consists of changes to the physical environment that help people to meet their basic needs and to be more productive. And the fifth capital asset is financial capital (F) denoting the financial resources that people use to achieve their livelihood objectives. It includes flows and stocks that contribute to consumption as well as production of livelihood resources. It includes the availability of cash or equivalents that enable people to adopt and change to different livelihood strategies.

Materials and Methods

The study area

Bariadi District was selected as the study area for the following reasons. First, it falls within the semi-arid areas of Tanzania where there are frequent food shortages due to uncertainty of rainfall and land degradation resulting from soil exhaustion. Second, the district's economy is dependent on crop production and livestock keeping, and according to NAPA (2007) these sectors are more vulnerable to CC&V impacts which have resulted from prolonged drought incidences and increase in temperature due to rainfall variability.

Bariadi District is located between latitudes 2°15′ and 3°10′ S of the equator and longitude 33°40′ to 35°10′ E of Greenwich. Bariadi District is among five districts forming the Simiyu Region, others being Maswa, Itilima, Meatu and Busega districts (Fig. 8.2). The district covers an area of 5484.68 km^2. Bariadi District is bordered by Busega District in the west, Mara Region in the north and Arusha Region in the east. In the south it is bordered by Itilima District. Administratively Bariadi District is made up of four divisions, 25 wards and 83 villages.

The study was conducted in two villages in the district, Ibulyu and Mahaha. Ibulyu village is located in Sakwe Ward, and 4 km from the district headquarters in Bariadi. The village is accessed through the Bariadi–Kasoli–Mwanza road. On the other hand, Mahaha village is located in Bunamhala Ward, 7 km from the district headquarters, along the Bariadi–Itilima–Meatu road.

According to the National Population Census of 2012, Bariadi District had a population of 422,916 people (201,718 males and 221,198 females) with an annual average growth rate of 1.8% (regional growth rate), and a population density of 77.1 persons/km^2 (URT, 2012). While the annual average growth rate of 1.8% is similar to the regional growth rate, the population density is relatively higher than the regional density of 63 persons/km^2. The district had 82,946 households with an average household size of 7.1 persons/household. The average household size was higher than the regional and national household sizes of 6.9 and 4.8 persons/household, respectively. However, Meatu and Itilima districts' household sizes were the highest; averaging 7.4 and 7.2 persons/household, respectively. In comparison with

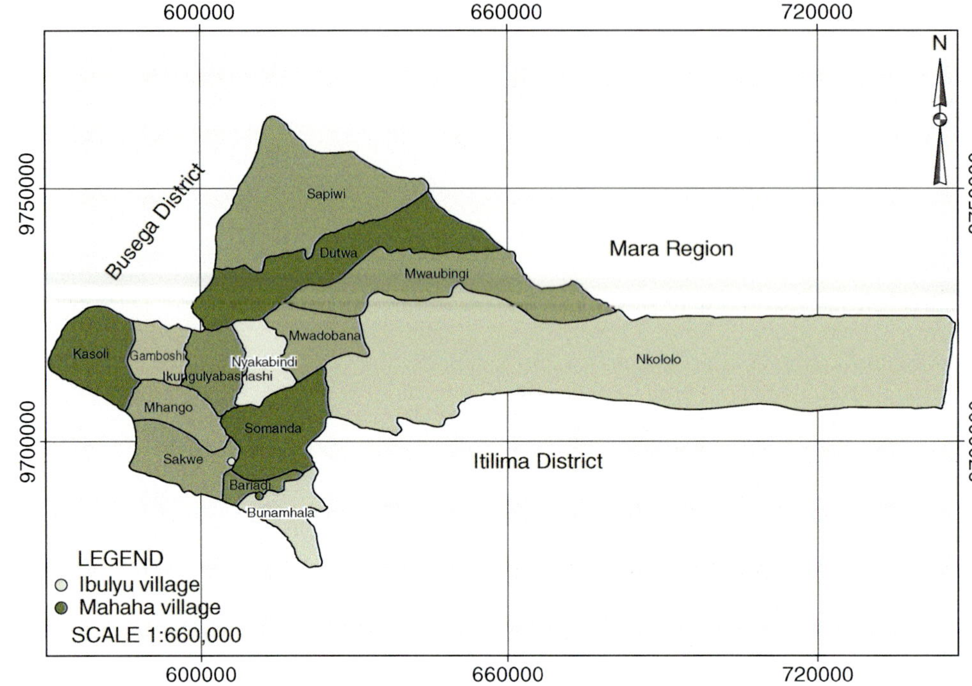

Fig. 8.2. Bariadi District showing ward boundaries and study area. (From BDC, 2010.)

other districts in the region, Bariadi District had the highest population, followed by Maswa District with 344,125 people. Busega District was the least populated district with 203,597 people. Itilima and Meatu districts had a population of 313,900 and 299,619, respectively.

Nevertheless, according to a 2010 District Demographic Survey Ibulyu village had a population of 7318 whereby 1046 were males, 1874 were females and 4398 were children below the age of 15 (BDC, 2010). The village had six hamlets (Madukani, Ngwagabate, Mhango, Mwabasabi, Lugongo and Ilomelo) and 755 households. Meanwhile, Mahaha village had a population of 5064 people whereby 1604 were males, 1732 females and 2268 were children below the age of 15. The village had 516 households and five hamlets, namely Mahaha, Nhigwa, Magalai, Mwansasi and Msanzu.

The Sukuma ethnic group mainly occupies Bariadi District. The Sukuma are Bantu-speaking people who live in north-western Tanzania around Lake Victoria (also known as Sukumaland). According to some sources, ancestors of the Sukuma were part of the extensive migrations of people speaking early forms of Bantu, in the first millennium AD. They, along with the Nyamwezi further south, seem to belong to the same group as the Bantu of western Uganda (Cory, 1954). Sukumaland proper was then inhabited by roughly 1,000,000 people who belonged to the central Eastern Bantu and comprised, with the Nyamwezi and related tribes, a group very closely related in language and customs (Cory, 1954, p. 1).

There is no tradition extant about the original inhabitants of the area. According to the Sukuma themselves, before the arrival of the Hamitic groups, the country was just bush, dense in some places and sparse in others. A European explorer described the Sukuma as driving their cattle into the forest during the dry season. The country was sparsely populated everywhere. The scattered human colonies consisted of a hundred odd people under the leadership of an *ntemi*, a name which only later became the title of a chief (Cory, 1954, p. 2). The Sukuma fought with the Taturu and Maasai in eastern Usukuma during the 19th century, and they have grown so large because of the migration of other ethnic groups joining the Sukuma (Homes and Austen, 1972, p. 378).

Today the Sukuma are the largest ethnic group in Tanzania, constituting about 13% of the total population. The Sukuma are both agriculturists and livestock keepers. However, with increased regional integration in Tanzania, the district is currently resided by different ethnic groups, especially in trading centres and in the district headquarters.

Bariadi District has a tropical type of climate with clearly distinguished rainy and dry seasons. The average temperature for the region is about 28°C and the minimum rainfall is 600 mm/year and the maximum 900 mm/year. The district has two seasons a year: the rainy season and the dry season. The rainy season usually starts between mid-October and December and ends in the second week of May. Normally, it has two peak seasons. The first peak lies between mid-October and December, while the second and longer one, falls between February and mid-May. As such, the whole rainy season covers a total of almost 6 months, with a dry spell, which usually occurs in January.

The dry season begins in mid-May and ends in mid-October. This is a period of about 5 months. The dry season is the worst period as the soils are hard to cultivate, pastures become very poor, and availability of water for domestic use and livestock become acute (Lyimo and Kangalawe, 2010).

The majority of residents in Bariadi District are mainly agropastoralists who rely on a combination of crop production and livestock keeping. Other sectors include wildlife and tourism activities in Maswa Game Reserve and Serengeti National Park, and business activities and small-scale industries (mainly cotton ginneries) which are estimated to contribute 10–20% of the district's gross domestic product (GDP) (BDC, 2010).

Crop production is the most important source of livelihood (BDC, 2010). About 80% of the entire population depends on agriculture for subsistence. Agriculture contributes approximately 80–95% of the district's GDP. Crop production is undertaken under communal and individually owned farms which are mainly small scale. Crop production is rain-fed and mainly for subsistence. The main agricultural tools used are the hand hoe, ox-ploughs, power tillers and a few tractors for cultivation. Crops produced include both food crops (maize, cassava, sorghum, beans, rice, chickpeas, groundnuts and sweet potatoes) and cash crops. For many years cotton formed the major cash crop in the district. However, currently there is a shift towards cultivation of maize and rice as cash crops due to cotton price fluctuations and increased costs of production. Vegetables and fruit gardening are practised through small-scale traditional irrigation.

Livestock keeping is the second most important source of livelihood in the district. The major livestock kept in Bariadi District are cattle (mainly stunted zebu type), goats, donkeys and sheep. Livestock are kept under free-range grazing. Previously, livestock were kept in individually or communally owned enclosures locally known as '*ngitiri*'. But with population increase and prolonged drought, pastures have become scarce thus forcing pastoralists to graze livestock in farms after harvesting, or in abandoned farms as well as along riverbanks and roads. Other pastoralists have sold their livestock and have migrated to other regions such as Morogoro, Rukwa, Mbeya, Lindi and Pwani looking for pastures.

The local breed of cattle (zebu type) are preferred because they have high rates of reproduction, higher resistance to diseases, and can survive the chronic shortages of pastures that affect the district, especially during the dry season. For the Sukuma, livestock have great value. They use cattle for subsistence, socio-economic security, paying of dowry and fines, and as a symbol of wealth and status. Cattle also serve as a medium of exchange and transportation, and they provide traction for cultivation. Moreover, cattle are an object of affection and supreme religious significance.

Research design

This study employed both qualitative and quantitative research designs. A quantitative design was used to capture information through a household survey where heads of households were interviewed as the representative of their households. A qualitative design was used to complement data that was collected through the household survey. The qualitative data was collected through focus group discussions (FGDs), key informant interviews and field observation. The study employed a cross-sectional design where data was collected at a single point without repetition from the representative sample (De Vaus, 1993). This was due to the nature of the study, time and financial constraints.

Sampling procedure

Purposive sampling was used in selecting villages, which suited the objectives of the study. The criteria used to select the villages included the presence of households which depended on agriculture and livestock keeping as the main source of livelihood. Key informants for the study were also purposively selected basing on their expertise, knowledge and experience about the environment and climate, CC&V, as well as how climatic variations have affected the social–ecological set up of the respective study area and key livelihood activities. Key informants in the study included: (i) those with experience of agriculture, livestock and water; (ii) representatives of the land, natural resource and environment departments; (iii) village agriculture and livestock extension workers; (iv) elders from the respective villages; and (v) the village chairman and the village executive officer.

Simple random sampling was used to determine the households to be interviewed during the household survey in both villages. The household formed the unit of the study for questionnaire administration, whereby heads of households were interviewed to represent their households. A household was identified as group of people who eat from a common pot, sharing the same house and maybe cultivating the same land (Njana, 1998). URT (2002) defined a household as the arrangement made by persons individually or in groups for providing themselves with food and other essentials for living or a group of people who live together and share expenses.

The sampling exercise for this study was conducted by dividing the village into their respective hamlets. Thereafter, households were randomly picked from each hamlet with the guidance of the hamlet leaders. Random sampling was applied because households in the village had no clear pattern of arrangement; also socio-economic activities in the village are similar in all households so it was easy to apply simple random sampling for household interviews.

The sample for FGDs was purposively selected to ensure that various groups in the village had equal and fair representation in the study. These groups included the youth, women, the disabled, widows, elders, farmers and pastoralists. The study involved all these groups to avoid domination by one group and/or gender, especially males who dominate most of the decision-making processes in agropastoral societies.

Sample size for the study

According to Bernard (1994), the adequacy of the sample size is assessed by how well the sample represents the whole population under study. Therefore, the more accurate a sample is required, the larger the sample must be. Also, Kothari (2004) argued that sampling design must result in a truly representative sample and be such that the results of the sample study can be applied, in general, for the universe with a reasonable level of confidence.

Based on these criteria, the study sampled 12% of the total households in each study village for household interviews. Thus, a total of 102 households were interviewed: 60 and 42 households from Ibulyu and Mahaha village, respectively. The selected sample size was considered to be adequate since the collection of required data was complemented by other methods used such as FGD, key informant interviews, field observation and secondary data review.

As outlined by Kothari (2004), the key informant interviews were based on selection of people with relevant knowledge, expertise and experience in relation to this study. A sample size of between 15 and 33 respondents should be adequate for key informant interviews. Therefore, the study adopted a total sample of 28 key informants, which provided adequate information required for the study. The sample for FGDs must be small enough for everyone to have their voices represented yet large enough to capture a range of voices (Onweugbuzie and Leech, 2007).

According to Liamputtong (2010) focus group interviews involve a group of 6–8 people who come from similar social and cultural backgrounds or who have similar experiences or concerns. Johnson and Christensen (2004) suggested a sample size for focus groups of between 6 and 12 persons, whereas Langford *et al.* (2002) and Morgan (1997) recommended 6–10 individuals. However, Krueger (2000) warned that more than 12 members in such groups tended to limit participants opportunity to share their views and observations. Despite this warning, this study undertook one FGD in each village, with a sample size of 14 participants in Ibulyu village and 13 participants in Mahaha village.

Data collection methods

Secondary data collection

Secondary data, relevant to the study, were obtained from different sources such as published and unpublished research papers. These data were obtained from the University of Dar es Salaam library, the Institute of Resource Assessment documentation centre, and Tanzania Meteorological Agency, the Ministry of Agriculture and Cooperatives, the Ministry of Livestock and Fisheries Development, Bariadi District Council, the National Bureau of Statistics and Internet searches. Such information included rainfall and temperature data, demographic characteristics, socio-economic activities, biophysical information and policies and programmes relevant to climate change. Results from these reviews have been used to support various arguments related to the study.

Primary data collection

QUALITATIVE DATA. Participatory rural appraisal (PRA) techniques were used for collecting in-depth knowledge of local people to CC&V. Such methods have the advantage of soliciting more information from local people, since they encourage participation and dialogue between local people and outsiders (researchers), as well as among local people themselves (Poffenberger *et al.*, 1992; Chambers, 1994; Mikkelsen, 1995). The PRA techniques used in this study included FGD, key informant interviews, wealth ranking and field observation.

Focus group discussion (FGD). According to Liamputtong (2011), focus group methodology is useful in exploring and examining what people think, how they think, and why they think the way they do about the issues of importance to them without pressuring them into making decisions or reaching a consensus. FGD builds up a collective and creative enthusiasm, which leads to sharing and discussing new ideas and concepts introduced by an outsider who is familiar with them (Chambers, 1994).

Two FGDs were carried out, one in each village of study. Checklists were used to provide guidance on the discussion; each discussion taking about 90 minutes. Key issues of the discussion were on community awareness and perception of climate change, indicators of CC&V, observed and perceived impacts as well as coping and adaptation strategies employed by the local community to avert climate change impacts.

Key informant interviews. A key informant interview is a qualitative in-depth interview with people who know what is going on in a community. Key informants are usually people who are knowledgeable, know the area well, and can provide insights about people living or working in the community. They are informative about the nature of the issue being investigated and can give recommendations for solutions (Carter and Beaulieu, 1992). The key informants must be purposively selected from the set of study participants to increase the likelihood that the key informants represent those of the other study members (Sankoff, 1971; Maxwell, 1996).

Eleven key informants from Ibulyu and five from Mahaha village were drawn for interviews guided by an open-ended interview schedule. Key informants from Ibulyu included the village chairman, village executive officer, one village council member and an elderly pastoralist who owned large herds of cattle in the village. Key informants in Mahaha village were the village chairman, village executive officer, one hamlet chairman, one vegetable grower/cultivator and a former Kilulu Cooperative Society clerk.

The main objective of conducting key informant interviews was to gather knowledge from prominent people as they possessed resourceful information from their area, which was deemed useful for this study. Also, key informants from government offices were interviewed to get the general knowledge concerning the study topic. They also provided statistical information and government plans, programmes and interventions for climate change which could not be easily obtained from the villages. These departments included agriculture and livestock, water, land, natural resources and environment.

Wealth ranking. Wealth ranking was then carried out to identify local criteria used to differentiate households on the basis of wealth and well-being (Mukherjee, 1992). Group interviews were carried out with local key informants who were asked to group households according to their wealth status by going through the list of household heads. To get everyone thinking along the same lines, participants were asked to define what was meant by the concept 'wealth', or what types of assets a wealthy person would have. It was revealed that among the Sukuma community wealth mostly means the size of farms owned by a household and livestock herds owned. However, other attributes were included such as: (i) income; (ii) type of house structures owned; (iii) access to farm inputs and implements; and (iv) food security. The household formed the unit of ranking and the ranking was done as described elsewhere by Mung'ong'o (2016).

Physical observations. Participatory physical observations were made to capture and cross-check issues raised by household interviews, FGDs and key informant interviews. Transect walks were carried out in each village to observe the effects of human activities on water sources, from the main water sources such as the Ibulyu and Senani rivers, to the swampy areas. During the transect walks photographs were taken to provide pictorial evidence on issues relevant to the study. Field observation when used accurately helps the researcher eliminate any subjectivity from the respondent's views (Kothari, 2004).

QUANTITATIVE DATA

Household questionnaire. A semi-structured questionnaire was administered in both villages. Pre-testing of the questionnaire for modification and improvements of the research questions preceded the household survey. In this study, the household heads were interviewed and in case the heads were absent their wives, spouses or other representatives were interviewed instead.

In each of the study villages a total sample of 12% of the households were randomly selected. Within the village the same proportion of households was selected from each sub-village. The questionnaire had four major sections reflecting the research objectives: (i) respondents' general information; (ii) community perception and indicators of CC&V; (iii) socio-ecological vulnerability and repercussion of such changes to community livelihood; and (iv) coping and adaptation strategies employed to avert CC&V impacts.

Data analysis and presentation

Quantitative data collected through household questionnaires were analysed by using Statistical Package for Social Sciences (SPSS) version 20.0. The question variables were first coded and then imported to the software for cross-tabulation, enabling the calculation of frequencies and percentages. Meanwhile, content data from FGDs, field observations and key informant interviews were analysed thematically and used to supplement data collected through household interviews. Microsoft Excel 2007 was used to analyse rainfall and temperature data. These data were summarized and presented using frequencies, histograms, charts and photographs.

Results and Discussion

Socio-economic characteristics of the respondents

The distribution of the socio-economic characteristics of the respondents in the study villages was described in terms of sex of respondents, marital status, age, size of household, education level and main economic activities. The socio-economic status of the community influences its resilience towards CC&V impacts.

Age and sex of respondents

The study involved 102 respondents from both sexes in both villages (60 respondents and 42 respondents for Ibulyu and Mahaha, respectively). Out of all the respondents in both villages, 62.7% were male and 37.3% were female, (Fig. 8.3). In Ibulyu village 73.3% of respondents were male and 26.7% were female, while in Mahaha village, 47.6% of respondents were male and 52.4% were female.

In this regard male respondents dominated the study because the focus of the study was to interview household heads. This finding reflects the patriarchy of African societies where often a man heads the household. In the households where female respondents were interviewed these were either single-parent families headed by women or they were interviewed in the absence of their husbands. Male dominance was also accentuated by women's reluctance to be interviewed; hence often their sons represented them.

Involvement of all genders in the study was crucial as livelihood resources access and control is always gendered (Rossi and Lambrou, 2008). Both women and men have common and distinct forms of knowledge and skills in resource use. This is due to differing responsibilities they have in production, reproduction and trade (Nelson and Stathers, 2009). Thus recognition of these gendered knowledge systems and skills may provide a rich resource for understanding climate change. Also, men and women were likely to have different options and 'safety

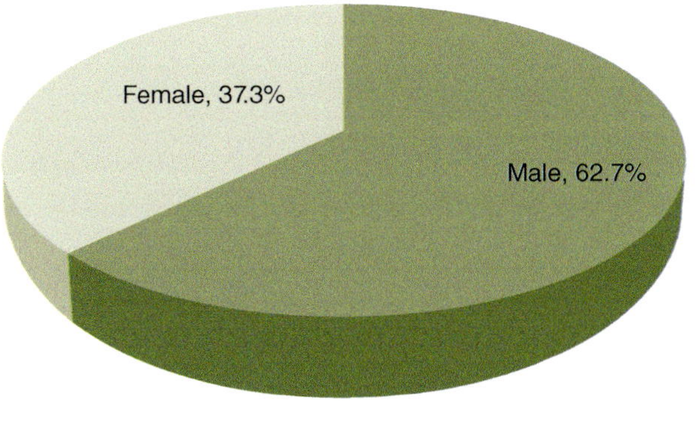

Fig. 8.3. Sex of the respondents. (From field survey, March 2014.)

nets' for adaptation to and coping with climate change (Rossi and Lambrou, 2008).

In respect of age of the respondents, the sampled population's age ranged from 28 years to 75 years, with an average age of 47 years. The majority of the respondents (87.9%) were in the working age group of 15–64 years while the rest (12.1%) of the respondents constituted the dependent group aged above 64 years. This observation corresponds well with the National Population and Housing Survey (URT, 2013) that showed that 53.1% of the population was aged between 15 and 64 years, while 44.6% and 3.3% were young people below 15 years and the elderly at more than 64 years, respectively. Such a population structure dominated by a young and an elderly population indicates that the respective authority needs to invest more in education for the young population and in the health sector for the elderly.

Dominance of an active working group in the study area indicates that the area has an abundant labour supply, which can easily be involved in various livelihood activities. Also, the youth and the active working population are more adaptive and willing to learn modern technologies than the elderly. The ability of this group to learn about and use new technologies means there is an opportunity to provide them with new technologies for adaptation to and mitigation of the impacts of CC&V.

In terms of age and sex most of the respondents were in the age group 35–49 years, and of these the study found that 53.1% and 52.6% were male and female respondents, respectively. The second most common age group was 20–34 years of which 26.4% were female and 12.5% were males. However, there were fewer females in age group 50–64 years (10.5% were females compared with 21.9% males) and in the age group above 64 years 10.5% were female and 12.5% were males (Table 8.1).

That the proportion of female respondents is high in the lower age group and declines towards the higher age groups is due to the fact that girls usually get married at a young age compared with males (females either marry after completing primary education or drop out of school and opt for marriage, while males enter marriage mainly in their late 20s). The decline of female respondents in the higher age group is also due to reluctance of elderly female members to be interviewed, thus asking their children to represent them.

Household size

Data from this study revealed that household size in the study area ranges between two and 19 people per household. Households with six to ten people were dominant (73.3%), whereas 26.3% of households had more than ten persons/household (Table 8.2).

The average household size was 8.4 people/household. This is relatively higher than the district average household size of 7.3 people/household, the regional average of 6.9 and the national household size of 4.8 people/household (URT, 2012). In rural areas such as Bariadi District large household sizes could supply more labour, especially when the majority of household members are in the working group. However, household sizes have negative implications on natural resources and environmental sustainability, especially when the community's livelihood activities are highly dependent on natural resources (Adekunle, 2009).

Table 8.1. Sex of the respondents by age group. (From field survey, March 2014.)

	Age group (%)				
Sex	20–34 years	35–49 years	50–64 years	>64 years	Total
Male	12.5	53.1	21.9	12.5	100
Female	26.4	52.6	10.5	10.5	100
Average	19.5	52.9	16.2	11.4	100

Table 8.2. Percentage of household sizes in Ibulyu and Mahaha villages. (From field survey, March 2014.)

No. of people/ household	Ibulyu (%)	Mahaha (%)
1–5	16.7	42.86
6–10	56.6	28.57
10+	26.7	28.57
Total	100	100

Marital status

The study found that 80.3% of the households, were headed by a pair of parents (mother and father), 15.7% of households were widowed, whereas 2.0% were single and/or separated. The analysis shows that households with paired parents were most efficient in the production activities due to adequate labour force and equal division of labour. They had more advantages for livelihood activities diversification as they were engaged in multiple activities such as crop production, livestock keeping and off-farm activities such as business. Divorced and single-headed households were mostly engaged in non-farm activities such as food vendoring, selling of the local brew and casual labour, especially during the agricultural season. Therefore, the majority of households being headed by paired parents were effectively organized socially and economically, and hence were more resilient to any change in the production system.

Education level

Education is a human capital asset, which is increasingly mentioned as necessary for livelihood diversification. The findings of this study indicated that primary education (Standard 1–7) was the highest level of education attained by most respondents (75.8%). On the other hand, some 24.2% of respondents did not attend school at all. This percentage was, however, lower than the average district literacy rate of 45% for people aged 5 years and above (URT, 2002). None of the respondents had more than primary education. In Ibulyu village 68.3% attained primary education, while 31.7% never attended formal education. For Mahaha village, 83.3% had primary education, while 16.7% had no formal education at all (Table 8.3).

Table 8.3. Respondents' level of education by village. (From field survey, March 2014.)

Education level	Ibulyu (%)	Mahaha (%)
No formal education	31.7	16.7
Primary education	68.3	83.3
Total	100	100

Furthermore, the study found that illiteracy was higher among females, whereby 21.1% had no formal education compared with 18.5% of the males. In terms of age group, it was found that respondents aged 50–64 years and above 64 years had higher rates of illiteracy at 22.1% and 50%, respectively. On the other hand, literacy was high among the 20–34 years age group with 88.9%, followed by 85.2% of the 35–49 years age group.

These findings show that the level of literacy of the respondents in the study villages was generally low which limited the level and ability of the community to innovate and adopt new ideas and technologies, especially those related to climate change adaptation and mitigation. An illiterate community is less innovative than a literate community thus affecting the dissemination of new adaptation technologies. As pointed out elsewhere by Adekunle (2009), the ability of an individual to adapt to change and accept new ideas and innovations, including climate change adaptation and mitigation technologies, is directly influenced by the level of education as literacy supports ease of access to agricultural extension services and new technologies.

Main socio-economic activities

There are various socio-economic activities that are carried out in the study area. These include agriculture, petty business and wage employment (Fig. 8.4). However, many of the respondents were engaged in more than one economic activity. Rain-fed crop production formed a major part of the livelihood as 100% of the respondents in both villages

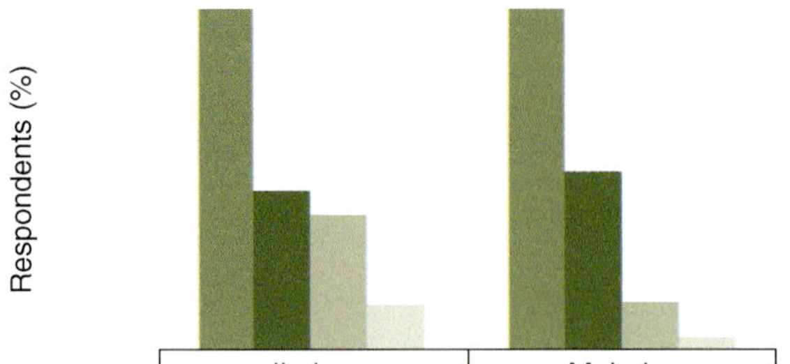

	Ibulyu	Mahaha
Crop production	100	100
Livestock keeping	46.7	52.4
Petty business	40	14.3
Wage employment	13.7	4.3

Fig. 8.4. Main economic activities in the study area. (From field survey, March 2014.)

were engaged in crop production. Rain-fed agriculture depends on the patterns of rainfall and its reliability. Smallholders, characterized by hand-hoe and ox-plough cultivation with limited use of agricultural inputs, dominate crop production. It was revealed by this study that crop production was declining over time mainly due to prolonged dry spells and shortage of rains. Major food crops grown in the study area included maize, rice, sweet potatoes, sorghum, cassava, beans, groundnuts and sunflower. Previously, cotton was produced as the main cash crop but currently due to price fluctuations and rising costs of production, food crops such as maize, rice, beans, groundnuts and sunflower were replacing cotton as cash crops.

Livestock keeping was the second most important socio-economic activity after crop production, with about 47% of respondents in Ibulyu and 53% of respondents in Mahaha village involved in livestock keeping. Most of the animals kept included local breeds of cattle, sheep and goats, which were kept under free grazing regimes. Other livestock kept included pigs and poultry. Petty business activities are now emerging as alternative sources of income complementing agriculture. Data from this study show that as many as 40% of respondents in Ibulyu and 14.3% of respondents in Mahaha were involved in different petty businesses, such as food and water vendoring, carpentry and masonry, selling and buying of crops, running kiosks, bicycle and motorcycle repairs and charging of mobile phones using solar power. The least important among the major sources of income was wage employment, which involved activities such as making bricks, carrying building sand, and cultivation and weeding during the farming season (13.7% and 4.3% of respondents in Ibulyu and Mahaha, respectively).

In recent years most of the youth are now engaged in bicycle transportation of passengers locally known as *dala dala* as well as motorcycle transport *boda boda*. Although the economy is based on the major crops grown, it was established that household involvement in an off-farm activity increased households' resilience, especially during the dry season when agricultural yields are very low.

With regards to petty business activity far fewer respondents in Mahaha village (4.3%) were involved in this activity than in Ibulyu (13.7% of respondents). This is due to the difference in the levels of development in the two villages. As Ibulyu is more urbanized than Mahaha village, it means that there are more petty business opportunities. Given that rain-fed farming and livestock keeping are the main economic activities in both villages this implies that CC&V will have significant impacts on the livelihoods of these communities. Nevertheless, Ibulyu villagers have more options for coping with CC&V impacts since they have more opportunities to diversify their economy easily than the Mahaha villagers.

Community knowledge, perceptions and indicators of CC&V

Local knowledge about climate

To assess local knowledge about climate the study used the same variables used by Lyimo and Kangalawe (2010) and Majule *et al.* (2013). These elements included rainfall, temperature, drought, wind and humidity. The study found that rainfall, temperature and drought were highly perceived by respondents as elements of climate (Fig. 8.7), and this was because community livelihoods are mainly influenced by the availability of reliable rainfall to support crop production and livestock keeping, while prolonged drought and increased temperature adversely affect the respondents' crops and livestock. Also, it was revealed that respondents had sufficient knowledge to identify their local climatic conditions.

Understanding climate change showed narrow variations among age groups. Rainfall, temperature and drought were ranked high as elements of climate. Among the age group 20–34 years, rainfall was ranked high by 43% of respondents in this age group, followed by drought with 29% of respondents, while temperature and humidity ranked low with 19% and 10% of respondents, respectively. Among the age group 35–49 years, rainfall was also given high priority by 40% of respondents in the age group, followed by temperature (28% of respondents), drought and humidity (24% and 8% of respondents, respectively). There was little variation between the groups 50–64 and above 64 years. Between them 31% and 35% of responses, respectively, ranked rainfall highly, followed by 31% and 24% for temperature, 24% and 26% for drought, as well as 12% and 18% for humidity. These findings portray that the community's perception of climate is based on the behaviour of elements which affect the respondents' livelihood activities and that there is a narrow variation among livelihood activities undertaken by each age group.

Furthermore, based on the level of education of the respondents, 54% of respondents without formal education identified rainfall as an element of climate; this is due to the fact that the community livelihoods were influenced by rainfall reliability. Only 15% of these respondents identified drought, while temperature and humidity ranked high with only 22.9% and 20% of these respondents, respectively. Contrary to that, respondents with primary education ranked rainfall and drought high (80% and 85% of these respondents, respectively), while 77% and 46% of these respondents identified temperature and humidity as elements of climate, respectively. This shows that indigenous knowledge still prevails in the society and influences community understanding and decision making in various livelihood aspects. Also, formal education supplemented and added extra knowledge to indigenous education. However, incidences of crop failure and livestock deaths were reported to result from rainfall scarcity and increased incidences of drought. Variables such as wind, flood and humidity ranked low, as they did not directly affect communities' daily livelihood activities.

Local perceptions and indicators of CC&V

Local perceptions and indicators of climate change were based on local experience over the past 20 years. Data from this study show that the local community is aware of their environment and hence has adequate experience of their local climate. The study revealed a growing concern among respondents that CC&V is already occurring. However, they associated climate change with variability in weather conditions, which is related to rainfall inconsistency and unpredictability over years rather than actual change. The variability was related to variations in agricultural seasons in a year. The majority of the respondents in Ibulyu and Mahaha villages (94% and 97%, respectively) attributed changing climatic conditions mainly to decreased rainfall amount and increased incidences of drought and temperature increase.

During the survey respondents gave multiple responses about the local indicators of CC&V, which were outlined and ranked according to their level of importance to community livelihoods (Table 8.4). Rainfall variability was most perceived by 100% and 95% of respondents in Ibulyu and

Table 8.4. Respondents' perceptions of local indicators of climate change and variability (CC&V). (From field survey, March 2014.)

Indicators	Ibulyu (% of respondents)	Mahaha (% of respondents)
Rainfall decrease	100	95
Temperature increase	94	97
Recurrent food shortage	77	92
Late rainfall	70	76
Decrease in crop production	60	57
Lengthening growing season	62	52
Decreasing number of livestock	20	43

Mahaha villages, respectively, as rainfall had become highly unpredictable: there had been delays in onset of rainfall and early cessation of rains. Temperature variability was ranked highly too, but after rainfall variability; it was claimed that there had been an increase in temperature and drought which had significantly affected agricultural activities. Other major concerns were related to indicators such as the lengthened growing season, late rainfall as well as recurrent food shortage resulting from decreased crop productivity and decrease in the number/death of livestock (Table 8.4).

However, responses on perception and interpretation of indicators of CC&V varied among male and female respondents; this was due to variations in roles and responsibilities. As argued by Nelson and Stathers (2009) the division of labour might expose household members to different experiences, which in turn affected their way of thinking, their perception of environmental stimuli and ultimately their responses.

Perceptions of rainfall variability

The study found that there has been variability in the trend and characteristics of rain over the past 20 years. Findings from both FGDs and key informant interviews correspond to those of the household survey, as they related rainfall variability with El Niño rains, recurrent drought and temperature increase. They indicated that there had been some years which they experienced heavy rainfall and some with erratic rains. The years with high rainfall were associated with increase in agricultural production and water availability, while the latter were associated with crop failure and livestock deaths.

However, it was revealed that they were experiencing more years with poor rainfall than in the past. One elderly key informant in Mahaha village claimed that in past years dry spells came after an interval of time and its trend could be predicted. Years ending with number 4 from 1974 onwards were believed to be bad years characterized by poor rainfall leading to poor agricultural harvests and livestock productivity, while in current years the trend was becoming uncertain and difficult to predict. Also, increase in occurrence of dry spells had shortened the rainy season as the length between rain days was becoming longer causing drying of crops. As a result farmers were reported to quit planting paddy in some years due to drying of swamp areas, especially in 1994, 2006 and 2013.

Furthermore, respondents indicated that rains were now concentrated in fewer months and particularly fewer days in those fewer months, after which there were many drier months which reduced the consistency of rainfall. Respondents in FGDs were able to recall years which had poor or no rainfall, characterized by drought incidences. They mentioned that the years in which they started witnessing climate variability were 1984, 1994 and 2006; rainfall was scarce in these years, resulting in crop failure and death of livestock. Similarly, 1997/98 was a bad year, where crops were affected by heavy rainfall and high blowing wind resulting from the El Niño rains.

Perception of temperature variability

The majority of the household interview respondents (93%) acknowledged that there had been temperature increases for the past 20 years. Most of the respondents perceived temperature variability through change in the heat and hot months in a year. Of the respondents, 97% reported that hot months were increasing due to the occurrence of prolonged dry spells in the wet months. Also, it was revealed that in the past, mornings were very cold, warming up by noon, with the evenings and the nights becoming cold again, except for the normal warmer months. However, in present years the coldness of the area was no longer the same as the normal cold times of the day had become warmer.

During key informant interviews and FGDs it was claimed that temperature increase resulting from drought incidences had led to crop failure and livestock deaths. 1974, 1984, 1994, 2006 and 2013 were noted by the local community as the most dry and hottest years, characterized by significant crop failure, water scarcity and death of livestock.

Impacts of CC&V on rural livelihoods

Community vulnerability to CC&V impacts

To assess households' vulnerability to climate change, the approach used followed the Intergovernmental Panel on Climate Change (IPCC) typology as presented in Kazembe (2009). Community vulnerability is defined as a function of biophysical and socio-economic factors prevailing in the community. The variables are selected to reflect the household's adaptive capability, sensitivity and exposure to climate change variability. Exposure relates to the degree of climate stress on a particular unit of analysis and is regarded as long-term climate change variability. Adaptive capability is considered as the ability of a system to adjust to actual or expected climate stress, or to cope with the consequences. Meanwhile, sensitivity is referred to as the degree to which a system will respond to changes in climate (O'Brien *et al.*, 2004).

Kelly and Adger (2000) argued that the extent to which the community is vulnerable to climate change depends on exposure and sensitivity to climate change and the ability to adapt to new conditions. The socio-economic well-being prevailing in the community influences the community's ability and capacity to adapt to CC&V. This study selected the livelihood assets (A) as an indicator of well-being through wealth ranking.

The household's livelihood assets reflect the extent of vulnerability and different adaptive capacity to the impacts of climate change (Lyimo and Kangalawe, 2010). Poor households are more vulnerable to CC&V due to their limited adaptive capacity caused by possession of fewer livelihood assets (Madaka, 2007). Through key informant interviews and FGDs based on locally accepted criteria and perception of wealth and poverty, three socio-economic groups were identified in terms of household wealth: (i) rich; (ii) medium wealth; and (iii) poor (Table 8.5).

The rich or the *nsabi* wealth group had more livelihood assets and diverse livelihood activities, including farm and non-farm activities. The *nsabi* were involved in livestock keeping, buying and selling crops, were government employees, owned and ran shops, and owned a tractor and ox-ploughs. It was also noted that in case of drought or food shortage the group were able to buy food or use their own reserve stocks. They could also hire casual labour for cultivation and to take care of their animals and move them to other places where there was water and pasture during drought periods. The group was reported to pursue cumulative

Table 8.5. Social stratification of households (%) in Ibulyu and Mahaha villages. (From field survey, March 2014.)

Name of village	Rich	Medium wealth	Poor	Total
Ibulyu	10	43	47	100
Mahaha	8	51	41	100
Average	9	47	44	100

strategies and, hence, they were relatively more adaptive to the impacts of climate change compared with the other two groups. The majority of the households falling in this group comprised government employees such as teachers, clinical officers at dispensaries, businessmen and extension officers and the village executive officers.

The medium wealth group comprised the majority of the households and they possessed more livelihood assets compared with the poor group but were not like the *nsabi* group. It was this concentration of wealth that made Cory to remark that 'These are poorer than rich men in Sukumaland' (Cory, 1954, p. 100). Households in this group had limited room for diversification compared with the rich group. They were involved in small businesses which did not involve much capital, such as selling of the local brew and keeping a few herds of cattle. Income from diversification to off-farm activities by this group complemented income from farming, especially during adverse climatic conditions.

On the other hand, the poor wealth group (the *nh'abe*) comprised the economically impoverished households. It included, among others, the disabled, the elderly, widows, divorcees and orphans. The *nh'abe* wealth group was reported to have very limited livelihood assets. This makes them highly vulnerable to the impact of climate change and this group was pursuing more coping strategies and little adaptation. Their livelihood diversification to off-farm activities mainly included being hired as labour for food projects.

During food shortages the poor group mostly depended on social capital (social networks): depending on aid from friends, relatives and neighbours as well as use of dried sweet potatoes (*michembe*). These households could neither afford to get a balanced diet nor three meals a day for the whole year. Heads of households of this group could not afford the education expenses for their children. The majority of the children dropped school at early stages of primary school and very few attained a full primary education. It was reported that these poor households could rarely afford medical expenses and mostly depended on traditional healers.

Comparing the villages, Table 8.5 shows that Mahaha village had more households in the medium wealth group (51%) than those in Ibulyu village (43%). By contrast, Ibulyu village led with the number of households in the poor group (47%) but also had a higher number of households falling in the rich group (10%). Generally, the stratification of the surveyed villages indicated that the poor and medium wealth groups comprised the largest number of households (91%). The percentage of households in the rich category was very low in both villages (9%), implying that these communities may be highly vulnerable to climate change. Through FGD it was revealed that inadequate or lack of income sources, limited productive farms, poor social networks, poor health conditions and lack of cattle were among the reasons that some households were more vulnerable to the impacts of climate change than others (Table 8.6).

As noted by Blaikie *et al.* (1994), households that had access to resources and social networks were less vulnerable. Although they could experience greater losses (in absolute terms) than the poor, it could be argued that resource-rich households were more resilient in that they were able to recover more quickly from a stress stimulus. The same was also found by Majule *et al.* (2007), whereby the richer people had greater sources of income and assets and were, therefore, less vulnerable compared with the poor.

Impacts of climate change on crop production

DECREASE IN CROP PRODUCTION. The study revealed that decrease in crop harvest was attributed to the changing climate. Both food and cash crops have been significantly affected by CC&V hence resulting in food insecurity and economic instability. The majority of respondents (69.5%) acknowledged that crop yield is decreasing. The decline or fluctuation in crop production was associated with occurrence of drought, increased crop pests and diseases, temperature increase and rainfall variability. FGDs revealed that

Table 8.6. Socio-economic grouping of households based on perception of wealth and poverty. (From field survey, March 2014.)

Rich (9%)	Medium wealth (47%)	Poor (44%)
Have farm of >10 acres	Have a farm of between 5 and 10 acres	Have farm of <4 acres
Have modern house made of cemented/burned bricks, corrugated iron sheet and use solar power	Have moderate house made of burned bricks, corrugated iron sheet	Muddy house roofed with grass/iron sheet and use wick lamp for lighting
Have large herds of cattle/goats more than ten animals	Owns small herds of cattle and goats mainly for cultivation	Only keep poultry
Are food secure	Partially food secure, sometimes have food shortage	Food insecure, relies on relatives and relief food
Own tractor, motorcycle, ox-plough	Owns bicycle, ox-plough and hand hoe	Owns hand hoe
Can afford children's school fees and medical services	Can partially afford school fees for children mainly to secondary level and medical services to district level	Can afford primary school fees for children
Owns shop, bicycle or motorcycle garage and mobile charging services	Owns kiosk, food venders, bicycle garage	Usually hired as casual labour

in the past farmers used to harvest 8–13 bags/acre in Ibulyu village and 10–12 bags/acre in Mahaha. To date crop productivity has declined to 3–4 bags/acre (about 1–2 bags/ha). It was further claimed that paddy and maize crop production have been seriously hit by climate change due to drying of swamp areas, which were used for cultivation of paddy.

Also, it was reported that farmers have quitted paddy cultivation due to the dryness of the swamps or they have harvested very little or nothing at all, especially during the dry years of 1994, 2006 and 2013, as well as in 1997/98 when El Niño rainfall seriously affected crops. FGDs and the household survey revealed that the study area has been repeatedly hit by spells of drought over the last 20 years, resulting in absolute crop failure and low crop yields and ultimately acute food shortage. The most notable incidences of drought occured in 1999 (locally named *tonga)*, which resulted in famine and hunger; and 2006 (locally named *balabala)*, during which the normal rainy months of October to December were dry, and there was no rainfall until February 2007. Despite naming these as the most noted dry years, respondents claimed that rains do not have the same strength compared with past years, and also the pattern and distribution is currently uneven.

Local experience shows that in the past there were two major seasons for cultivation: the first season used to be from October to December while the second, and major maize-planting season, was between March and May. However, with the changes in the weather patterns the two seasons are no longer as distinct as they used to be and production has become single seasoned. On the other hand, sunflower, sweet tomato and sorghum production were reported to have increased. This is because these crops require less water compared with maize and rice. Tomatoes and other vegetables were said to have moderately increased because of increased demand. The production of such vegetables is expanding into the valley bottoms, especially in the dry season, taking advantage of the residual moisture in the soil and ease to irrigate such crops. The trends of major crops produced in the study area are presented in Table 8.7.

Table 8.7 shows that many of the respondents did not know the productivity trends of many of the crops grown in the area. Reasons behind this included the difficulty in separating productivity change associated with climate change and change

Table 8.7. Respondents' perceptions of crop productivity trends in the study area. (From field survey, March 2014.)

	Respondents' responses (%)							
	Increasing		Decreasing		Fluctuating		Do not know	
Crop	Ibulyu	Mahaha	Ibulyu	Mahaha	Ibulyu	Mahaha	Ibulyu	Mahaha
Maize	6.7	33.3	33.3	66.7	60.0	–	–	–
Sorghum	13.3	–	3.3	–	13.3	–	70.0	100.0
Finger millet	–	–	–	4.8	–	4.8	100.0	90.6
Rice	6.7	–	23.3	28.6	33.3	19	36.7	52.4
Beans	3.3	–	16.7	9.5	36.7	–	43.3	90.5
Groundnuts	13.3	38.1	23.3	23.8	26.7	4.8	36.7	33.3
Sunflower	43.3	23.8	10.0	4.8	26.7	9.5	20.0	61.5
Cotton	6.7	–	30.0	81	43.3	9.5	20.0	9.5

associated with non-climatic factors such as declining soil fertility and use of agricultural inputs and implements. Also, some respondents did not grow a particular crop or they had abandoned growing a particular crop due to various factors, including climate change.

CHANGE IN CROP VARIETIES. Increase in outbreak of diseases and pests, drought and increased temperature have significantly affected crop production in the study area. The most affected crops are those which require reliable rainfall, especially maize, rice and beans. Due to these impacts farmers have opted to plant crop hybrids which withstand long periods of moisture stress as well as pests and diseases. The majority of respondents (94%) reported using improved crop varieties, which they claimed produced better yields; even during moisture stress they claimed to get a harvest compared with the indigenous varieties which die even at maturity. Few respondents (6%) from the study villages claimed that they did not see any difference between using the local and improved varieties, besides some farmers claimed they could not afford to buy the required inputs and implements due to financial capital constraints.

CHANGE IN CROPPING PRACTICE AND SEASON. Poor crop yield was also associated with changes in the farming calendar in the study area. The majority of respondents (83%) revealed that changes in the farming calendar were linked to rainfall unreliability, especially in the last 20 years. In the past, rainfall seasons started in October–December (OND) for the first peak and the second came in March–May (MAM). In recent years rainfall had become unpredictable, sometimes starting early in September and ceasing early or, if continued, doing so in relatively low intensity and with uneven distribution. Unreliable rainfall had affected the growing season and had limited farmers' timing of the right time to plant. Thus, the farmers had become fearful to plant because they are not sure whether the rains would continue and last for a long period or would cease early and affect the germination and growth of the crops. Those who decided to plant were not sure if they would harvest or crops would fail due to drought and temperature increase.

According to the FGDs, there had also been change in cropping patterns due to unreliable rains. For instance, land preparation, application of agricultural inputs such as fertilizers, pest control and weeding of different crops in the villages had also changed. Such change in cropping patterns, however, had no direct correlation to climate change. The use of fertilizers, for example, was pushed by changes in soil characteristics and structure (Table 8.8). Due to decrease in soil fertility farmers were using nitrogen fertilizers and manure to boost/improve soil fertility. There had also been changes in the weeding period, especially with maize and

Table 8.8. Respondents reported changes in cropping practice in the study area. (From field survey, March 2014.)

Change in farming practice	Ibulyu (%)	Mahaha (%)
Land preparation	50.0	14.3
Use of agricultural inputs	66.7	61.9
Pest control	70.0	90.5
Weeding	83.3	95.2

cotton. In order to increase crop yield also there was increase in the use of pesticides to control crop pests and diseases.

OUTBREAK OF CROP PESTS AND DISEASES. The majority of respondents in the two villages perceived that incidences of crop pests and diseases had increased over the past 20 years and that the pests had become more prevalent with time (Table 8.9).

Such incidences were locally perceived to have considerably affected the productivity of crops produced. It was revealed that increased prevalence of crop diseases and pests is a result of increased temperature and decline or unreliability of rainfall. This is because pests are mostly favoured by high temperature conditions rather than cooler conditions. The prevalent crop pests and diseases are cotton justice, American bollworms, bag flies and aphids. It was observed that pests and diseases have caused detrimental effects to maize crops as well as cotton flowers.

Impacts of CC&V on livestock keeping

The types of livestock kept have not changed much with CC&V but the numbers seem to have been influenced considerably. The majority of respondents were of the opinion that livestock, particularly the grazers, were declining in numbers. Among the factors for the decline were shortages of quality pastures and drinking water. It was reported that in previous years every household had livestock, but now only 49.5% of respondents were involved in livestock keeping. FGDs revealed that previously the richest herdsman could own more than 500 head of cattle, while at the time of the study the wealthiest herdsman was reported to have around 60–120 cattle. Climate change has put livestock at risk, with animals dying alongside pastures that are deteriorating in condition and drying water sources.

Table 8.9. Respondents' perceptions of the status of crop pests and diseases over the past 20 years. (From field survey, March 2014.)

Pests and disease trend	Respondents' responses (%)	
	Ibulyu	Mahaha
Increasing	50.0	61.9
Decreasing	6.7	14.3
No change	40.0	19.0
Do not know	3.3	4.8
Total	100.0	100.0

The majority of livestock keepers depend on rainfall to recharge water for their livestock, so during the dry season livestock keepers had to walk longer distance in search of water and pasture which in most cases ended in resource use conflicts. It was revealed that water availability for livestock was increasingly becoming difficult. The main groundwater sources such as the Ibulyu River in Ibulyu village, and the Senani and Ndoba rivers in Mahaha have declined/dried up due to the prolonged drought. During the dry season livestock keepers dug temporary wells along the dried rivers to get water for their livestock. Widespread livestock mortalities were reported due to the lack of water and pasture, especially in prolonged dry years such as 1994, 1999 and 2006. This had diminished pastoralist herds faster than the animals could be replaced, putting pastoralist livelihoods under increasing pressure.

Also, it was claimed that increase in livestock pests and disease is related to climate change. Livestock diseases have led to poor livestock health, reduced milk production and even deaths. It was revealed that the number of cattle is declining due to outbreak of livestock diseases. The major diseases which affect livestock are the tickborne diseases, namely East Coast fever (ECF), anaplasmosis, babesiosis, black quarter, contagious bovine pleuropneumonia (CBPP) and heartwater. Livestock keepers argued that most of the livestock pests such as ticks (*kupe*), flies and aphids (*utitiri*) are common during dry periods compared with rainy seasons. ECF, locally known as *Ndigana kali*, was indicated as the most dangerous disease, reported to have killed cattle in the previous years. Newcastle disease, fowl pox, fowl typhoid, coccidiosis and fowl cholera were said to affect poultry in the study area. However, the increase in livestock pests and diseases was also associated with non-climatic factors such as: (i) inadequate vaccination; (ii) poor handling of livestock sheds; (iii) inadequate and dysfunctional cattle dips; and (iv) lack of sustainable livestock-keeping knowledge among herdsman.

Impacts of CC&V on natural resources

The increase in water scarcity was also associated with deteriorating quality, as one of critical climate change impacts on water resources. The study found that in recent years, water scarcity is increasing and as a result there has been early drying of water courses and wells. As rainfall intensity reduces and as it lasts for a shorter period of time, the rivers and wells are becoming seasonal as they dry out after a short period the rain has ended or remain with only a low amount of water. Through FGD it was claimed that permanent rivers such as the Ibulyu, Senani and Ndoba have turned into seasonal rivers as they dry out early in May or June or remain with a low amount of water, especially during the normal dry months after the rains have ended. Also, during the rainy season water intensity in rivers has declined. Decline in water intensity or drying of rivers was associated by increased temperature resulting from prolonged incidences of drought, which increases the rate of surface water evaporation.

Coping with impacts of CC&V

The rural livelihoods context, including climate-related trends and shocks, together with people's capital asset base varies over time and space resulting in a wide range of coping and adapting strategies. According to the IPCC (2001), adaptation to current CC&V often produces benefits as well as forming the basis for coping with future anticipated climate change. The study has established that people respond differently to climatic events in their communities based on their adaptive capacity.[1] This capacity depends on the socio-economic grouping to which one belongs. Thus people in different socio-economic groups cope and adapt to weather variations and extremes differently. Majule *et al.* (2007) noted that the well-off households are cushioned by their resource endowments hence may not feel the pinch of drought or famine whereas the poor have limited capacity due to limited resources endowment. Coping and adaptation strategies include those concerning: (i) crop production; and (ii) livestock keeping.

Coping and adaptation strategies: crop production

The main household coping strategies in the surveyed villages are: (i) buying food (100% of households); (ii) getting assistance from relatives (52%); (iii) getting relief food (92.2%); (iv) selling livestock to buy food (74.5%); (v) selling other household assets to buy food (49%); (vi) working for food (86.3%); (vii) borrowing food (78.4%); (viii) reducing the number of meals (68.6%); (ix) migration to other places (76.5%); and (x) storage of dried food (locally known as *michembe*) and vegetables (89%). Other strategies mentioned by most of the respondents are the improvement of veterinary services, improvement of access to agricultural extension services and improvement of infrastructure and transport. Such production improvements are perceived to succeed through increased effort and engagement in production activities.

The study found that to cope with food shortages households buy supplementary food from retailers, but households which cannot afford to buy food for a long period, especially the poor group, sell household assets and livestock so as to buy food for their family. They also depend on their relatives who can assist them as well as getting relief food from the government (it was reported that the government and other agencies provided food during food shortages in 1998, 2006 and 2013). Furthermore, other households borrow food from sellers, neighbours or relatives and have to pay this back during the next harvest. In addition to that, households sometimes had to reduce the number of meals per day from three or two to one meal; drinking porridge in the morning precedes this. Casual labour has been important to poor households as the poor provide casual labour on farms, for grazing livestock and other activities to earn their livelihoods. In a critical food shortage situation, household goods and assets such as land and livestock are sold to enable the household's chances of long-term survival. Storage of dried food has been useful during food shortages, especially dry periods; dried food and vegetables stay edible for up to 3 years.

On the other hand, various adaptation strategies were devised by the respondents. These strategies corresponded well with those identified elsewhere by Lyimo and Kangalawe (2010) and Majule *et al.* (2013). The most popular strategies in the study area included: (i) planting drought-resistant crop varieties (94.1% of households); (ii) livelihood diversification (45.1%); (iii) movement to new resource areas (90.2%); (iv) exploitation of social relations and rural–urban linkages (76.5% and 80.4%, respectively); (v) seasonal or permanent migration (86.3%); (vi) proper timing of farm operations (96.1%); (vii) planting of fast-maturing and high-yielding varieties (94.1% and 92.2%, respectively); (viii) adoption of crop rotation and mixed cropping (34.9% and 82.4%, respectively); and (ix) rainwater harvesting (96.1%). Other minor agricultural adaptation strategies included farming along the river valleys and wetlands, application of fertilizers and pesticides and change in weeding patterns.

In this study it was revealed that farmers practised a combination of these adaptation strategies, as a single strategy was inadequate in adapting to the impact of CC&V. Several strategies when combined together were likely to be more effective than a single strategy. However, due to inadequate financial capital in the study area it was revealed that not all farmers were able to engage in all of the above-outlined adaptation strategies.

PLANTING DROUGHT-RESISTANT CROP VARIETIES. Prolonged drought and inadequate rains are among major aspects affecting crop performance in the study area, which necessitates the need for using drought-tolerant crop varieties. It was revealed that the indigenous crop varieties required high amounts of moisture/water for germination and growth. Hence as water became scarce it resulted into crop failure. The most popular indigenous crop varieties were *Gembe*, *Nchanana* and *Kakuyu*. Due to failure of indigenous varieties, farmers had resorted to planting drought-resistant crop varieties, for both food and cash. Most planted improved maize seed varieties from PANA and SEEDCO companies such as Nyati, Pundamilia, Tumbili and Simba, while cotton seeds were supplied by Quton Company. Other drought-resistant crops grown in the study area included sweet potatoes, which is planted by the majority of the households, as well as sorghum and finger millet planted by a small proportion of the households.

TIMING OF FARM OPERATIONS, PLANTING FAST-MATURING AND HIGH-YIELDING CROP VARIETIES. From FGDs it was revealed that early planting of crops is no longer the same as in previous years because of late rainfall onset, so crops planted early may be affected by a prolonged dry spell. Therefore, change in onset of rainfall had forced farmers to set their planting time after the first onset of the rains. Together with this, farmers were embarking on planting fast-maturing crop varieties. In the study area the respondents claimed that these varieties matured between 60 and 90 days after germination which fitted very well with the shortened period of rainfall. Moreover, the crop varieties produced higher yields compared with the long-maturing indigenous varieties. It was, for example, claimed that 1 acre of a maize crop could produce between eight and 12 bags compared with the highest yield for indigenous varieties which was six bags/acre (Mongi *et al.*, 2010).

ADOPTION OF MIXED CROPPING. Mixed cropping involves planting more than one crop in the same field. Cereal, legume and nut crops are planted together. This type of farming is common in all villages in the study area. Mixed farming is used as a strategy to ensure that if one crop fails because of drought others can survive due to differences in crop cycles, rooting depths and water requirements (Tumbo, 2007). The traditionally mixed crops in the study area include maize, sunflowers, beans and cowpeas. According to the FGDs and key informant interviews, maize, cotton and sunflowers were occasionally mixed in the same plot.

The reasons for mixed cropping were land scarcity and climate change, which had led to a decline in soil fertility and moisture levels. The key informants claimed that mixed cropping was vital in avoiding and reducing loss of all crops in case of harsh climatic conditions. If some crops are affected a few of them could survive. It was further revealed, for example, that sorghum and sunflowers could better survive harsh conditions than maize. Moreover, if cereals were mixed with groundnuts or a legume crop the latter helped to improve soil fertility by nitrogen fixing and retension of soil moisture. Due to variations in maturing of these crops, the land was also prevented from remaining bare after harvesting cereal crops.

From the FGDs it was clear that farmers had adequate knowledge of the type of varieties to plant in the same field, drought tolerance and maturity variation among the crops. According to these researchers, farmers had field knowledge on the advantages of mixing crops with varying attributes in terms of maturity period (e.g. maize and beans), drought tolerance (maize and sorghum), input requirements (cereals and legumes) and end-uses of the products (e.g. maize as food and sunflowers for cash).

PRACTISING CROP ROTATION. Crop rotation involves change of crops planted on the same piece of land in successive years. This practice serves as a strategy for adaptation to CC&V. Respondents in FGDs claimed crop rotation was a highly viable strategy for adaptation to CC&V impacts. They asserted that cotton and maize crops had normally been rotated every 2 years. Cotton helped to maintain soil moisture and restore fertility for the maize crop. Scientifically, the cotton plant has taproots which absorb nutrients from deep down in the soil unlike the maize crop which has shallow roots (Majule and Mwalyosi, 2005; Majule *et al.*, 2013). Despite the importance of crop rotation in moisture and fertility stress conditions, few respondents were practising crop rotation in the study area due to land scarcity. Also, the majority of respondents lacked knowledge about crop rotation.

MIGRATION TO OTHER AREAS IN SEARCH OF ARABLE LAND. Migration is an important adaptation strategy in times of climate change as observed by Yanda and Mung'ong'o (1999). These authors indicated that migration in rural areas was more often than not a movement of labour (seasonal or permanently) from one agricultural area to another area, or across sectors. Migration between and within urban and rural areas is central to a household's ability to ensure livelihood security during food shortage.

In the study area it was revealed that people were migrating to other areas in search of suitable arable land, either permanently or temporarily (as they return home after harvesting). The migration trend was increasing due to land scarcity and decline in agricultural yields in the home lands. Most farmers were migrating to other regions, mainly Morogoro, Mbeya, Rukwa and Coast Regions. Others were migrating to urban centres such as Bariadi, Shinyanga, Kahama, Mwanza, Musoma and other major cities in the prospect of getting paid jobs. There were also a few people who were migrating to work in mining centres. Migration was dominated by men, leaving women and children at home. Young girls were also migrating to a lesser extent and were employed as housekeepers, barmaids, hotel workers and food vendors, mainly in the regional capital Bariadi.

RURAL–URBAN LINKAGES AND SOCIAL RELATIONS. A rural–urban linkage is among the vital adaptation strategies in the study area. It entails taking agricultural (food and cash crops) and livestock products (cattle, goats, sheep and chicken) as well as forest products (e.g. charcoal and firewood) to town for sale. This is especially during the market (*minada*) days at Bariadi town, which is usually on a Thursday each week. Also, due to reliable transport and short distances to Bariadi town centre, the study villages purchased many of their household necessities in town. This linkage enhanced communities' capacity to withstand impacts related to climate change through availability of a market for their products and access to various services from Bariadi town at affordable prices.

Social relations through remittances from relatives and migrant family members play an important role in household well-being during food shortage and difficult periods (Gwambene, 2007). People who received remittances in the study villages tended to be less affected by shocks in terms of access to food and social services. Relatives living in urban areas helped their rural counterparts especially during hard times. Remittances included money, food, clothes, assisting in paying medical and education bills and other assistance depending on the actual requirements of their relatives. However, respondents claimed that such remittances had not always been adequate enough to rescue them from sustained economic hardship. This was because most of their relatives in the urban areas were engaged in casual labour jobs, where their earnings were very little and not enough to support their families and relatives back home.

MOVEMENT TO KEY RESOURCE AREAS AND RAINWATER HARVESTING. Increased incidences of drought and dry spells adversely affect soil moisture and fertility and lead to high costs of cultivation in the normal farmland. In response to such incidences farmers in the study area had moved to cultivate swamps, wetlands and along river courses. Most of the swamps and wetlands in both villages had been converted to farmland for rice cultivation and vegetable gardens. Conversion of these areas to farmland has resulted in a decline of surface water and drying up of wetlands and swamps, which are the main source of water in the villages.

Traditional knowledge related to water harvesting and use has been highlighted as one of the most important adaptation requirements by Barron and Okwach (2005). Rainwater harvesting has been practised as a major strategy to overcome long-distance tracking in search of water for domestic use. In most villages availability of fresh water from the drilled boreholes has been declining. Some wells are discharging a low amount of water and others have completely dried up. This is particularly the case during the dry periods from June to September, where women and children have to walk long distances to neighbouring villages or hamlets to fetch water, where there are long queues of people waiting to collect water. Therefore, in the study villages rainwater was collected and stored to be used during the dry season. However, due to financial capital constraints, villagers were not able to buy or construct large water facilities which could store large volumes of water to be used for long periods during the dry times. The commonly used equipment included water drums and jelly canes. Very few had large water tanks. It was revealed that after the rains had ended the household could use the harvested water for an average of 2 weeks and a few for up to 1 month.

LIVELIHOOD DIVERSIFICATION. Livelihood diversification strategies including integration of livestock, crop production and non-farm activities are crucial to enhance the adaptive capacity of local communities (Lyimo and Kangalawe, 2010). Diversification of the rural non-farm economy is an alternative to declining farm productivity due to CC&V and reduces overdependence on natural resources. Due to the decline in agricultural productivity in the study area local communities had resorted to engaging in other livelihood activities to complement the already struggling agricultural sector. These livelihood activities included petty business such as running kiosks, food vendors, brick making, carpentry and masonry, bicycle and motorcycle repairs, transportation by using bicycles or motorcycles as well as mobile phone charging using solar power.

CHANGE IN FARM INPUTS AND IMPLEMENTS. Increase in use of agricultural inputs results in increased agricultural productivity. Limited access and availability of agricultural inputs have contributed to the decline in agricultural production in Shinyanga Region (Lyimo and Kangalawe, 2010). In the study villages as a result of declining soil fertility and outbreak of crop pests and diseases, farmers were applying fertilizers and pesticides. Nitrogen fertilizer (N_2) and manure were widely used, especially on maize farms. These inputs helped to improve soil fertility and hence also increase production. Use of nitrogen fertilizer improves crop production and hence increases food security (NAPA, 2007; Galloway *et al.*, 2008).

Nitrogen fertilizers in the study villages were obtained at Bariadi town, Shinyanga or Mwanza city. They were also obtained through government subsidies where farmers got fertilizers at half price. However, due to limited financial capital the majority of farmers used manure instead of fertilizer. Also, increase in crop pests and diseases, which ultimately reduced crop harvest, forced farmers to use pesticides to control destructive pests and diseases. To a large extent the use of pesticides helped farmers to control destructive pests and diseases affecting crops. The most commonly used pesticides included traditional firewood ashes (*majivu*) and industrially made pesticides such as Acteric Super, Herarete and Phenom.

The study found that 53.5% of the respondents were changing from using the hand hoe for cultivation to ox-plough (32.7% were using ox-ploughs whereas only 4.3% used tractors[2] for cultivation). Only a few reported using a combination of both implements. Respondents explained that they had shifted from using the hand hoe for cultivation so as to improve productivity, reduce extensive labour costs and reduce the time needed for cultivation. Also, they claimed that ox-ploughs and tractors broke large tracts of soil which helped crop root penetration. This corresponds well with findings from (Sims and Kienzle, 2006) who observed that the use of ox-ploughs and the tractor involved deep soil turning thus increasing nutrient mixing, conserving soil moisture and increasing water retention from runoffs. Soil turning increased infiltration, aeration and microbial activity, which aided quick plant growth and increased yields.

Coping and adaptation strategies: livestock keeping

CC&V affects the quality and amount of forage from grasslands and can cause an increase in livestock pests and diseases. Severe drought directly affects water resources and forage availability for livestock leading to the loss of large numbers of livestock in most pastoral areas (Lyimo and Kangalawe, 2010; UNECA, 2011). Livestock keepers have a vast wealth of knowledge on how to adjust to and cope with climatic variations (Mwandosya *et al.*, 1998). In arid and semi-arid areas, pastoralists have developed various strategies to cope and adapt to CC&V impacts. Such strategies are: (i) migration in search of pasture and water in neighbouring areas; (ii) livestock species diversification; (iii) switching between capital assets; (iv) distributing livestock among relatives and friends in other areas to ensure they are not all wiped out in a disaster; and (v) reciprocity in resource use (Orindi *et al.*, 2006). Similar coping and adaptation strategies were outlined in the study area. They included emphasis on small stock (35.2% of households), distribution of livestock to different places (38%), migration of livestock keepers to different places (83%), livestock vaccination and treatment (63%), livestock enclosure relocation (21%), digging of temporary water wells (79%), grazing on farmland after harvest (98%) as well as introduction of improved livestock varieties and zero grazing (31%).

MIGRATION OF LIVESTOCK KEEPERS TO SEARCH FOR PASTURE AND WATER. As observed by Lyimo and Kangalawe (2010) in their study in Shinyanga District, deteriorating pasture condition and drying water sources had forced pastoralists to migrate to other regions searching for water and pasture. In the study area it was revealed that pastoralists who owned large herds were migrating to other regions where they expected to have more reliable pasture and plenty of water for their livestock. The majority of the migrating pastoralists were seasonal, although a few had moved permanently alongside their families. Regions to which they were moving to included Morogoro, (particulary Kilosa, Mvomero and Kilombero Districts), Mbeya (in Mbarali, Chunya and Momba Districts – formerly part of Mbozi District), Rukwa, Katavi, as well as the Coastal Regions of Lindi, Mtwara and Pwani. However, Isaya (2009) and Mombo (2011) found that the movement of these communities was responsible for the outbreak of land use conflicts between farmers and pastoralists in different parts of the country.

ALLOCATION OF LIVESTOCK PASTURE ENCLOSURES. Traditional silvo-pastoral practices (locally known as *Ngitiri*, *Ronjo* or *Mlimbiko* in Shinyanga, Arusha and Kilimanjaro, and Dodoma, respectively) were used by pastoralists to alleviate shortage of dry-season fodder supply. In these practices, either a farm was enclosed and pasture was left to regenerate or a rangeland was subdivided into pasture zones and some zones were reserved for dry season grazing only. Such systems relied on customary laws and institutions that were respected by all community members (Orindi *et al*., 2006). In the study area the *Ngitiri* practices were introduced in the 1990s as part of the Hifadh Arhi Shinyanga (HASHI) programme in Shinyanga Region. This system emphasized that pastoralists should allocate part of their communal or household land for livestock pasture depending on the herd size they possessed. Currently, the enclosure system is practised only marginally. Pastures in these enclosures are local grasses. There is no expatriate influence in the management of the *Ngitiri*[3] thus the quality of pastures are poor and they are small in size ranging from 0.25 acre to 1 acre, hence they cannot feed large livestock herds.

DISTRIBUTION OF LIVESTOCK TO DIFFERENT PLACES AND EMPHASIS ON SMALL LIVESTOCK. Distribution of livestock to relatives, neighbours or children was outlined as one of the adaptation strategies to CC&V in the study area. Pastoralists with large herds of livestock were reported to distribute livestock to other places and retain a herd of manageable size. It was revealed that outbreak of livestock pests and diseases, poor pasture and water scarcity had forced livestock keepers to distribute their herds. This practice was claimed to be important as it reduced the risk of livestock death due to outbreak of livestock diseases (Mung'ong'o, 1995). If livestock deaths occurred they could be replaced by livestock that had been distributed to other places. Remaining with small and manageable herds required a low amount of pasture and water while labour, vaccination and treatment costs were reduced as well.

EMPHASIS ON SMALL STOCK AND POULTRY. Due to increased scarcity of water and pasture for livestock, livestock keepers started to keep small stock such as goats, sheep and pigs. This was because small stock required lower amounts of input in terms of pasture, water and labour power compared with large livestock like cattle. They claimed that goats and sheep can even feed on tree and banana leaves whereas pig feeds are obtained cheaply from milling machines, cotton ginneries as well as human food remains. Therefore, this makes small stock less vulnerable to climate change impacts, specifically vulnerability due to pasture and water scarcity. Silitshena and McLeod (1998) noted that during severe droughts cattle populations declined due to lack of water and grazing while for goats, the trend was in the opposite direction.

However, though cattle are more vulnerable to CC&V compared with small livestock, cattle are a major asset among the Sukuma community. Cattle are an asset of prestige, a source of labour especially for cultivation, and a symbol of wealth/power. Therefore, the majority of respondents were reluctant to switch to small animals. It was revealed that during the hard times, one head of cattle could be sold for up to

TSh500,000 while the maximum price for a goat could only be TSh100,000. Therefore, if a household required a large amount of money it had to sell a large number of goats compared with cattle. On the other hand, the market for sheep is too minimal hence farmers did not keep them extensively. Regarding pigs, the respondents claimed that they were easily attacked by disease, especially flu'. Also, religious beliefs inhibited some respondents from keeping pigs, especially Muslems and Seventh Day Adventists which is among the main religious domain in the study area.

It was also noted that the majority of households had adopted poultry keeping, specifically chickens, to supplement other economic sectors. Poultry keeping was emerging as an alternative income-generating sector in the study area. This was due to readily available markets for eggs and chicken meat in Bariadi town and Lagangabilili Centre in Itilima District, which are easily accessible from the study villages.

OTHER LIVESTOCK-KEEPING ADAPTATION STRATEGIES. Another livestock-keeping adaptation strategy was grazing of livestock on farm plots after crop harvests. During this time farmers allowed pastoralists to feed their livestock with crop residues and weeds. This method was widely practised in all study villages. Respondents claimed that this practice helped livestock keepers in that they did not have to move long distances in search of pastures during the dry period. However, despite the importance of this strategy to livestock keeping it was said to have negative consequences for crop production, specifically when using hand hoes for tilling the soil, as trampling by livestock made the soil hard and compacted and hence difficult to plough. It also limited water and moisture penetration into the soil.

Also, vaccination and treatment of livestock had been adopted as a means of reducing livestock pests and diseases. In Ibulyu village there is a newly constructed cattle dip, which is used for livestock dipping. However, there was no dip in Mahaha village at the time of this study. Nevertheless, the majority of the pastoralists reported that they individually treat their livestock as they claimed that the cattle dip did not function regularly. Vaccination of livestock and poultry was carried out to prevent outbreaks of seasonal diseases. Such vaccination and treatment was done by the livestock extension officer from the village and the district office.

Government and non-governmental initiatives

In order to increase the adaptive capacity of the local community and consequently enhance resilience of the society against CC&V impacts, external interventions and assistance based on policy, materials, technologies as well as markets for agricultural and livestock products are imperative. The interventions might be from the government or from non-governmental organizations (NGOs). Based on this, the study attempted to determine the government and NGOs' interventions to assist the community in coping and adapting to CC&V. The study found that 67% of the respondents in Ibulyu village said they had received external assistance while 73% of respondents in Mahaha village acknowledged the same.

In terms of the form of external assistance received, an average of 73% of respondents in both villages reported that these interventions had been in the form of materials, whereas 19.3% noted that interventions had varied depending on the situation. The remaining 7.7% said they they had received a mixture of support. The material support provided was mainly food during food shortages, while other supports included: (i) digging of water boreholes; (ii) training on alternative income-generating activities such as poultry and pig production; (iii) establishment of women's entrepreneurship groups; and (iv) training on sustainable agricultural practices. Other assistance from the government was formalization of village land and provision of a Certificate of Customary Right of Occupancy (CCRO) to owners in Ibulyu village in 2010. It was reported that the cost of issuing the CCRO was fully covered by the government.

On the other hand, respondents were asked if they were satisfied with the assistance they were getting from the government and other NGOs; 78% of the respondents in Ibulyu and 83% in Mahaha village were not satisfied. The reasons behind dissatisfaction in both villages included agriculture and livestock officers not regularly visiting them to advise and assist them. The officers only visited them when there was a major problem, or when they were ordered to do so by their employers. Also, the DASIPs programme did not include their villages therefore they felt isolated. For Mahaha village their land was not adjudicated thus they did not get CCROs, which could be used as collateral for securing loans. Also, they claimed that agricultural inputs and implements were getting to them very late; sometimes after the cultivation season had begun.

Conclusion

This study set out to deepen our understanding of the socio-ecological resilience of agropastoral communities to CC&V impacts in Bariadi District, Tanzania. Specifically the study intended: (i) to examine the local perceptions, indicators and trends of CC&V; (ii) to determine the impacts and assess household vulnerability to CC&V; and (iii) to examine the coping and adaptation strategies used by agropastoralists in relation to CC&V impacts in the district.

From the discussion above, it is clear that the study area is already experiencing CC&V. The local communities are aware that over the past 20 years their local climate has constantly been changing. They associate changing climate with changes in the intensity, pattern and distribution of climatic elements which are directly affecting their daily livelihood activities, especially rainfall. The study found that over the period of study, the area has been experiencing a decrease in rainfall intensity, unpredictable rainfall onset, early rainfall cessation as well as increase in temperature and frequent occurrence of dry spells in the wet months. Also, bad years characterized by erratic rainfall have been increasing.

It was generally observed that CC&V had negatively affected the farming system in the study area. Crop production and livestock keeping, which are the main sources of income in the study area, are rainfall dependent, hence alterations in rainfall intensity, onset and distribution has significantly affected these sectors thus exposing the community to food insecurity. Also, the ecological setting of the area has significantly been altered to the extent that it cannot provide the required services and products such as water, fertile soil and soil moisture, which are crucial for plants and crop growth as well as livestock sustainance.

Observed changes in the production system have negatively affected community resilience hence increasing their vulnerability. This is because the local community mostly depends on rain-fed crop production and livestock keeping as their main sources of income. It was observed that poor households, widows, children, disabled people and the elderly are more vulnerable to CC&V impacts. The vulnerability of these groups was mainly attributed to the ownership of few resources and less productive farms, limited labour sources, limited social networks and limited options for economic diversification.

In order to respond to climatic stresses short-term coping and long-term adaptation strategies have been devised by the local community. These responses have been both on- and off-farm options such as change in crop varieties, change in farming calendar as well as changing types of farm implements and inputs, migration of livestock keepers, doing casual labours, business, food vendors, as well as bicycle and motorcycle transportation. Also, there have been interventions from government and private agencies to combat climate change and assist the community, especially during food scarcity.

Other interventions have included land formalization in Ibulyu village to enable the local community access to loans, construction of water wells, establishment of District Agricultural Sector Investment Project (DASIP) and District Agriculture Development Plans (DADPs) to improve agricultural production, as well as establishment of an environmental management programme under the Lake Victoria Environmental Management Project (LVEMP). However, these interventions have not reduced climate change stressors such as temperature increase and drought, which have led to crop failure, decrease in productivity of both crops and livestock. Also, some of the adaptation strategies practised are responsible for further environmental degradation such as cultivation of vegetables, maize and beans along the river courses which has accelerated river erosion and sedimentation. It was also observed that rice farms in the wetlands and swampy areas are resulting in the destruction and drying of water sources.

Notes

[1] Adaptive capacity is the ability of a system to adjust to climate change (including climate variability and extremes), to moderate potential damages, to take advantage of opportunities, or to cope with the consequences (IPCC, 2001).

[2] Through the Agriculture First (*Kilimo Kwanza*) slogan the government provided each village with one tractor to reduce production costs.

[3] *Ngitiri* is a Sukuma word which refers to the land set aside for livestock grazing and tree planting for firewood and other uses.

References

Adekunle, V.J. (2009) Contribution of agroforestry practice in Ondo State, Nigeria, to environmental sustainability and sustainable agriculture production. *Afrika Focus* 22(2), 27–40.

Bariadi District Council (BDC) (2010) *Bariadi District Socio-Economic Profile, Simiyu Region*. District Executive Director's Office, Bariadi, Tanzania.

Barron J, Okwach G (2005) Run-off water harvesting for dry spell mitigation in maize (Zea mays L.): Results from on-farm research in semi-arid Kenya. Agric Water Manag 74: 1–21.

Berkes, F., Colding, J. and Folke, C. (2003) *Navigating social-ecological systems: Building resilience for complexity and change*. Cambridge University Press, Cambridge.

Blaikie, P., Cannon, T., Davis, I. and Wisner, B. (1994) *At Risk: Natural Hazards, People's Vulnerability, and Disasters*. Routledge, London.

Carpenter, S.R. (2001) Alternate states of ecosystems: evidence and its implications. In: Press, M.C., Huntly, N. and Levin, S. (eds) *Ecology: Achievement and Challenge*. Blackwell, London, pp. 357–383.

Chambers, R. (1994) The origin and practice of participatory rural appraisal. *World Development* 22(8), 953–969.

Cory, H. (1954) *The Indigenous Political System of the Sukuma and Proposals for Political Reform*. The Eagle Press, London.

De Vaus, D.A. (1993) *Survey in Social Research*. Westview Press, London.

DFID (1999) Sustainable Livelihoods Guidance Sheets. Department for International Development, London.

Galloway, J. *et al*. (2008) Transformation of the Nitrogen Cycle: Recent Trends, Questions, and Potential Solutions. *Science* 320, 889–92.

Gwambene, B. (2007) Climate change and variability adaptation strategies and implications on land resources in Rungwe District, Tanzania. Dar es Salaam University Press, Dar es Salaam.

Homes, C.F. and Austen, R.A. (1972) The pre-colonial Sukuma. *Journal of World History* 14, 378.

Intergovernmental Panel on Climate Change (IPCC) (2001) *Climate Change 2001: the Scientific Basis. Contribution of the Working Group I to the Third Assessment Report of the Intergovernmental Panel on Climate Change* [Houghthon, J.T., Ding, Y., Griggs, D.J., Noguer, M., van der Linden, P.J., Dai, X., Maskell, K. and Johnson, C.A. (eds)]. Cambridge University Press, Cambridge.

Intergovernmental Panel on Climate Change (IPCC) (2007) *Climate Change 2007: Impacts, Adaptation and Vulnerability. Contribution of Working Group II to the Fourth Assessment Report of the Intergovernmental Panel on Climate Change*. [Parry, M.L., Canziani, O.F., Palutikof, J.P., van der Linden, P.J. and Hanson, C.E. (eds)]. Cambridge University Press, Cambridge.

Isaya, I. (2009) Fact Finding Mission Report on the prevailing land dispute at Namwawala village in Kilombero District, Morogoro Region. Haki Ardhi (Land Rights Research and Resources Institute) and Legal and Human Rights Centre (LHRC), Dar es Salaam, Tanzania.

Johnson, B. and Christensen, L. (2004) *Educational Research: Quantitative, Qualitative and Mixed Approaches*, 2nd edn. Allyn & Bacon, Needham Heights, Massachusetts.

Kazembe, L.N. (2009) Modelling individual fertility levels in Malawian women: A spatial semiparametric regression model. *Statistical Methods and Applications* 18(2): 237–255.

Kelly, P.M. and Adger, W.N. (2000) Theory and practice in assessing vulnerability to climate change and facilitating adaptation. *Climatic Change* 47, 325–352.

Kothari, C. (2004) *Research Methodology: Methods and Techniques*. New Age International Limited, New Delhi.

Krueger, R.A. (2000) *Focus Groups: a Practical Guide for Applied Research*, 3rd edn. Sage, Thousand Oaks, California.

Langford, B.E., Schoenfeld, G. and Izzo, G. (2002) Nominal grouping sessions vs. focus groups. *Qualitative Market Research* 5, 58–70.

Lema, A.M. and Majule, A.E. (2009) Impacts of climate change, variability and adaptation trategies on agriculture in semi-arid areas of Tanzania: the case of Manyoni District in Singida Region, Tanzania. *African Journal of Environmental Science and Technology* 3(8), 206–218.

Liamputtong, P. (2011) Focus Group Methodology: Principles and Practice. SAGE, London.

Lugenja, M., Meena, H.E. and Stephenson, M. (2006) Climate change impacts on livelihoods in Tanzania and adaptation options: experience of floods and drought in Rufiji. CEEST Foundation, Dar es Salaam, Tanzania.

Lyimo, J.G. and Kangalawe, R.Y.M. (2010) Vulnerability and adaptive strategies to the impact of climate change and variability. The case of rural households in semi-arid Tanzania. *Environmental Economics* 1(2), 89–97.

Madaka, T. (2007) Community vulnerability and adaptation to the impacts of climate variability and extremes on Simiyu Wetlands: the case of Simiyu Wetlands, Lake Victoria Basin. MSc. (NARAM) dissertation, University of Dares Salaam, Dares Salaam, Tanzania.

Majule, A.E. and Mwalyosi, R.B.B. (2005) Enhancing agricultural productivity through sustainable irrigation. A case of Vinyungu Farming System in selected zones of Southern Highland, Tanzania. In: Sosovele, H., Boesen, J. and Maganga, F. (eds) *Social and Environmental Impacts of Irrigation Farming in Tanzania: Selected Cases*. Dar esSalaam University Press, Dar es Salaam, Tanzania.

Majule, A.E., Gibson, R. and Chiwatakwenda, A. (2007) Climate change adaptations in low potential area of Tanzania: local perceptions, vulnerability, current adaptations and in future strategies in Maluga village, Iramba, Singida. *Climate Change and Adaptation in Africa (CCAA)-Tanzania, Malawi Working Paper* No. 1.

Majule, A.E., Kauzeni, A.S. and Mujwahuzi, M. (2013) Exploring opportunities for climate change adaptation in semi-arid areas of Tanzania: a case of Nzega District in Tabora Region. *African Journal of Environmental Science and Technology* 7(8), 758–769.

Maxwell, J.A. (1996) *Qualitative Research Design*. Sage, Thousand Oaks, California.

Mikkelsen, B. (1995) *Methods for Development Work and Research. A Guide for Practitioners*. Sage, Thousand Oaks, California.

Mombo, F. (2011) Ratification of the Ramsar Convention and sustainable wetlands management: situation analysis of the Kilombero Valley wetlands in Tanzania. *Journal of Agricultural Extension and Rural Development* 3(9), 153–164, 2 September.

Mongi, H., Majule, A.E. and Lyimo, J.G. (2010) Vulnerability assessment of rain-fed agriculture to climate change and variability in semi-arid, Tanzania. *African Journal of Environmental Science and Technology* 4(6), 371–381.

Morgan, D.L. (1997) *Focus Groups as Qualitative Research*, 2nd edn. Qualitative Research Methods Series 16. Sage, Thousand Oaks, California.

Mukherjee, N. (1992) Villagers' perceptions of rural poverty through the mapping methods of PRA [Participatory Learning and Action]. Special Issue: Wealth ranking. *RRA Notes* 15, 21–26.

Mung'ong'o, C.G. (1995) Social processes and ecology in the Kondoa Irangi Hills, Central Tanzania. Meddelanden Series B 93. Department of Human Geography, Stockholm University, Stockholm.

Mung'ong'o, C.G. (2016) Tale of three villages: a study of social stratification in three pastoral communities in Simanjiro District, northern Tanzania. In: Yanda, P.Z. and Mung'ong'o, C.G. (eds) *Pastoralism and Climate Change in East Africa*. Mkuki na Nyota, Dar es Salaam, Tanzania, p. 227.

Mung'ong'o, C.G. and Mwamfupe, D. (2003) Poverty and changing livelihoods of migrant Maasai pastoralists in Morogoro and Kilosa Districts, Tanzania. REPOA, Research Paper No.03.5. Mkuki na Nyota Publishers, Tanzania.

Mwandosya, MJ, Nyenzi BS, and Luhanga, ML (1998). T*he Assessment of Vulnerability and Adaptation to Climate Change Impacts in Tanzania*. Centre for Energy, Environment, Science and Technology (CEEST), Dar-es-Salaam, Tanzania.

NAPA (2007) National Adaptation Programme of Action (NAPA). Division of Environment, Vice President's Office, Dar-es-Salaam, Tanzania.

Nelson, V. and Stathers, T. (2009) Resilience, power, culture, and climate: a case study from semi-arid Tanzania, and new research directions. *Gender & Development* 17(1), 81–94.

Njana, R.N. (1998) Prospects of local people's involvement in management of Forest Reserves. A case study of the North Mahiwa-Kisara Catchment Forest Reserve, Morogoro, Tanzania.

MSc. dissertation, Sokoine University of Agriculture, Morogoro, Tanzania.

O'Brien, K., Leichenko, R. and Kelkar, U. (2004) Mapping vulnerability to multiple stressors: climate change and globalization in India. *Global Environmental Change* 140, 303–313.

Onwuegbuzie, A.J. and Leech, N.L. (2007) A call for qualitative power analyses. *Quality & Quantity. International Journal of Methodology* 41, 105–121.

Orindi V.A., Kibona, E. and Moindi, D.M. (2006) Addressing climate change in the development process in Tanzania. *African Centre for Technology Studies (ACTS), Ecopolicy Series* No. 18. Acts Press, Nairobi.

Poffenberger, M., MacGean, B. and Gadgil, M. (1992) *Field Methods Manual: Diagnostic Tools for Supporting Joint Forest Management Systems*. Joint Forest Support Management Programme. Society for the Promotion of Wasteland Management, New Delhi.

Rossi, A. and Lambrou, Y. (2008) *Gender and Equity Issues in Liquid Biofuels Production: Minimizing the Risks to Maximize the Opportunities*. Food and Agriculture Organisation of the United Nations (FAO), Rome.

Sankoff, G. (1971) Quantitative aspects of sharing and variability in a cognitive model. *Ethnology* 10, 389–408.

Silitshena, R.M.K. and McLeod, G. (1998) *Botswana: a Physical, Social and Economic Geography*. Longman, Gaborone, Botswana.

Sims, B.G. and Kienzle, C. (2006) Farm power and mechanization for small farms in sub-Saharan Africa. FAO, Rome.

United Republic of Tanzania (URT) (2012a) *National Population Census 2012*. Tanzania National Bureau of Statistics, Dar es Salaam, Tanzania.

United Republic of Tanzania (URT) (2012b) National Population Census 2012. Tanzania National Bureau of Statistics, Dar es Salaam, Tanzania.

Yanda, PZ and Mung'ong'o, CG (1999) Farming systems of Kasulu District, western Tanzania. Research Paper No. 45. Institute of Resource Assessment, Dar-es-Salaam, Tanzania.

Yanda, PZ, and Mubaya, CP (2011) Managing a changing climate in Africa: Local level vulnerabilities and adaptation experiences. Mkuki Na Nyota Publishers, Dar-es-Salaam, Tanzania.

9 Natural Resource Use Conflicts in a Changing Climate: The Case of the Wetlands of Kilombero and Kilosa Districts in Tanzania

Emma T. Liwenga* and Florian Silangwa

Institute of Resource Assessment, University of Dar es Salaam, Dar es Salaam, Tanzania

Abstract

In Tanzania studies on conflicts between subsistence farmers and agropastoralists/pastoralists as well communities versus conservationists have been frequently reported. These include studies on water use conflicts in the Pangan River Basin. Agropastoralists have also been evicted as a result of these conflicts. The overriding tendency has thus been centered on discussion about what keeps pastoralists and crop cultivators apart rather than what keeps them together. Although conflicts occur mainly as a result of resource competition, some conflicts constitute only one aspect of a much more complex interaction between the two groups. Little has been documented in examining the relative contribution of climatic and non-climatic factors in aggravating these conflicts and how this triggers various dimensions of conflicts among livelihood groups. This study examined the relative contribution of climatic and non-climatic factors as causes of natural resource conflict, based on a study conducted in Kilosa and Kilombero districts in south-central Tanzania. The study further examined conflict resolution mechanisms in terms of adaptive capacity to address these issues in a changing climate and in particular explored the roles of institutions in conflict resolution, and the associated challenges and opportunities involved. The study was conducted through a household survey, focus group discussions and expert interviews with Kilosa and Kilombero district land planning officers. The aim was to understand the challenges of land use planning and how those transformed into conflicts between users. The results of the interviews in the two districts revealed that there were land conflicts related to boundary disputes between and among villages, investors against villages or communities, and government agencies versus villages or investors. Unlike Kilosa, Kilombero District had more land conflicts involving villages and government institutions. There is a complicated land conflict in 25 villages against the Ramsar site area.

Introduction

General overview

Conflicts over natural resources such as land, water and forests are ever-present throughout the world. People everywhere compete for natural resources to enhance their livelihoods (Ayling and Kelly, 1997). However, the dimensions, level and intensity of conflicts vary greatly from place to place based on a number of factors. Conflicts over natural resources may have class dimensions, putting those who own the resource against those who own nothing but whose work makes the resource productive. Political dimensions may dominate where the state has a keen interest in public good, such as

*liwenga99@gmail.com

conservation, or in maintaining the political alliance it needs in order to remain in power (Suliman, 1999). The intensity of conflict may vary enormously from confusion and frustration among members of a community over poorly communicated development policies to violent clashes between groups over resource ownership rights and responsibility.

It is evident that Africa is facing a number of climate shocks that are intensifying poverty, inequality and the disruption of livelihoods. Indeed, concern in the early 1990s over the negative impact of climate change strengthened fears that environmental degradation and demographic pressures would displace millions of people in the developing world and lead to huge social upheaval. Sub-Saharan Africa is among the most vulnerable to the negative effects of climate change, and faces the greatest challenges of adaptation.

There is a general consensus that climate variability and change exacerbate the scarcity of natural resources on the African continent, where the majority of people depend on land, water and the oceans for their livelihood. This scarcity happens through sudden climate events or through slower changes and variability, such as changes in temperature and rainfall. Increases in extreme or sudden events – such as flooding or prolonged droughts – reduce the availability of arable land, water, food and fish stocks. Slower, insidious changes and variability in temperature and rainfall patterns – that cause desertification, water and land shortages – are no less hazardous, as they place long-term stresses on already vulnerable communities.

Resource-based conflicts in Tanzania

In Tanzania, studies on conflicts between subsistence farmers and agropastoralists/pastoralists as well communities versus conservationists have been frequently reported. These include a study on water conflicts in Pangani (Mbonile, 2006) and conflicts over land use (Mbonile and Mwamfupe, 1997). Agropastoralists have also been evicted as a result of these conflicts such as on the Usangu Plains in Mbeya (Matee and Martin, 2006). The overriding tendency has thus been centered on discussion on what keeps pastoralists and crop cultivators apart rather than what keeps them together. Surely conflicts occur mainly as a result of resource competition. None the less they constitute only one aspect of a much more complex interaction between the two groups. Little has been documented about the relative contribution of climatic and non-climatic factors in aggravating these conflicts and how this triggers various dimensions of conflicts among livelihood groups. Despite the various efforts to address the natural-resource-based conflict, these continue, on a diverse scale. This chapter examines the relative contribution of climatic and non-climatic factors as causes of natural resource conflict, based on a study conducted in Kilosa and Kilombero districts in Tanzania. The chapter further examines conflict resolution mechanisms in terms of adaptive capacity to address these issues in a changing climate.

The Intergovernmental Panel on Climate Change (IPCC) (2007) has indicated that climate change and variability have the potential to impact negatively on water availability, and access to and demand for water in most countries, but particularly in Africa. It is now also conventional wisdom that climate change has already affected, and will continue to affect, human security (including ecological sustainability), which may result in human rights abuses and intrastate and interstate conflict and wars.

Resource users include small-scale farmers, pastoralists, large-scale landowners, and private companies (e.g. forestry industries, mining, hydropower and agribusiness agencies), but it is mainly the last two mentioned that increasingly influence natural resource management decisions. Some people may use resources in ways that undermine the livelihood of others. Power differences between groups can be considerable and what is at stake may, for some, be a matter of survival. The resulting conflicts often lead to chaotic and wasteful deployment of manpower and the depletion of the very natural resources on which livelihoods are based.

The use of natural resources is susceptible to conflicts for a number of reasons.

First, natural resources are embedded in an environment or interconnected space where actions by one individual or group may generate an effect far from the site. Second, natural resources are also embedded in a shared social space where complex and unequal relations are established among a wide range of social actors such as agro-export producers, small-scale farmers, ethnic minorities and government agencies. As in other fields with political dimensions, those actors with greatest access to power are also able to control and influence natural resource decisions in their favour (Buckles and Rusnak, 1999). Third, natural resources are subject to increased scarcity due to rapid environmental change, increasing demand, and their unequal distribution (Homer-Dixon and Blitt, 1998).

Recently, natural resource conflict, especially in the wetland areas of Tanzania, has drawn considerable attention because natural resource disputes often have devastating effects on the sustainability of people's livelihoods. This has been the case in the wetland areas of Kilombero and Kilosa districts, in the Morogoro Region. Moreover, almost all these conflicts typically involve disputes such as those between farmers and pastoralists in Kilosa and local communities over access to land and forest products in protected areas. The effects of natural resource conflicts are multidimensional and the crisis is more serious when it comes to developing countries (Gebremariam, 2011). It is in this context that this chapter examines the relative contribution of climatic and non-climatic factors as causes of natural resource conflict in Kilosa and Kilombero districts and considers the historical, social, ecological and institutional dimensions among the livelihood groups.

Conceptual and theoretical framework

Figure 9.1 shows the conceptual framework on which this study was based, indicating the link between climate change, resource scarcity and resource use conflicts. As the current study assumed that both climatic and non-climatic factors could be the cause of resource use conflict, both types of factor were explored. As such this conceptual framework was integrated with the UK Department for International Development's (DFID) Sustainable Livelihoods Framework (1999) in order to contextualize other factors that affect livelihood systems, in terms of shocks, stress and seasonality.

The study further examined conflict resolution mechanisms in terms of adaptive capacity to address these issues in a changing climate and in particular explored the roles of institutions in conflict resolution, and the associated challenges and opportunities involved.

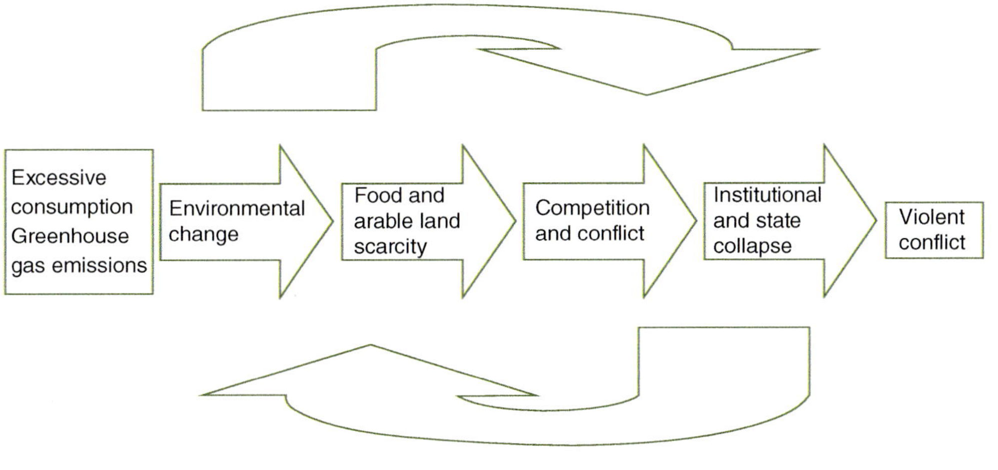

Fig. 9.1. Conceptual framework indicating the link between climate change, resource scarcity and resource use conflicts. (From Brown *et al.*, 2007.)

Methodology

Location of the study area

Using the Rufiji Basin, as a case study, two districts Kilombero and Kilosa (Fig. 9.2) were selected for detailed study. Both districts are endowed with a great diversity of natural resources. Kilombero District is located on the western side of Morogoro Region. It borders with Kilosa and Morogoro Rural District to the north-east, Mufindi and Njombe to the south-west and Kilolo District to the north, all in the Iringa Region, Ulanga District to the south-east (along the Kilombero River) and Songea Rural District of Ruvuma Region to the south. The Kilombero District is situated in Morogoro Region in eastern Tanzania and lies between longitudes 35°563' E and 37°797' E and latitudes 7°654' S and 10°023' S (Fig. 9.2). Kilosa District Council is one of six local government authorities in

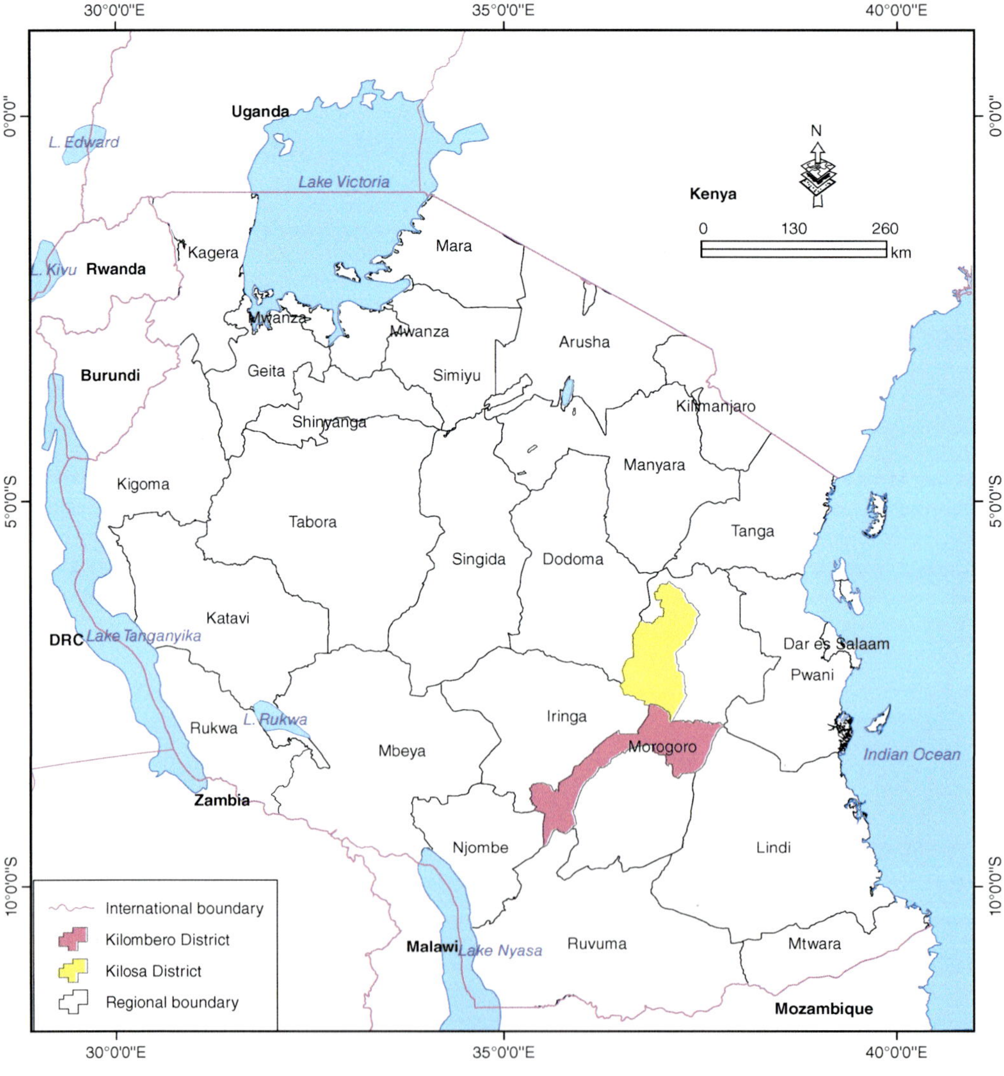

Fig. 9.2. Map showing the location of Kilosa District and Kilombero District in Tanzania. (From Geographic Information System (GIS) Laboratory, Institute of Resource Assessment, University of Dar es Salaam, 2017.)

Morogoro Region. The district lies between longitude 36°030' E and 37°030' E and latitude 5°55' S and 7°53' S. Located on the western part of the region, the council forms 20% of the area of the entire region.

Agroclimatic conditions

Generally, the districts have high temperatures (hot weather conditions) and have bimodal rainfall patterns. Short rains begin towards the end of November and end in January or February. Long rains usually start in March and end in May or June. The average temperature in the districts range from 26°C to 32°C. The average rainfall ranges from 1200 mm/year to 1600 mm/year. Kilombero and Kilosa districts experience seasonal flooding which causes some parts of the area to be inaccessible during the long rainy season. Flooded areas are mostly fertile due to frequent flooding. A diversity of crops are grown in each zone including sugarcane, paddy, maize, cassava, vegetable and fruits (oranges, mangoes and pawpaw).

Data collection

The study was conducted through a household survey, focus group discussions and expert interviews with Kilosa and Kilombero district land local government officials.

Discussion of the Findings

Overview of the existing natural resource use conflicts

This section analyses the nature and types of resource use conflicts in Kilosa and Kilombero districts. Due to heavy dependency on water and land, resource use conflicts in Morogoro Region are primarily long-standing rural phenomena rooted in the livelihood systems of both pastoralists and farmers in the study villages. Ninety-nine per cent of all respondents in the two districts reported having experienced conflicts related to land and water resources. However, it was revealed during participatory resource appraisal (PRA) sessions that land and water resources are either owned by the central government as conservation or protected areas managed by district councils and the National Parks Authority or owned by individuals (private investors, farmers or groups of pastoralists) from within and outside the region.

Based on discussions with various stakeholders, it was found that conflicts between village communities and investors in the district were to a large extent experienced in a low-lying fertile wetland which was free from pests and animal diseases. Furthermore, conflicts are concentrated along the rivers where crop farming is practised throughout the year. Respondents in key informant interviews reported the presence of human–wildlife conflicts (especially elephants versus farmers) in both districts. Generally, the findings show that land and water are the most important resources reported to be the major cause of conflicts in the area. While land resource was the main source of conflicts in Kilosa District, water resource conflict was more pronounced in Kilombero District with some cases of land conflicts in some villages.

Water resource conflicts on the other hand are highly concentrated along rivers where both farmers and herders depend on the river for irrigation and water for the livestock, respectively. This difference is attributed to the fact that Kilombero District is more involved in irrigated farming along the Kilombero River and its tributaries, thus there is friction between and among farmers over irrigation water and farmers versus herders when cattle are taken across crop fields to the rivers for water. A detailed account of these conflicts is given in the following sections.

Land use conflicts in the study area

Kilombero and Kilosa districts form one of the resource-rich and very fertile lowland wetlands in Tanzania. Interviews with key informants in both districts said that the lowland wetlands are also crop and animal pest free and land in this area has vast potential for large-scale agricultural mechanization and livestock production. It was further reported that since the colonial period, the land and water resources and other attributes of this area have attracted investors and other migrants from all over the country and beyond. As a result there has been a large influx of migrants, particularly livestock keepers.

Land use conflicts in Kilosa District

The study conducted interviews with the Kilosa District Land Planning Officer to understand the challenges of land use planning and how that transformed into conflicts between users. The results of the interviews revealed that throughout the district, there are land conflicts related to boundary disputes between and among villages, investors against villages or communities, government agencies versus villages or investors. The discussion with officials revealed that from July to September 2013 there were 83 conflicts in six villages, namely Kilangali, Malui, Madudu, Miyombo, Kivungu and Magomeni. Also, 53 other conflicts were also registered in five villages namely Tindiga, Miyombo, Kivungu, Chanzuru and Kilangalima from October to December 2015. Analysis of responses to the household survey indicated that the two main sources of conflicts in Kilosa District were perceived to be crop damage by livestock (58%) and boundary disputes (34%) (Fig. 9.3). Over 83% of respondents felt that land conflicts have been increasing.

Land use conflicts in Kilombero District

In Kilombero District, there was a similar picture of land use conflict to that of Kilosa District. It was found that land conflicts in Kilombero District have been caused by either lack of or failure of land use plans at the village level. Figure 9.4 shows the number of cases of land use conflict in Kilombero District over the period 2007–2016. Unlike Kilosa, Kilombero District has more land conflicts involving villages and government institutions. There is a complicated land conflict in 23 villages against the Ramsar site area (a Ramsar site is land listed as a Wetland of International Importance under the Ramsar Convention on Wetlands of International Importance). Villages that are involved in this land conflict are Mahutanga, Miwangani, Namwawala, Ihenga, Kalenga, Ikwambi, Ijia, Isago, Luvilikila, Mkangawalo, Kidete, Ikule, Merera, Msita, Chisano, Ngwasi, Kalengakelu, Msolwa-Mlimba, Ngalimila, Ipinde, Utengule, Iduindembo and Ngombo.

It was also reported during FGDs that there has been a tendency of rich people, mainly herders, from outside the district illegally grabbing village lands by bribing village authorities. This has resulted in major conflicts between herders and farmers over land. In Kiberege village within the 1000 ha of land allocated for livestock keeping, a single pastoralist has monopolized 460 ha since 1992. Only 540 ha were left for other herders, which have now been taken by farmers. During the land use planning in 2012, this land was considered as grazing land for all herders in the village, thus leading to conflicts over land with the owner. The incident culminated into fighting between farmers and herders as well as failure to implement the land use plan established in 2012. The number of pastoralists is going up rapidly with time hence increasing the number and

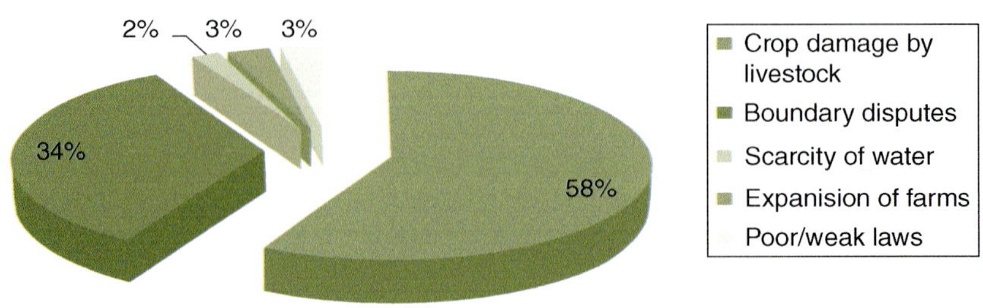

Fig. 9.3. Source of conflicts in Kilosa District. (From household survey, 2017.)

level of land conflicts. For example, in 1995 Kiberege village had eight pastoralists, but by 2016 there were 36 known herders with over 1681 cattle. In addition there were 41 large-scale farmers who bought between 50 ha and 460 ha each between 1996 and 1998 so that a total of about 9120 ha of land is owned by these farmers. This has left small-scale farmers without land, and hence has resulted in conflicts. Data obtained from Kilombero District Land and Housing Tribunal, shows an increasing trend in the number of land conflict cases reported each year (Figure 9.4, Land disputes reported at Kilombero Land and Housing Tribunal).

According to interviews with officials at the District Land Committee there are many more cases of land conflicts that are not reported. Most of these land conflicts are reported and resolved by the village land committee. Of the 99 villages in Kilombero, 56 villages have village land committees. Although results from the household survey indicate that conflicts over forest resources were reported only in Kilombero District, key informant interviews at Kilosa District revealed that there are cases where herders get into conflict with government authorities as result of invading forest reserves for fodder, especially during drought conditions. According to the Kilosa District Natural Resource Officer, when this study was conducted more than a quarter of the cattle available in the district were already in protected areas, and every day a number of herders could be forced out from these forests. Conflicts over forests that have been reported in Kilombero District are found in Mkasu village and the neighbouring village of Mpanga over Kikande Kande or Nakandubwa forest reserve. The forest is a source of income through timber and charcoal making, and also through money earned from Selous Game Reserve since part of the forest forms part of a Wildlife Management Area (WMA). Mkasu village is also in conflict with Udizungwa Conservation Area Authority over a 5 km extension of the conservation area into the village land.

Drivers of resource use conflicts in the study area

Natural resource conflicts, particularly land and water resources conflicts, are closely linked as land and water are highly desired resources by communities and individuals. In Tanzania, land, in combination with inequitable access to resources, resource degradation and demographic pressures, has been a key driver in violent conflict. A further problem is the lack and disrespect of land use plans, which are undergoing changes related to modernization and globalization. Lack of

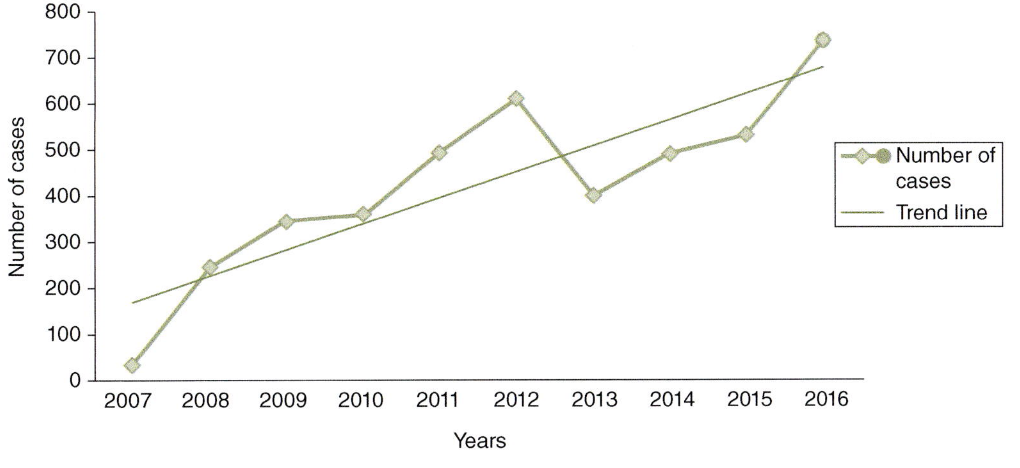

Fig. 9.4. Land disputes reported at Kilombero Land and Housing Tribunal 2007–2016. (From fieldwork, 2017.)

appropriate policy, weak land institutions and conflict resolution mechanisms contribute to the continued existence of conflicts in the region. Drivers of natural resource conflicts in the study area include the following: (i) increased population due to influx of people in the area; (ii) climate change and variability; (iii) absence or disrespect of land use plans; (iv) presence of undeveloped farms; (v) poor governance; (vi) unregulated resettlement; and (vii) ineffective dispute resolution systems.

Influx of people in the area

Due to the agricultural land which is suitable for cultivation and the plentiful green grass which attracts the pastoralists, coupled with the abundance of rivers and streams leaving most part of Kilosa and Kilombero districts green almost throughout the year, the two districts witness an influx of subsistence farmers and pastoralists from different parts of Tanzania such as Arusha, Dodoma, Singida, Mwanza, Shinyanga and Tanga.

The procedure for receiving migrants in the two districts has been prepared by the district councils. The procedures clearly state that 'no entry of cattle in Kilosa District is allowed without a written permission of the district council authority', however, mainly due to poor governance, the leaders of the village authorities do not implement these procedures. In the FGD it was learnt that the pastoralists tend to collect money by selling part of their livestock in order to influence the village leaders and get access to land without informing the district authority. It was also observed that the agropastoralist (the Sukuma and Gogo) and pastoralist (Maasai and Barbaig) migrants in Kilosa and Kilombero have a strong network with relatives in their areas of origin, and therefore once they manage to get access to land, they tend to invite their relatives without informing the village and district authorities.

Climate change and variability

Climate change and variation are expected to be one of the causes of natural resource conflicts. During periods of drought when water becomes scarce, competition for access and use of water sources will intensify. As a result, conflicts are more likely to occur as a struggle to utilize this scarce resource increases. Water-use conflicts occur when one water source is used for more than one use by different users and the uses are not complementary. Conflicts over water access and use may also arise from the uneven distribution of available water between uses and users. During the rainy season, when water is plentiful, members of one village can access water sources at another nearby village. However, as water becomes scarce owing to prolonged drought, villages may prohibit members from another village from using water sources located within their jurisdiction. In this study water conflict caused by climate change and variability was experienced in Kiberege Ward in Kilombero District. Discussions with different stakeholders in Kilosa and Kilombero suggested that the climate has changed over recent decades resulting in prolonged droughts, unpredictable rainfall, floods and new outbreaks of animal and human disease. Drought-induced migration is linked to land use conflicts.

Absence or disrespect of land use plans

The absence or disrespect of land use plans was reported to be one of the causes of land use conflict in Kilosa and Kilombero districts. The discussion with the districts' land planning officers and town planners revealed that some of the village land use plans have been designed as a solution to the conflict over land resources. According the planners, in 2007 39 villages in Kilosa District were involved in establishing land use plans under the support of Tanzania Forestry Conservation Group (TFCG) (18 villages), World Wide Fund for Nature (WWF) (seven villages), MKURABITA (two villages), KASTAN Mining (one village), Molan (two villages) and the National Land Use Planning Commission (NLUPC) (seven villages). Only seven of these villages (those sponsored by WWF) managed to complete the process and have land use plans. In Kilombero District, a total of 27 villages have no land use plans, while 15 have land use plans but their village boundaries have not been surveyed and mapped.

It was further noted that in most of the villages in Kilosa District, the land use plans were not implemented because the village leadership refused to allocate land for livestock keeping, despite the presence of a large number of pastoralists in the villages. It was further learnt that 32 villages in Kilosa did not get even get to the end of the process because 23 of these villages failed to complete the process due to presence of land and boundary disputes among the villages. The other nine villages failed to complete because they did not agree on allocating grazing land to livestock keepers for the reason that livestock keepers were few and the livestock could graze their crops.

It was further elaborated that when the land use planning process was near to completion in 2010, some wards and villages were divided into two or three villages for political reasons to win the interest of voters in the 2010 general election. The divisions of wards and villages made the whole process of village land use plans which were already in place and those which were in preparation to be a waste of efforts and resources. Such villages include Magubike, Ruaha and Kisanga. The process of dividing the villages for political reasons became even more rampant in the 2015 election. New villages included Ulaya Kibaoni, Mvumi and Tindiga A and B. Besides these divisions, the land survey process for the divisions was not participatory. Discussions with village leaders confirmed that in many villages, such as Tindiga A and B, leaders were not involved in the survey, which makes it difficult for them to identify the boundaries of their villages.

Presence of undeveloped farms

Presence of a large number of undeveloped farms with land exceeding 50 acres was reported to be another cause of conflict in Kilosa and Kilombero districts. Discussion with district land planning officers revealed that in Kilosa there are more than 183 undeveloped farms with legal ownership but these have never been used for more than 15–20 years. The presence of these farms has provided room for encroachment by farmers and herders. In the course of using these farms, the two groups have found themselves in conflict. Farmers are blaming livestock keepers for grazing on their farms, while at the same the livestock keepers complained about farmers' interference in grazing land.

The presence of the undeveloped National Park has created room for illegal immigrant pastoralists and farmers to create a village within the National Park that has been involved in conflicts. The immigrant farmers in Tindinga village believe that the National Parks Authority (TANAPA) has taken part of their village. This has a created a conflict between Tindiga A village and the Eleto Group which own a piece of land within the National Park. People were born and grew up on these large land plots. It was reported by villagers in Tindiga that some of this land which has remained unused was owned since 1929 and the land ownership has been changing from one to another without making any developments on this land. With increasing population caused by natural increase and influx of people, the requirement for land has also increased, and as the people knew the history of farms which were abandoned they have been forced to encroach on these farms, but trying to force them out means conflict.

Poor governance

Poor governance in issues related to land administration is a major problem in many developing countries. In terms of small-scale petty corruption, a recent international survey found that over one in five people reported having paid a bribe when dealing with land services, placing land as the third most corrupt sector (Transparency International, 2010–2011). This challenge was also reported to contribute to land conflict in Kilosa and Kilombero. The villagers in Tindiga, for instance, complained about delays in addressing the conflict by the respective authorities. Villagers explained that the lack of taking immediate action when farmers reported destruction of crops by herders had aggravated the problems.

It was narrated that the respective authorities could take 2–4 days to arrive at the site and they do not take any measures against the herders because by the time they arrive they found that the herders had moved to another area. The situation is different when the farmers decide to take action by killing the livestock for destroying their crops: the respective authorities tend to arrive at the site on time taking strong measures with the farmers including taking them into custody. The double standard in handling these conflicts has aggravated the conflict between farmers and pastoralists.

Unregulated resettlement

Kilosa District is home to people from other regions such as Singida, Mwanza, Shinyanga, Tabora and Dodoma. This unregulated influx has resulted in an increase in the number of pastoralists and farmers in Kilosa District villages. Pastoralists from Ihefu who were evicted without proper plans for resettlement made their way to Kilosa adding to the number of people on the same land. Problems were associated with granting of the entrance permits by the gatekeepers who are the village leaders. It was explained that some village leaders would allow the herds to be taken into their villages without consideration of existing land use plans.

Ineffective dispute resolution systems

Land disputes often exist in such large numbers that they overwhelm the capacity of the existing dispute resolution systems, whether this is court-based or customary. In turn, this means that conflicts may take a very long time to resolve or be seen as impossible to finalize. This can drive frustration that may be expressed through violence. Furthermore, dispute resolution systems may be difficult to access or use for reasons of distance, prohibitive fees, documentation requirements, procedural knowledge or illiteracy. Some dispute resolution procedures may be especially challenging to access by women or marginalized ethnic groups.

Major actors in resource use conflicts

Findings indicate that in both study districts and all study villages, over 59% of conflicts involved farmers groups against herders. At least 11% of all conflicts are between farmers themselves while conflicts among herders themselves accounted for 9%. Conflicts between individuals were reported to account for 8% of conflicts and over 7% involved village communities and government authorities.

Analysis of household surveys and results from PRA sessions in all villages indicated that most of the conflicts in the study districts are violent (Fig. 9.5). The overall picture of analysis indicated that 87% of all conflicts are violent. However, discussion with district officials revealed that while conflicts occur almost everywhere in Kilosa District, some of these conflicts such as those involving organizations and government agencies are at an early stage and no physical fighting has been involved.

Farmer–pastoralist conflicts

As pointed out by Matzko (1985), the pastoralists view crops as grass and they argue that when cattle consume crops such as rice and maize, they fatten and produce plenty of milk for their families for sale. Farmer–pastoralist conflicts are the most dominant in both districts of Kilosa and Kilombero. The conflicts involve two different groups of people with different socio-economic backgrounds. Principally Kilosa and Kilombero districts are endowed with good geographical factors that support both pastoralism and agricultural activities; the existence of fertile soils, rivers that flow throughout the year and presence of grazing areas and valleys that are green throughout the year constitute the reason for the influx of people who come to take advantage of these conditions. This has resulted in frequent emergence of conflicts between farmers and pastoralists, which have often resulted in injuries and deaths since the use of firearms during fighting is increasingly becoming

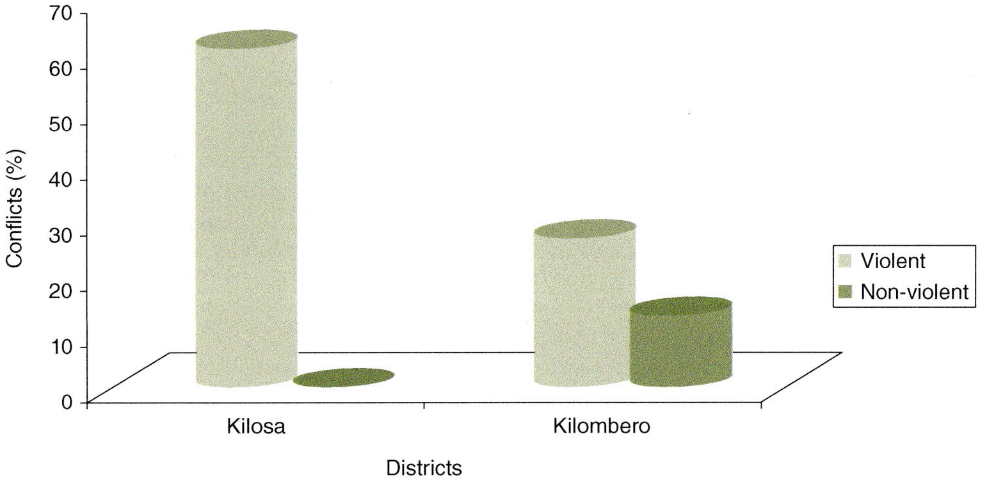

Fig. 9.5. Magnitude of natural resource conflicts in the study area. (From fieldwork, 2017.)

common. Conflicts occur especially during the dry season or when rainfall is not sufficient. Moreover, for the subsistence farmers the valleys are suitable for irrigation and dry season irrigation competing for these resources therefore results in clashes in many areas in these districts.

Villages–institutions conflicts

Conflict between villages and institutions is another type of conflict, which was observed in the two districts. Villagers reported that there is an increasing incidence of conflicts between government or private institutions with village governments over boundary disputes. As stated earlier in the chapter, in Kilombero District there is a complicated land conflict between the Ramsar site and 23 villages. Discussion with district land planning officers revealed the disagreement between the two parties emanates from the fact that the boundaries of the Ramsar site were set in the 1970s when the population in these areas was low. However, with increasing population due to migration, there is a need to revisit the boundaries in order to reduce conflicts between the Ramsar site and the adjacent villages. Other reported conflicts involving villages and institutions were between Utengule and Iduindembo villages against the hunting block, Ikule against Tanzania People's Defence Force and Bwawani village against Kiberege prison.

Pastoralists versus agropastoralists

Conflicts involving livestock keepers is another type of conflict we observed in our case study area. Conflicts among herders were reported to occur between the Maasai pastoralists in Mabwegere village against the Sukuma in Mambegwa village who are agropastoralists. Based on the discussion with village leaders in Mabwegere, it was revealed that the conflicts between the two groups is mainly caused by encroachment of the Sukuma agropastoralists in Mabwegere village which was designated by the district authority as a pastoralists' village. The Sukuma as agropastoralists need land for grazing as well as for cultivation and they have needed more land to perform these activities. It was reported that there is a tendency of people from one tribe to steal cattle from the other tribe and often such incidences result in clashes between the respective tribes.

Investors and village governments

According to the Kilombero District Executive Officer, for many years the conception of investment in the district meant so much land being taken without developing it. This provides a founding explanation of the recurrent land uses and the conflicts involved in Kilosa and Kilombero districts. It is this historical account that has also shaped the land management and planning complications in the two districts.

Other conflicts in the district include conflicts between investors and villagers, conflicts between national parks and dwellers and conflicts between two neighbouring villages. There is a frequent conflict in Magomeni Ward in Kilosa between investors and villagers. In addition, the occupation of some wetland areas by foreign investors for large-scale farming was also named by the respondents as one of the main sources of conflict in the district. Lack of specific regulations and systems as to how pastoralists could access water for the cattle has also forced the pastoralists to move across farmland and destroy crops.

Implications of natural resource conflicts on livelihoods

Natural resource disputes undermine economic growth in Kilosa and Kilombero districts as well as the economic well-being of communities in a number ways. These include: (i) loss of life, injuries and destruction of properties; (ii) disruption of livelihood activities; (iii) impacts on land and water resources; (iv) decreased agricultural productivity and food insecurity; and (v) impacts on management issues and communities' relationships.

Loss of life, injuries and destruction of properties

During the FGDs with district officials in Kilosa and Kilombero, it was revealed that natural resource conflicts have resulted in loss of life, injuries and destruction of properties in the study area. In explaining the impact of natural resource conflicts in Kilosa District, officials cited an example of land conflict between Mabwegere and Mambegwa villages, which occurred in 2008. In this conflict one pastoralist and seven farmers lost their lives. Furthermore, in 2014, 38 houses were burnt. In 2015, one pastoralist and two farmers lost their lives. Similar to this, it was reported that during one of the clashes between Mkindo and Kambala villages in 2014, about six people lost their lives, most of them being males. Such reported incidences have a significant effect on the households concerned, which immediately become female-headed, and hence this increases the burden of family care on the women.

Disruption of livelihood activities

Disruption of the usual way of life has been one of the major impacts of these conflicts in the two districts. Traditionally in farming societies in these districts and in most parts of Tanzania, women play a pivotal role in feeding their families, so with the recurring conflicts in the two districts women find it difficult to play this critical role as there are times when they have to stay at home in fear of being ambushed in their fields. The situation is more or less similar to men who sometimes are compelled to spend their time fighting and this is done at the expense of productive activities such as farming and attending livestock. Moreover so many people in communities have been displaced internally and generally this undermines the entire economy of the societies involved. Such conflicts have now resulted in breakdown of societal values and are changing established family roles as women have been recently reported to assume additional responsibilities for the family.

Pastoralist–farmer conflicts have also had negative implications on the provision of social services such as education and health services in villages. For instance, the conflict which occurred in 2008 in Mabwegere led to closure of the school for about 2 weeks because neither students nor teachers were found on the school premises as they all ran away to save their lives in the bushes. Children belonging to the affected household also did not go to school as they were all camped at Msowero village. Apart from closing of schools, villagers reported that patients who sustained grievous injuries overwhelmed health centres and dispensaries. There were also those who sought asylum for their lives at both Msowero and Mabwegere villages. The district dispatched health practitioners to attend to the women who had taken refuge at a Lutheran church in Mabwegere village.

Impacts on land and water resources

Destruction of social norms such as communal and individual land ownership and the displacement of water sources has brought about loss of land as a means of livelihood. The findings show that about 62% and 75% respondents in Kilosa and Kilombero districts, respectively, had no permanent land for agricultural activities. For example, the expansion of villages and the need for the Udzungwa Mountains National Park and Ramsar site for conservation purposes has resulted in a reduction of land available for agriculture, rituals and graves in several villages, including Mkasu and Msita.

Furthermore, a discussion with communities and field observations revealed that the increased human population together with the expansion of irrigation farms and increased livestock in Kilombero District has created a scarcity of water and pastureland especially during the dry season. The expansion of farms in villages such as Nyamwezi and Mkasu has reduced the potential grazing land and livestock routes, which has resulted in pastoralists encroaching on to farms, leading to destruction of crops and irrigation infrastructure, and hence frequent conflicts between farmers and livestock keepers.

Decreased agricultural productivity and food insecurity

Natural resources conflicts such as those based on land and water resources have several effects on livelihoods, including leading to low food production for households. Natural resource conflicts may lead to reduction in output and income of crop farmers as a result of the destruction of crops by cattle. It may also reduce a farmer's savings and the ability to repay credit, as well as affecting the food security and economic welfare of urban dwellers who depend on these farmers for their food supply.

The findings in the study area have shown that land conflicts between farmers and pastoralists, which have involved a number of killings, have discouraged farmers from being involved in agricultural activities. Natural resource conflicts have also resulted in destruction of crops, killing of livestock and the farmer's displacement, which all contribute to food shortage in the study area. In the farmer's community, the women who remain behind are especially affected as they stop going to the distant farms for fear of attack from the pastoralists. Such displaced farmers have become a source of liability to other farmers from whom they have to beg for food for themselves and their families. This has created a vicious cycle of poverty in such communities (Tor *et al*., 2009).

From the findings, about 53% and 80% of the household respondents in Kilosa and Kilombero districts, respectively, reported having suffered food shortage. Sometimes the conflict situations resulted in hunger and famine as reported by 42.6% and 17.5% of the household respondents in Kilosa and Kilombero, respectively. Very few household respondents in Kilosa District reported on the lack of a balanced diet. However, the effects of conflict on food security can be clearly understood by considering changes in different sources of income at the level of the household. Such consideration may focus on crop sales, production for a household's own consumption and livestock production. This is due to the fact that household income controls consumption of food in any community.

Regarding pastoral communities, resource use conflicts have limited livestock mobility which allows rational use of grazing areas within the district. Respondents in Kilosa (Mabwegere village) said they had lost the corridors and access to grazing land on the other sides of the village. While in Kilombero District, the pastoralists from Bwawani village had lost access to water sources (Kiberege River) in Mkasu village. According to Odhiambo *et al*. (2012), such insecurity to access resources puts pastoral systems under pressure and later reduces its productivity, meaning that the land use system fails to respond to ecological and climatic variability, resulting in ecological degradation.

Impacts on management issues and communities' relationships

Natural resource conflict with communities over forest reserves and the Ramsar site in Kilombero has undermined the management and conservation prospects of such valuable natural resources. It was observed that the existing forest or land use conflicts in Msita and Mkasu villages, for instance, have largely undermined the sustainability of conservation efforts made by the surrounding communities. Villagers around the forest have withdrawn themselves from involvement in conservation. Hence the land use conflict has resulted in poor collaboration between the villagers and government agencies, when such collaboration is necessary for effective management of resources.

Conflict resolution mechanisms in natural resource management

Management of natural resources basically involves the management of conflicts over scarce resources, in this case land and water resource, between agriculturists and livestock keepers, village government and institutions, villages and investors as well as pastoralists and agropastoralists. In Tanzania, the National Land Policy acknowledges the growing number of land and resource tenure conflicts (National Land Policy, 1997). The policy statement, with regard to land conflicts, suggests the need for establishing a land dispute settlement machinery by strengthening the existing quasi-judicial bodies who deal with such disputes. The dispute settlement mechanism is known as the councils of village elders (*Mbaraza Ya Wazee Ya Ardhi*). Land conflict resolution mechanisms are structured so that the councils of village elders have primary jurisdiction in all land matters, including settling disputes over individualization of tenure.

In order to understand the effectiveness of the conflict resolution mechanisms, respondents were requested to express their views with regard to the effectiveness and presence of justice in resolving natural resource conflicts in their areas. The findings show that 77.3% and 88.3% of the respondents in Kilosa and Kilombero districts, respectively, were dissatisfied with the level of justice in conflict resolution and all the government bodies involved in natural resource conflict resolution. In Tindiga village, issues associated with poor governance were reported when it came to resolving conflicts between farmers and herders. Local communities had concerns that the authorities do not deal with villagers' complaints effectively as they underestimate the damage to crops/ farms or the level of injury/loss of livestock (and damage to property such as houses) and sometimes this leads to little or non-payment of compensation.

Conclusion and Recommendations

It is evident that the impacts of changing climatic conditions on the availability of natural resources, coupled with factors such as population growth due to influx of people, weak governance and land tenure challenges, have led to increased competition over scarce natural resources – most notably fertile land and water – and this has resulted in tensions and conflicts between communities and livelihood groups. It has been further established that disputes between different actors (including farmers, pastoralists, investors and conservationists) demonstrates that uncontrolled movement of people, land grabbing and poor management of natural resources like land and water and the failure to implement on a participatory basis management plans of natural resources is a major cause of these disputes.

Addressing conflict is a prerequisite for sustainable natural resource management. If not addressed in an effective and timely manner, natural resource conflicts can adversely affect community livelihoods and result in resource degradation. The study has estblished several strategies that local communities, resource users and public officials use to manage and to resolve conflicts. These include: (i) the use of informal institutions such as elders; and (ii) formal methods which involve the village land committee, village council, police forces and courts.

The authors of this chapter recommend that the genesis of the conflict should be addressed with the seriousness it deserves, not just to offer quick-fix, short-term solutions, but to come up with more sustainable solutions to the problem. In order to achieve this, it is important to take into consideration the interests and way of life of both the communities involved in a dispute and see how they can be harnessed for the purposes of peace and tranquility. The pastoralist communities are nomadic in nature and they pay a great deal of attention to their livestock, while farming communities too pay a lot of attention to their farming activities. Therefore, they should be given the same attention when there is a need for resolution.

The law enforcement mechanism, where there are frameworks, is another issue that must be taken seriously by the leadership in the two districts, especially with regard to how people are admitted in the villages and the capacity of allocated areas to accommodate livestock. Proper land allocation should be followed by authorizing bodies and the villages' village councils and village assemblies should be involved in management of village lands as per the Village Land Act. A participatory land use planning approach may be useful, instead of creating villages for pastoralists only or allocating certain areas for farming. This is because even in some of the pastoral villages there is evidence that they practise both farming and livestock keeping.

In relation to policy issues, and in particular policy responses related to migration as a result of climate variability impacts, the best solution is to strengthen the adaptive capacity of villagers who are vulnerable to climate change impacts. In this respect a multidimensional approach is crucial in order to highlight the importance of the role of individuals and the government at all levels in learning for adaptive management. As suggested by the International Organization for Migration (IOM) discussion paper on environment and migration (IOM, 2007), the principles for effective management of migration caused by climate change impacts should include: (i) proactive policy and early action; (ii) comprehensive and coherent polices that have sufficient budgetary support for long-term planning; (iii) bilateral and regional cooperation; and (iv) multi-stakeholder partnerships involving public and private service actors, non-governmental and inter-governmental organizations, trade unions, individual migrants and diaspora associations. Effective management of migration caused by climate change should not assume that climate-change induced migration is part of the problem; it may indeed be part of the solution.

References

Ayling, R. and Kelly, K. (1997) Dealing with conflict: natural resources and dispute resolution. *Commonwealth Forestry Review* 76(3), 182–185.

Balama, C., Augustino, S., Eriksen, S. and Makonda, F.B.S. (2016) Climate change perceptions and adaptation strategies by forest adjacent communities in Kilombero District, Tanzania. *Journal of Environment and Earth Science* 6(10), 39–53.

Brown, O., Hammill, A. and McLeman, R. (2007) Climate change as the 'new' security threat: implications for Africa. *International Affairs* 83(6), 1141–1154.

Buckles, D. and Rusnak, G. (1999) Cultivating Peace: Conflict and Collaboration in Natural Resource Management (Ottawa). In: Buckles, D., (ed) *Cultivating Peace: Conflict and Collaboration in Natural Resource Management*, World Bank Institute, Washington DC, pp. 1–10.

Department for International Development (DFID) (1999) Sustainable Livelihoods Guidance Sheets Framework. Available at: http://www.livelihoodscentre.org/documents/20720/100145/Sustainable+livelihoods+guidance+sheets/8f35b59f-8207-43fc-8b99-df75d3000e86 (accessed 8 January 2016).

Gebremariam, A.G. (2011) Multidimensional approaches to local water conflicts: a study based on Afar Region of Ethiopia. PhD thesis, Loughborough University, Loughborough, UK.

Homer-Dixon, T. and Blitt, J. (eds) (1998) *Ecoviolence: Links Among Environment, Population, and Security*. Rowman & Littlefield Publishers, Lanham, Maryland.

Intergovernmental Panel on Climate Change (IPCC) (2007) *Climate Change 2007: Impacts, Adaptation and Vulnerability. Contribution of*

Working Group II to the Fourth Assessment Report of the Intergovernmental Panel on Climate Change. [Parry, M.L., Canziani, O.F., Palutikof, J.P., van der Linden, P.J. and Hanson, C.E. (eds)]. Cambridge University Press, Cambridge.

Kant, S. and Cooke, R. (1999) Jabalpur District, Madhya Pradesh, India: Minimising Conflict in Joint Forest Management. In: Buckles, D. (Ed). *Cultivating Peace: Conflict and Collaboration in Natural Resource Management*, pp. 81–100. International Development Research Center, World Bank Institute, Washington DC.

Mattee, A.Z. and Martin, N.S. (2006) Ambivalence and Contradiction - A review of the policy environment in Tanzania in relation to pastoralism. Issue paper no. 140. IIED, London. pp. 48.

Mbonile M. (2006) Migration and Intensification of Water Conflicts in the Pangani Basin, Tanzania. *Habitat International* 29 (1), 41–67.

Mbonile, M.J. and Mwamfupe, D.G. (1997) In-migrants and their impact on land management: a case of Usangu Plains, Tanzania. *Tanzanian Journal of Population Studies and Development* 4(1), 37–53.

National Land Policy (1997) *National Land Policy*. The Ministry of Lands and Human Settlements, Dar-es-salam, Tanzania.

Odhiambo, M.O., Abate, F., Kimani, M.J. and Bataringaya, D.T. (2012) *Impact of Conflict on Pastoral Communities' Resilience in the Horn of Africa: Case Studies from Ethiopia, Kenya and Uganda*. RECONCILE/ Food and Agriculture Organization of the United Nations (FAO), Rome.

Suliman, M. (1999) The Nuba Mountains of Sudan: resource access, violent conflict, and identity. In: Buckles, D. (ed.) *Cultivating Peace: Conflict and Collaboration in Natural Resource Management*. International Development Research Centre, Ottawa, pp. 205–220.

Transparency International (2010–2011) Global corruption barometer. Available at: https://www.transparency.org/gcb201011(accessed 6 May 2020).

10 The Role of Ecosystem Services in Enhancing Climate Change Resilience of Local Communities: The Case of Ngarambe-Tapika Wildlife Management Area, Rufiji District, Tanzania

Richard J.M. Katondo* **and Agnes M.S. Nyomora**

College of Natural and Applied Sciences, University of Dar es Salaam, Dar es Salaam, Tanzania

Abstract

This study examined the role of ecosystem services in enhancing climate change resilience of local communities in Ngarambe-Tapika Wildlife Management Area (WMA). The study aimed to identify forms of ecosystem services that can be gained from conservation of a WMA in relation to climate change adaptation. The design for this study adopted both a quantitative and a qualitative research approach. The study was undertaken in Ngarambe-Tapika WMA located between latitude 39° S and 39°30' S and between longitude 12°30' E and 13° E. It is located alongside the north-eastern border of the Selous Game Reserve. The area is also the home of local people whose lifestyles and livelihoods are intricately tied to the biological diversity and the functioning of this natural system. Purposive sampling was employed in selecting respondents for the household questionnaire, focus group discussions and key informant interviews. This study found that income obtained from Ngarambe-Tapika ecosystems by the communities were invested in material welfare and livelihoods that enhance resilience to climate change, primarily social services (54.9%) such as construction of houses, dispensaries and rehabilitation of the primary schools, and some of the money was spent on electricity provision for the community and energy for light and water pumps. Other benefits included employment (16.5%), protection from dangerous and problematic wildlife (14.3%) and petty business (14.3%). Generally, in Ngarambe-Tapika WMA there is a need to emphasize conservation awareness and extension programmes which advocate sustainable utilization of wildlife resources, and adopt an integrated approach of climate-smart agriculture to address the challenges related to food insecurity and climate change and variability. The latter would enable increased agricultural productivity to support equitable increases in farm incomes, improve food security and build resilience of agricultural and food security systems to adapt to climate change and variability.

Introduction

The majority of local communities in developing countries depend on ecosystem services for their well-being (Millennium Ecosystem Assessment, 2005). Such ecosystem services go beyond the provision of material welfare and livelihoods to include security, resilience, social relations, health, production of food, fuelwood timber and non-timber forest products and regeneration of soil fertility (Kikoti, 2009; Costanza *et al.*, 1997). Local

*rjmkatondo@gmail.com

communities often rely directly on such services, products or land from nearby natural ecosystems to meet their livelihoods (Kikoti, 2009) and to provide insurance and coping mechanisms to increase flexibility and spread or reduce risks of major natural hazards and disasters, such as floods, drought (Mansourian *et al.*, 2006) and surprises (Millennium Ecosystem Assessment, 2005). Hence, maintaining natural ecosystems can contribute to physical protection against climate change impacts (Mansourian *et al.*, 2006).

Climate in East and Southern Africa is warming at a faster rate than global models have predicted it. Temperature rises of 0.25–0.5°C per decade have been recorded in Zambia (Chidumayo *et al.*, 2011). In Tanzania, predictions show that the mean daily temperature will rise by 3–5°C throughout the country while the mean annual temperature will rise by 2–4°C (URT, 2007). The rise in temperature is expected to affect the lifestyles of most local communities who depend on ecosystems for their livelihoods.

In East Africa, food security is not guaranteed through the markets but through seasonal rain-fed subsistence and dry farming (Mwebaza, 2009). Food security in the region is, therefore, intricately tied to climate variability and the predictability of planting and harvesting seasons. Minor changes in climate will thus have a disastrous impact on lives and livelihoods of poor people living in marginal environments with low agricultural productivity (UNEP, 2007). They will also have serious impacts on communities like forest dwellers and hunter-gatherers (Mwebaza, 2009). As a result of this dependency, any impact that climate change has on natural systems will directly threaten the livelihoods, food intake and health of the populations (New Economic Foundation, 2006).

For centuries wildlife resources as one form of ecosystem services outcomes have been utilized not only for subsistence, but also for commercial purposes. However, as the human population has expanded, wildlife resources have become subjected to severe pressure, which threatens their very existence and sustainability (Wilfred, 2010). In Tanzania community-based natural resource management (CBNRM) in the wildlife management areas (WMAs) was introduced outside the core protected areas in order to lay the basis for sustainable management and utilization of wildlife resources at the grassroots level (Wilfred, 2010). Protecting natural habitats indirectly has helped to boost the national economy revenues from tourism, beekeeping products, fishing and improving infrastructures in remote areas (Mansourian *et al.*, 2006).

This study explored the links between ecosystem services and resilience of local communities to the impacts of climate change. The aim of this chapter is to delineate forms of ecosystem services gained by local communities from protecting the ecosystem. It also outlines the benefits accrued from conservation of the WMA ecosystem that would help the communities to adapt to climate change impacts. This study was expected to raise awareness of local communities, policy makers, the donor community, non-governmental organizations (NGOs) supporting WMAs, and the Ministry of Natural Resources and Tourism of the potential of a new direction for WMAs and changes that are needed as far as climate change adaptation is concerned.

Materials and Methods

The study area

Ngarambe-Tapika WMA (767 km^2) is located between latitude 39° south and 39°30' south and between longitude 12°30' east and 13° east. It is located alongside the north-eastern border of the Selous Game Reserve (Fig. 10.1). The area is also the home of local people whose lifestyles and livelihoods are intricately tied to the biological diversity and the functioning of this natural ecosystem (Formo, 2010). The study area involved the two villages of Ngarambe and Tapika, which are located in the Rufiji District in Tanzania. By 2012 the villages had a population of 2453 persons, of which 1253 were males and 1200 females with an average household ratio of 4.3 (URT, 2013).

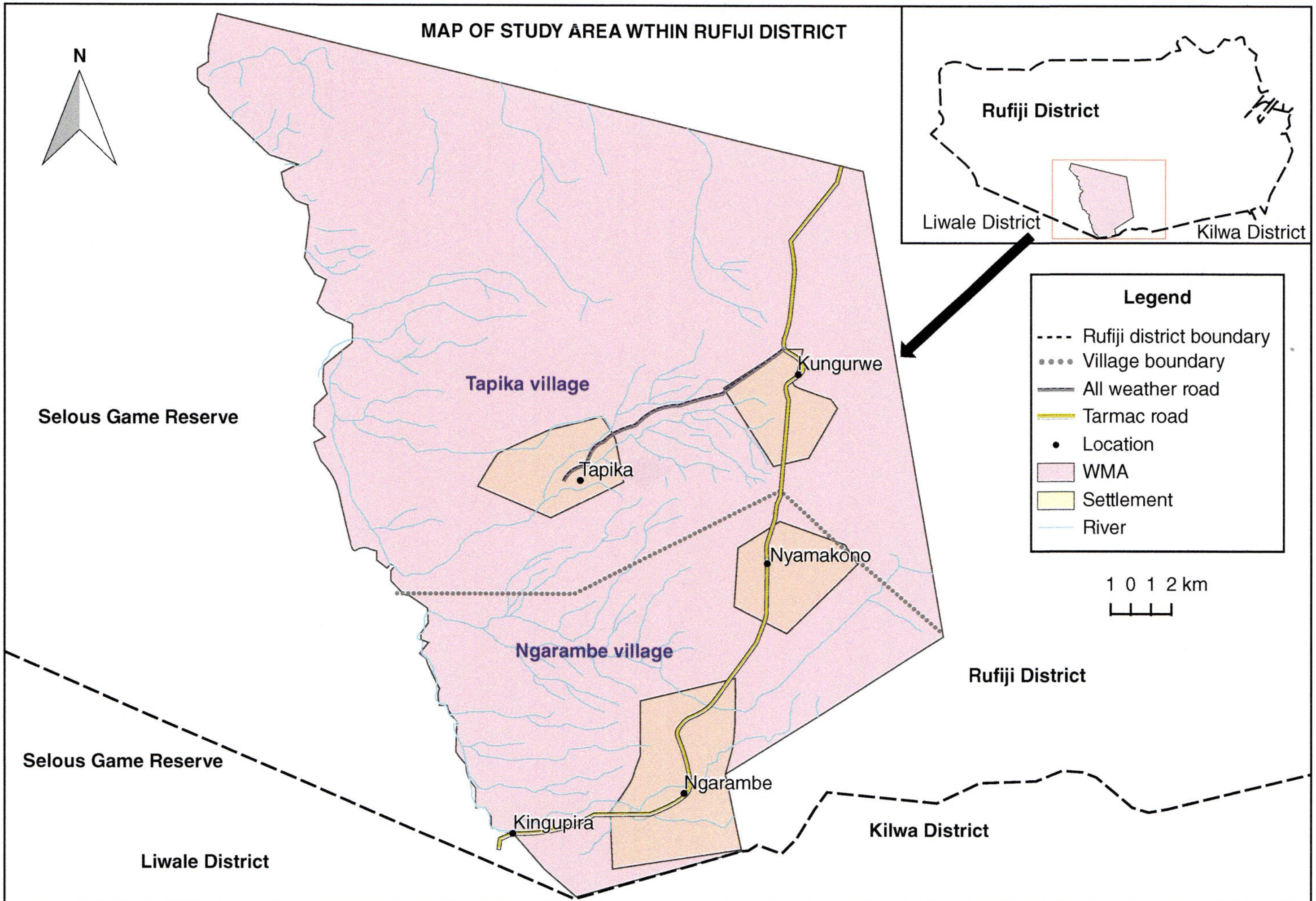

Fig. 10.1. Map of Ngarambe-Tapika Wildlife Management Area (WMA) in Rufiji District. (From Geographic Information System (GIS) Laboratory, Institute of Resource Assessment, University of Dar es Salaam, 2016.)

Research design

The design for this study adopted a mixed method research approach. Purposive sampling was employed in selecting respondents for a household questionnaire, focus group discussions (FGDs) and key informant interviews. This sampling method was used to ensure equitable representation of all members in the two villages. The sample size depended on the population size of each village and the nature of stakeholders related to the study. For the household survey a sample of 10% of the 553 households from the two villages was used as proposed by Shaghude (2004) and Kothari (2004).

Data collection techniques

Both primary and secondary data were collected in October 2014 using a structured and semi-structured questionnaire. Also, FGDs of five to eight respondents were held in the two communities. Interviews of key informants were also held with a few community elders, village government council members, Village Natural Resource Committee members and village game scouts to capture background information from people with long-term experience of the area. Direct observation using four transect walks were carried out at selected areas within the villages of Ngarambe and Tapika and within the WMA to witness physical destruction of crops by wild animals, water sources, farming practices affected during the drought and firewood collection by women, and harvesting of poles and timber. Quantitative data from the household survey was subjected to statistical analysis where frequencies and measures of central tendency were particularly determined. In order to compare the relations among variables cross tabulations were used. The study used Statistical Package for Social Sciences (SPSS) and Microsoft Excel software to analyse survey data. Qualitative data was described and presented based on predetermined themes, categories and patterns into which data from the field was obtained.

Results and Discussion

Social–economic activities

The main social–economic activities in Ngarambe and Tapika villages were agriculture-related activities (crop farming 98.2% of respondents, poultry keeping 17.6% and petty businesses 7.1%). Formal employees accounted for only 8.9% of respondents, but most respondents still practised farming (Fig. 10.2). Most villagers practised rain-fed

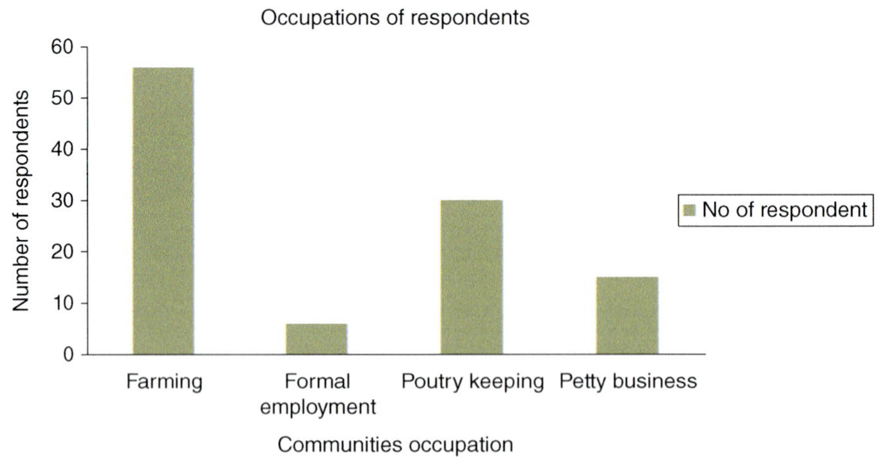

Fig. 10.2. Occupations of respondents. Some respondents have more than one occupation. (From survey data, 2014.)

subsistence farming for food crops, namely maize, millet, rice and cassava. Cash crops grown in the area were rice and sesame. Data from FGDs indicated that currently sesame had surpassed rice as a cash crop, providing more income to the community.

Ecosystem services obtained by the local communities

The results of this study showed that the communities around the WMA obtained more than one ecosystem service from the WMA. Figure 10.3 shows that the communities obtained water (100% of responses), energy (100%), agricultural land (98.2%), game meat (91.1%), thatching material (67.9%), non-timber products (32.1%), timber (26.8%) and medicine (17.9%) from the Ngarambe-Tapika ecosystem.

The communities agreed to use surface water and groundwater, including tap water, wells and rivers, for domestic use (drinking, washing utensils, building houses, etc.) from the ecosystem. Before the WMA was set up they obtained water from shallow and unprotected wells, rivers and surface runoff for domestic use, which was not safe for their health. Currently, as a benefit of ecosystem services accrued from tourist hunting (i.e. tourists who pay to go hunting) in the WMA that the communities protect, they were using tap water from dip-protected wells.

From the multiple responses the communities agreed that the ecosystem was their major source of energy. While 42.9% of the respondents used electricity from WMA generators for lighting, 28.6% used kerosene and 7.1% used solar energy. The villages were given the electricity generator to enable economic development and a better quality of life in the villages and reduce over-dependence on the ecosystem for wood energy. The power system in Ngarambe consisted of a 60 kVA generator and a grid system with underground cables. It was observed that a number of businesses along the main road, the schoolteachers' houses, the village government offices, the dispensary, the mosque, the wood factory machines, the edible oil pressing and refinery machines and the water pump were using electricity from the generator.

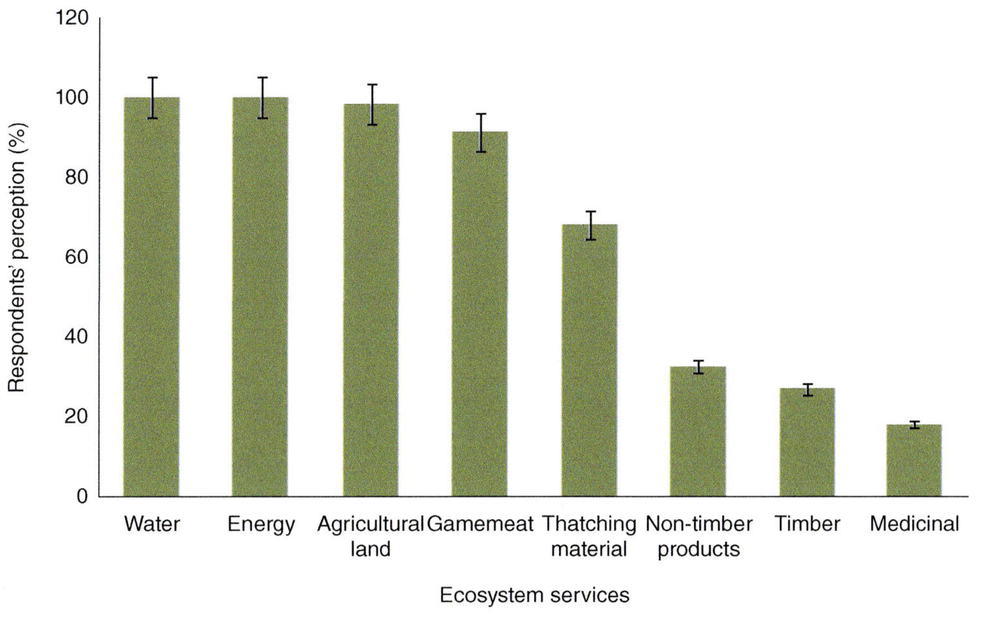

Fig. 10.3. Respondents' perceptions (%) when asked about the benefits the local community gains from ecosystem services. Error bars represent standard error of the mean. (From survey data, 2014.)

Figure 10.4 shows the proportions of each type of fuel used to provide energy which is used for cooking and income generation. All in all, fuelwood still contributed 100% while charcoal contributed only 46%. However, such fuelwood was collected mainly from farm fields, surrounding bushes, woodlands and to a certain extent from the WMA or forest reserve. Women generally collected the fuelwood, although men occasionally assisted. The woodlands were not managed and there were no restrictions in using them for firewood. Due to the increasing pressure on farmlands, availability of dry wood for fuel was at present decreasing. Although fuelwood harvest was not permitted in the WMA, the lack of clear-cut boundaries on the ground to distinguish between public land and the WMA made people take advantage of harvesting fuelwood from the WMA.

The data from this study confirmed that about 98.2% of all respondents obtained agricultural land as an ecosystem service by living near to the Ngarambe-Tapika WMA, which contributed to higher yield crop productivity. This was due to the good soil fertility and better environment for pollination agents in the reserve. Data from the FGDs acknowledged that the ecosystem attracted much rainfall which was important to farmers and prevented moisture loss from the soil and the rise in surface temperature as the WMA provided good land cover. These observations are confirmed by Locatelli (2016) who indicated that the agricultural system in Rufiji had been changing gradually through time; its evolution through time being influenced by the interplay between natural, ecological and anthropogenic factors. The natural factors included rainfall, floods and droughts. Other ecological factors included issues such as invasion by locusts, other pests and weeds, while anthropogenic factors included demographic changes, the political economy and changes in lifestyle. The FGDs also mentioned some of the above as factors that influenced the agricultural system in Ngarambe-Tapika.

Moreover, about 91.1% of the respondents agreed that they obtained bush meat from the WMA. Bush meat was obtained legally through local hunting licenses and was sold/supplied in villages' wild meat shops at a relatively cheap price of TSh800/kg at least twice a week during the hunting season (July–December) from 2003 to 2012. It was revealed that before the establishment of the WMA, bush meat was obtained through poaching. After the establishment of WMA, the Mungata Authorized Association (AA) in collaboration with the Ministry of Natural Resources and Tourism agreed on the legal provision of bush meat so as to improve nutrition and generate income for the villagers. It was clear from the FGDs that the provision of bush meat by the WMA had decreased the incidence of poaching in the Ngarambe-Tapika WMA, and the surrounding ecosystems.

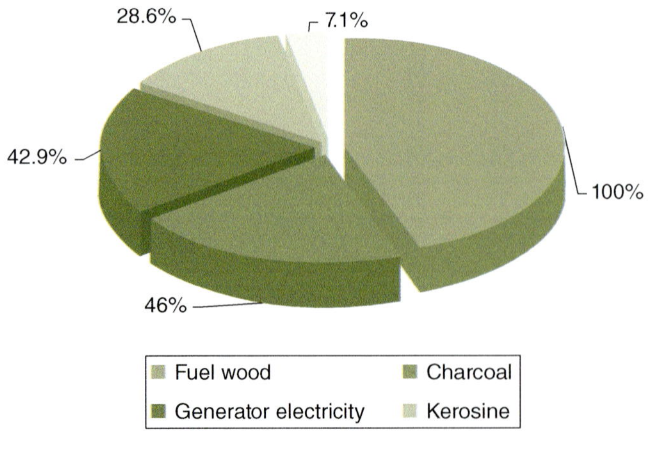

Fig. 10.4. Respondents' perceptions (%) of different types of fuel used for energy use. Some respondents use more than one type of fuel. (From survey data, 2014.)

It was also clear that Ngarambe and Tapika village communities utilized forests as an alternative source of income. The data from this study show, for example, that 26.8% of the respondents indicated that the communities obtained timber, and 67.9% of the respondents reported they get building and construction materials from the forests, mainly vertical poles and lattices, for walls, doors and windows including frames, roof frames and thatching grasses and fibres. Some 32.1% confirmed participating in mat making using raffia (*Raphia farinifera*), palm tree (*Borassus aethiopum*) leaves and sedge papyrus (*Cyperus papyrus*) harvested in the WMA. This is traditional work mainly carried out by women in Ngarambe-Tapika. Most of this artwork was used for beds, mattresses, seats, packaging materials and broom making in the area. These wares were sold in the villages and in nearby towns like Utete, Ikwiriri and other major urban centres such as Dar es Salaam. Money obtained was used to address household demands, including climate-related challenges (Rufiji District Council, 2012). Timber and non-timber forest products, such as firewood, wild fruits, mushrooms and fodder, also contributed to livelihood diversification, an adapting and anticipatory strategy that reduced the sensitivity of households and communities to climate variations.

Benefits and investment from ecosystem services

Benefits from wildlife utilization

Due to the positioning of Ngarambe-Tapika WMA adjacent to the Selous Game Reserve, the WMA is one of the greatest potential commercial areas for tourist hunting in Tanzania. At the time of this study tourist hunting was already at an advanced stage of generating revenue in the area. Revenue from tourist hunting, including local hunting (bush meat sold in Ngarambe and Tapika villages) had increased and reached over TSh142,671,763 million during the period 2003–2012 (Shemdoe and Jacob, 2012). As shown in Fig. 10.5 the highest collection of revenue was in the 2009/10 hunting season whereby TSh44,369,395 was collected from tourists, local resident's hunting and sale of bush meat in villages. This was followed by the season 2007/08 whereby TSh43,818,050 were collected. The least amount of revenue collected was for the season 2005/06 with only TSh5,295,000 was collected (Fig. 10.5). The reasons for fluctuation in income suggested by the FGDs included climate-related incidents (drought or floods) whereby wildlife moved into or away from Ngarambe-Tapika WMA and Selous Game Reserve in search of water and pasture. Other reasons

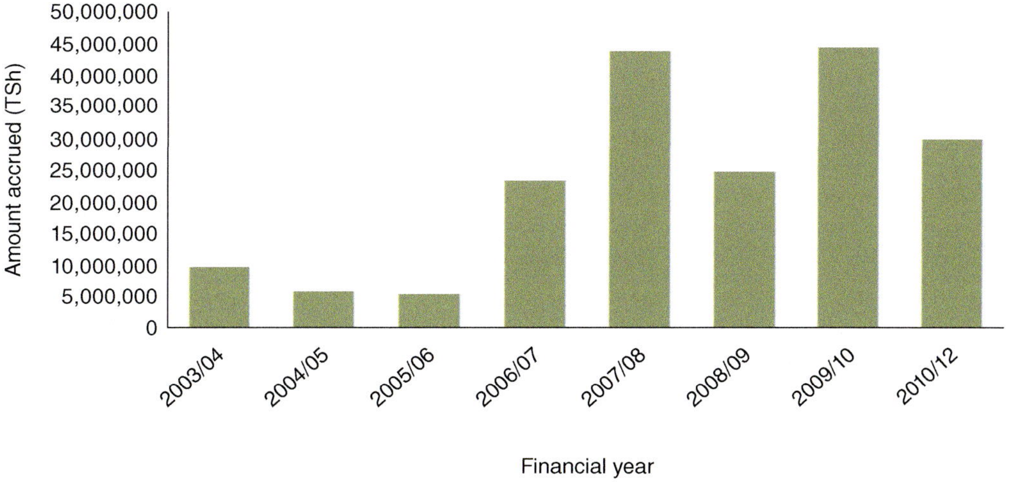

Fig. 10.5. Revenue collection in Ngarambe-Tapika WMA, 2003–2012. (From survey data, 2014.)

included fluctuations in quota allocation as well as the presence of tourist hunters.

About 50% of all revenue collected from tourist hunting was returned to the member villages and 50% went to the Community Based Organization for management and conservation purposes. All revenue passed from the Mungata AA to the villages went through a standard village council allocation and village assembly approval process to be used to invest in community social services. By the time of this study, most investments had been allocated for: (i) education support (e.g. timber from WMA for pupil desks, construction of classrooms, and money for secondary student support); (ii) tap water rehabilitation; (iii) fuel for the electricity generator; and (iv) a contribution to the construction of a village dispensary and houses for the health staff. Expenditure by the WMA itself included employment of ten village game scouts who carried out anti-poaching patrols. Data from the FGDs confirmed that investments from benefits accrued from the ecosystem services had accelerated the improvement of the livelihoods of the communities in the villages.

Other benefits from the use of ecosystem services

In general, this study has so far indicated that material welfare on livelihoods were mostly through social services (54%), formal employment (18%), protection from dangerous and problem wildlife (14%) and petty trading (14.3%). Moreover, Ngarambe-Tapika communities benefited for many years from the conservation of natural ecosystems protected against human-dominated ecosystems (agricultural land) and from the exploitation of biodiversity, especially wildlife.

In addition to generating revenue, WMA development also provided some local opportunities to sell local goods, furniture, chicken and handicrafts to visitors. During the FGDs the Ngarambe-Tapika communities also confirmed they obtained other benefits from the use of the ecosystem and assistances from other stakeholders of conservation within the WMA. The stakeholders included hunting companies, the Selous Game Reserve and NGOs such as the World Wide Fund for Nature (WWF)-Tanzania. Such assistance included: (i) security from dangerous and problematic wildlife during the severe drought; (ii) donations like food supplies during periods of flood and drought; (iii) provision of roofing materials; (iv) provision of vocational training grants for secondary and college students; (v) provision of uniforms and tents for the village game scouts; and (vi) building of mosques in both Ngarambe and Tapika villages.

The results of household interviews on the use/investment of revenue collected from the utilization of the ecosystem indicated that the largest portion was used for health services (22%) (both for traditional healers and a village dispensary), followed by other social needs (21%), house construction (20%), investing for education (15%), investing for agriculture (14%) and buying food (8%) (Fig. 10.6). The respondents agreed mostly to use money from ecosystem services for treatment of diseases such as malaria and diarrhoea in the village dispensary. This was because it was reported by the village health assistants and village elders, for example, that during the past few years the cases related to malaria were increasing due to increase in temperatures in the area.

Though most houses were made of poles and mud, some money collected was invested to build permanent houses with bricks and corrugated iron sheets. The use of iron sheets was increasing, especially among youth working as game scouts and other employees in hunting companies. Carpenters in the village wood factory agreed to improve their lives by selling house furniture to visitors in and out of the village.

Some of the community members used the revenue accrued in paying for tractors, paying for labour and other investments in agriculture and buying improved seed varieties. Farmers were asked about the use of fertilizer for more productivity but the answer was that no money was used since their land was 'still fertile'. As one FGD respondent said: 'Why use fertilizer when our land is virgin? If we use fertilizer we will turn this land to unproductive land.'

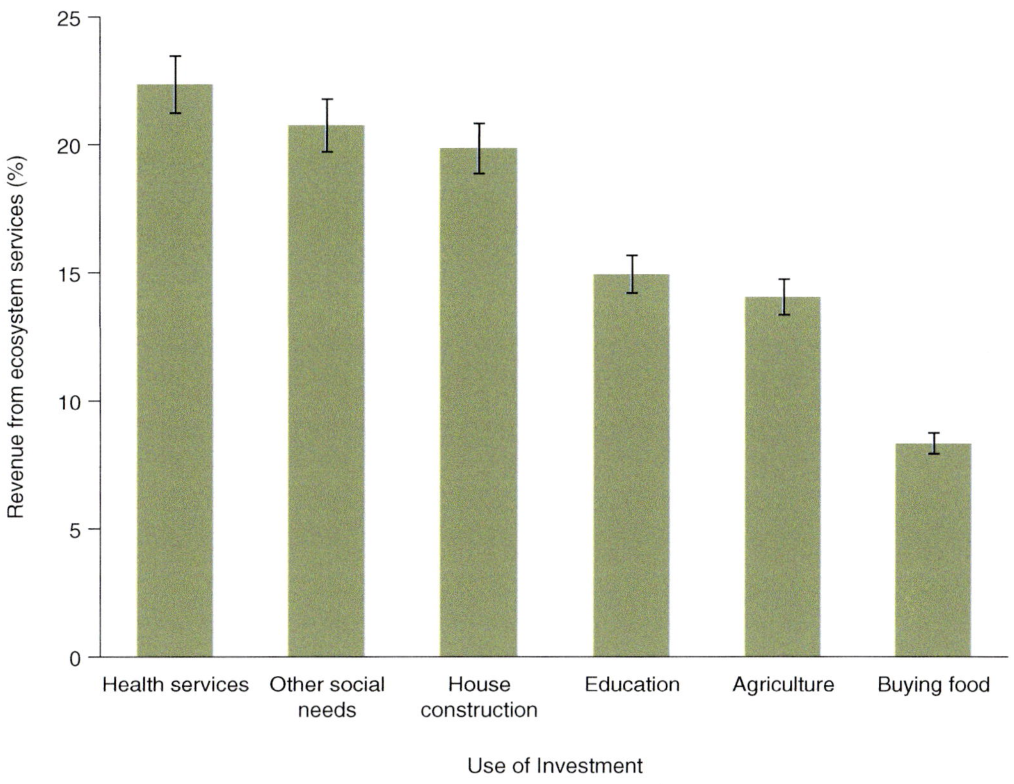

Fig. 10.6. Investments of revenue from ecosystem services. Error bars represent standard error of the mean. (From survey data, 2014.)

By the time of this study there was seasonal influx of people opening up new land for sesame production due to such fertility of the land.

Investments in education in the villages were varied. Of respondents, 15% said they allocated their money for education predominantly for buying school needs and paying school fees for secondary school students. They agreed that the assistance for their children was supported by hunting companies in their WMA. During the time of this study, more than 30 students had been sponsored outside the Rufiji District for secondary and college studies. The CBO also lent money to village government councils to support parents who failed to support/sponsor their children for studies, which would be refunded after harvest.

Regarding the literacy level of respondents, it was generally observed that about 62.5% of respondents had attained primary education, 19.6% had not attained any formal education, 16.1% had attained secondary education while only 1.8% had attained tertiary education. During the interviews it was revealed by the respondents that the main reasons for this were: (i) lack of awareness of the importance of education during the past years; (ii) low economic status of the resident farmers; and (iii) remoteness of the area.

Related observations were reported from other ecosystem services studies in Tanzania where the majority of farming communities who had attained low levels of education had hindered them from using state-of-the-art technology in their farming activities (Juma, 2011). Other observers have argued that it was easier for educated persons who attained secondary education and beyond to diversify their livelihood and adapt to the impact of drought and changing climate as they could engage in a variety of non-farming economic activities (Pauline *et al.*, 2017) such as beekeeping, fish farming and other tourism-related activities.

Investments in buying food were minimal. Only 8% of the respondents agreed that some money obtained was used to buy food in the study area during food shortage. Also, the FGDs confirmed that during severe drought, which accelerated hunger in the communities, they received some support from the government and the hunting companies who supplied foodstuffs to the local communities. Also, income obtained was devoted to social amenities such as buying transport facilities (e.g. bicycles and motorcycles) which helped people in food searching and transporting it from far off centres and towns like Utete. Other communities spent time selling labour to other farms, the Selous Game Reserve, and Kichi Forest Reserve boundary clearing. Similar observations have been documented whereby poor households were seen to spend most of the growing season working on other people's farms and engaging in off-farm activities so as to sustain their families (Pauline *et al.*, 2017).

Conclusion

Ecosystems play a key role in improving adaptive capacity and enhancing resilience of human societies to climate change. This was especially verified from Ngarambe-Tapika WMA ecosystem during this study. The diversity of functions and services provided by the Ngarambe-Tapika WMA ecosystem, such as the provision of wood and non-wood forest products, soil fertility, water regulation and conservation of biodiversity, gave the WMA a potentially significant role in adaptation approaches.

An ecosystem encompasses a variety of existing life forms, the ecological roles they perform and the genetic diversity they contain. It is the key to enhance resilience and adaptation of local community plant and animal species to climate change. The continued loss and degradation of biodiversity in Ngarambe-Tapika will weaken the ability of the ecosystem to respond to climate change and support the community.

In Ngarambe-Tapika villages the WMA revenue is accrued from selling of ecosystem services in the form of cash and food crop production, renting out agricultural land, using bush meat, water, energy, thatching material, timber and non-timber products from the forest. The communities benefited from the tourist hunting money, which was primarily spent on construction of their own houses, dispensaries, rehabilitation of the primary school, and government staff houses in the villages (for the primary school teacher and health staff). In addition, some of the money was spent on generators both in Tapika and in Ngarambe, which provided electricity for the communities and energy for the water pumps.

It is important for the local communities not only to possess the Ngarambe-Tapika ecosystem in their area, but also to have access and control over the ecosystem's use. The way in which the Ngarambe-Tapika ecosystem is used will determine the communities' sustainability over time, and consequently their vulnerability to various hazards.

It was observed that increasing local skills and local production contributes towards strengthening the resilience of the local economy, enhancing wealth and growth within the region. Diversifying skills via education and income-earning opportunities increases the options that people have to fall back on in times of climate change challenges and disasters. Actions at the community level such as conserving and managing WMAs is evidence that much can be achieved in terms of increasing the resilience and adaptive capacity of communities to cope with climate change.

For sustainability of the Ngarambe-Tapika ecosystem the communities, which depend entirely on subsistance farming, should be encouraged to practise and adopt an integrated approach of climate-smart agriculture (CSA) to address challenges related to food insecurity and climate change and variability. This would enable increased agricultural productivity to support equitable increases in farm incomes, improve food security and build resilience of agricultural and food security systems to adapt to climate change and variability.

References

Chidumayo, E., Okali, D., Kowero, G. and Larwanou, M. (eds) (2011) *Climate Change and African Forest and Wildlife Resources*. African Forest Forum, Nairobi.

Costanza, R., D'Arge, R., De Groot, R., Farber, S., Grasso, M., *et al.* (1997) The value of the world's ecosystem services and natural capital. *Nature* 387, 253–259. Available at: https://www.esd.ornl.gov/benefits_conference/nature_paper.pdf (accessed 28 April 2014).

Formo, R.K. (2010) The power and subjectivation. The political ecology of Tanzania's wildlife management areas. *International Environmental Studies and Development Studies*. Available at: https://nmbu.brage.unit.no/nmbu-xmlui/handle/11250/187699 (accessed 28 April 2014).

Juma, A. (2011) Farmers' adaptive strategies to the impact of drought: a case study of Iramba District in Singida Region. MA dissertation, Institute of Development Studies, University of Dar es Salaam, Dar es Salaam, Tanzania.

Kikoti, Z. (2009) Livelihoods and ecosystem services around protected areas. A case study from Ugalla ecosystem, Tabora, Tanzania. MSc. thesis, University of Klagenfurt, Klagenfurt am Wörthersee, Austria.

Kothari, C. (2004) *Research Methodology: Methods and Techniques*. New Age International, New Delhi.

Locatelli, B. (2016) Ecosystem services and climate change. In: Potschin, M., Haines-Young, R., Fish, R. and Turner, R.K. (eds) *Routledge Handbook of Ecosystem Services*. Routledge, London, pp. 481–490.

Mansourian, S., Belokurov, A. and Stephenson, P.J. (2006) The role of forest protected areas in adaptation to climate change. *Unasylva* 231/232(60), 63–69.

Millennium Ecosystem Assessment (2005) *Ecosystems and Human Well-being: Biodiversity Synthesis*. World Resources Institute, Washington, DC.

Mwebaza, R. (2009) The impact of climate change in eastern Africa. In: *Environmental Governance and Climate Change in Africa: Legal Perspectives*. Available: https://oldsite.issafrica.org/uploads/Mono167.pdf (accessed 28 April 2014).

New Economic Foundation (2006) Africa – Up in Smoke, 2. The second report on Africa and global warming from the Working Group on Climate Change and Development. Available at: https://www.preventionweb.net/files/1833_VL102100.pdf (accessed 28 April 2014).

Pauline, N.M., Vogel, C., Grab, S. and Liwenga, E.T. (2017) Smallholder farmers in the Great Ruaha River sub-Basin of Tanzania: coping or adapting to rainfall variability? *Climate and Development.* 9(3), 217–230. doi:10.1080/17565529.2016.1184607.

Rufiji District Council (2012) *The Authorized Association of Ngarambe-Tapika WMA, General Resources Management Plan, 2012–2022*. Rufiji District Council, Rufiji, Tanzania.

Shaghude, Y.W. (2004) Coastal impacts of water abstraction and impoundment in Africa: the case of Rufiji River. Final Report submitted to START, March 2004. Available at: http://www.oceandocs.org/bitstream/1834/187/1/coastal.pdf (accessed 25 January 2015).

Shemdoe, S.R. and Jacob, K. (2012) Understanding community based adaptation strategies to climate variability in fishing communities of Rufiji River basin in Tanzania. *African Journal of History and Culture* 4(2), 17–26.

United Nations Environment Programme (UNEP) (2007) *Biodiversity and Climate Change*. International Day for Biological Diversity. Convention on Biological Diversity. Available at: http://www.cbd.int/doc/bioday/2007/ibd-2007- booklet-01-en.pdf (accessed 28 April 2014).

United Republic of Tanzania (URT) (2007) *Wildlife Policy of Tanzania*. Ministry of Natural Resources and Tourism, Wildlife Division, Government of Tanzania, Dar es Salaam, Tanzania.

United Republic of Tanzania (URT) (2013) *2012 Population and Housing Census*. National Bureau of Statistics, Ministry of Finance, Government of Tanzania, Dar es Salaam, Tanzania.

Wilfred, P. (2010) Towards sustainable Wildlife Management Areas in Tanzania. Conservation letter. *Journal of Tropical Conservation Science* 3(1), 103–116.

11 Effectiveness of Existing Climate Smart Agricultural Practices in Tehuledere District, North-eastern Ethiopia

Adera S. Wassie[1] and Noah M. Pauline[2]*

[1]College of Agriculture, Wolaita Sodo University, Wolaita Sodo, Ethiopia; [2]Centre for Climate Change Studies, University of Dar es Salaam, Dar es Salaam, Tanzania

Abstract

This study sought to understand the effectiveness of existing climate smart agricultural (CSA) practices used by smallholder farmers in Tehuledere District, north-eastern Ethiopia, using empirical data collected over 3 months in 2016. Both qualitative and quantitative data were collected from primary and secondary sources, including historical climatic records and perceptions, indicatiors of climate change, impacts and responses. The study employed focus group discussions and a household questionnaire survey during data collection. Findings reveal that the majority of participants are aware of climate change and variability. The indicators of climate change include drought, off-season rainfall, too little and/or too much rainfall, and high temperature. The impact of climate variability observed by participants include: (i) decreasing crop yields and livestock production; (ii) increasing pests and disease; and (iii) decreasing water quantity and quality. The adaptation measures embraced to mitigate negative climate change effects include: (i) changing crop varieties; (ii) integrating livestock and crop production; and (iii) soil and water conservation practices. As far as CSA practices were found to be viable and effective response measures, support from innovative polices and strategies should be emphasized so as to address the barriers and bring about widespread adoption.

Introduction

Global warming has been established as one of the major concerns for the international community which is jeopardizing sustainability (Niang *et al.*, 2014). It is evident that climate change impacts on economic, social and human living conditions as well as the environmental ecosystem (WWF, 2006; Branca *et al.*, 2011; Stabinsky, 2014). For example, in eastern Africa, particularly countries in the Nile Basin, the temperature has increased from 0.2°C to 0.3°C per decade over the last 60 years (Deressa *et al.*, 2009). A case study on climate change vulnerability in Ethiopia, Kenya and Tanzania also shows an expected increase in temperatures of about 1.1–3.1°C, 1.0–2.8°C and 2–4°C, respectively, by 2060 (Brohan *et al.*, 2006).

The observed rising temperatures, increasing frequency of drought and unpredictable rainfall pose the greatest vulnerability for agriculture in East Africa (Adhikari *et al.*, 2015). On average 60% of the population depend on predominantly rain-fed agriculture

*noah.makula@gmail.com

run by small-scale farmers which contributes about 34% of the gross domestic product (GDP) and almost 60% of export income (Niang *et al.*, 2014). In Ethiopia, the proportion of people who depend on agriculture is 85% and 45% of the GDP is generated from this sector (FDRE, 2007).

Despite the large contribution of agriculture to development and food security in Africa, the sector remains highly vulnerable to climate change due to the low adaptive capacity of the system. A number of reasons have been established to explain this low adaptive capacity of African agriculture to climate change. As Niang *et al.* (2014) pointed out, some of the constraints in Africa include: (i) economic marginalization of the continent; (ii) financial constraints; (iii) national priority that pays more attention to poverty reduction than environmental sustainability; (iv) low research capacity and technical skills; and (v) insufficient or lack of institutions that deal with adaptation.

This implies that development policies should be set in such a way that policies and programmes can tackle the negative impacts of climate change with the aim of reducing poverty and food insecurity through contextual adaptation development strategies (Yanda and Mubaya, 2011). By contrast, agriculture emits about 14% of greenhouse gas which contributes to climate change (IPCC, 2007). Carbon sequestration in the soil and reduction of emission to the atmosphere is a result of climate science advancement (Niggli *et al.*, 2009).

Therefore, responding to climate change is important for attaining food security. At the same time, agricultural mitigation can minimize atmospheric greenhouse gas concentrations and slow climate change itself (Verhagen *et al.*, 2014). This can be attained by applying the concept of climate smart agriculture (CSA) that comprises a set of location-specific agricultural practices to reduce greenhouse gas emission and future global warming by integrating agricultural production with sustainable development (FAO, 2010; Branca *et al.*, 2011). CSA is considered as mediation of climate change through mitigation (IFPRI, 2014).

The Food and Agriculture Organization of the United Nations (FAO) developed the concept of 'climate-smart' agriculture in 2010 to improve agricultural adaptation systems to climate change (FAO, 2010). CSA is a holistic approach to attaining food security in changing climatic conditions while mitigating climate change with a contribution for other development agenda (Verhagen *et al.*, 2014). According to the original definition, 'climate-smart' agriculture aimed at 'sustainably increasing productivity, resilience (adaptation), reducing/removing greenhouse gases (mitigation), and enhancing achievement of national food security and development goals' (FAO, 2010, p. 2).

CSA strives to attain three major goals: (i) increasing food production; (ii) building resilience of the agriculture system to climate change impacts; and (iii) reducing agricultural greenhouse gas emission through promoting efficient use of land, water, soil and other environmental resources (FAO, 2010). CSA includes a number of technological policies and institutional interventions. It deals with seeds, water, energy and nutrients and some risk-averting interventions that increase the resilience and stability of agriculture and thus helps farmers adapt to and reduce the risk of climate change (Branca *et al.*, 2011; IFPRI, 2014). It helps to achieve long-term agricultural development concerns by integrating short-term priorities with other development goals, transforming and reorienting agricultural systems to adapt to climate change and variability (Verhagen *et al.*, 2014). Though CSA was developed in 2010 by the FAO, its effectiveness in Ethiopia is not fully assessed.

By interacting with smallholder farmers who use a variety of CSA practices (e.g. soil and water conservation, organic fertilization of the soil, crop diversification, weather forecasts and early warning of extreme events) we seek to understand the effectiveness of such practices in reducing the negative impacts of climate change in the region. Specifically, the study sought to: (i) assess how farmers perceive and respond to climate change; (ii) establish existing CSA practices; and (iii) analyse the evidence for effectiveness of such CSA practices.

Methodology

The study area

The field survey was carried out in Tehuledere District, South Wollo Zone Amhara National Regional State, Ethiopia (Fig. 11.1). The district was selected based on agroclimatic conditions since it is highly vulnerable to climate change, which may adversely affect the sustainability of the agricultural production system and the food security of the area.

Data collection

A two-stage sampling technique was applied to select the sample households. In the first stage, a purposive sampling method was employed to identify three representative kebeles from the district considering topography and agroecology. Tehuledere District has 24 kebeles with a total of 28,780 households with an average of 1199 households in each kebele (Central Statistical Authority, 2007). The second stage was a random sampling technique to select a total of 150 households from the sample kebeles. Therefore, 150 respondents from the three pre-selected kebeles (50 each) were selected due to time and financial limitations.

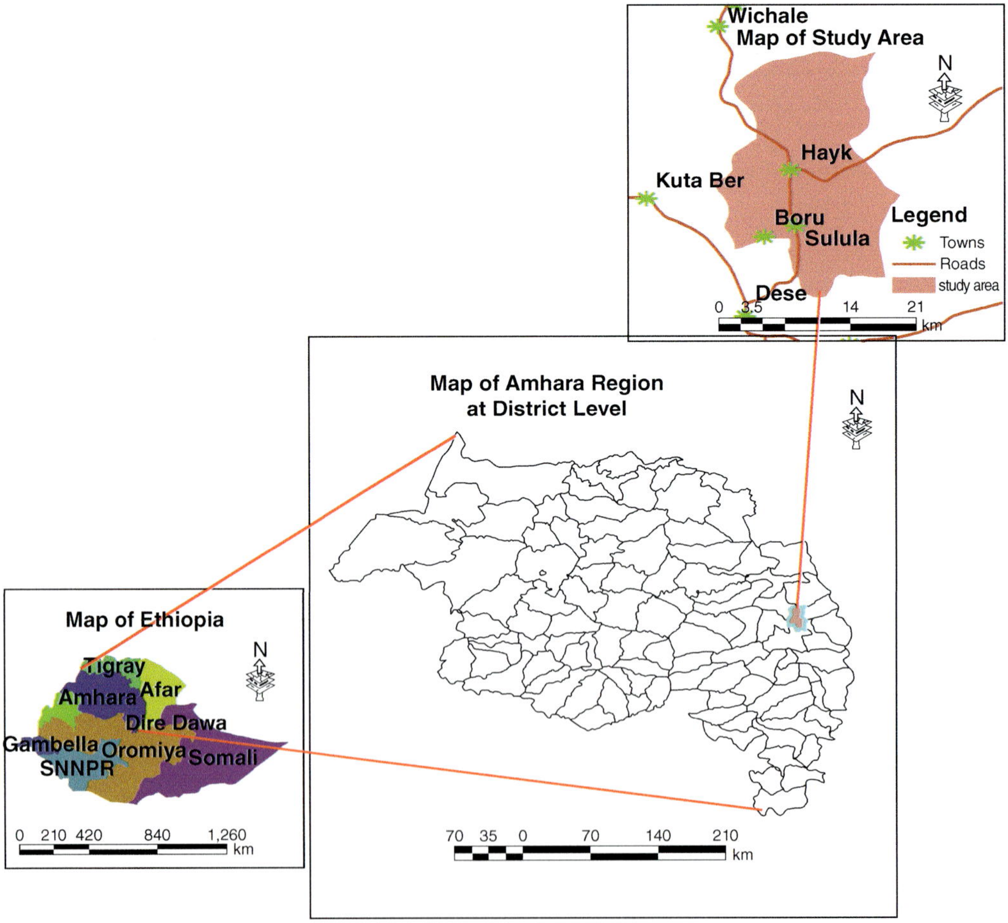

Fig. 11.1. Map of Ethiopia showing the location of the study area. (From Wassie and Pauline, 2018.)

Both secondary and primary data were collected in this study. Secondary data was obtained from different resources in the literature. Different documents related to the study were explored for in-depth understanding of how far the problem had been studied. Secondary data also gave an insight into what had not been covered about the research problem and what gaps there were to fill. It was also helpful to identify areas of focus. Primary data was collected from the field survey in the study area. The following methods were employed to collect primary data from the field: (i) household questionnaire survey; (ii) key informant interviews; (iii) focus group discussions; and (iv) field observations.

In the study area, about 150 questionnaires containing open- and close-ended questions were administered to the respondents. The semi-structured interview questions were designed to allow the interviewer to probe beyond the answers and respondents provided their views and understanding about the problem being investigated on the ground. The questionnaire was designed to gather qualitative and quantitative data pertaining to: (i) climate change variability; (ii) indicators of CSA effectiveness; (iii) existing CSA practices; and (iv) other aspects of households (including perception of the local community to CSA, access to climate-smart technology and information like weather forecasts and the extension service).

Two key informants from the zone and district agricultural office, who have an agricultural background, were interviewed for in-depth clarification of CSA preferences and effectiveness in the study area. In addition, one agricultural development agent in each sample kebele was interviewed to understand the extension service about climate-smart technologies and weather information in response to climate change trends, impacts and the existing stressors.

Open-ended questions were provided to the group, allowing the participants to share their views, experiences and understanding about the problems in the study area. Focus group discussions were undertaken by selecting six to eight participants from each kebele in the study area and interacting with them for 1 h. The participants were provided with information regarding the objectives and the purpose of the study before the discussions. Adequate representation was ensured by incorporating women, men, young and elderly members of the community as participants.

Counterchecking the information gathered by household survey and focus group discussion through field observation was important, particularly in summarizing reports about people's perceptions that were being studied. During field observation, photographs were taken to document actual daily activity of the community with regard to CSA practices. For example, agricultural diversification such as an agroforestry system or a mixed farming system (involving both crop and livestock production), rainwater harvesting practices as well as soil and water conservation practices (e.g. terracing) were observed and documented.

Data processing and analysis

The qualitative information obtained from focus group discussions and key informant interviews (including effectiveness of CSA) was analysed through the method of content analysis. Quantitative information obtained from household interviews (including existing CSA) was cleaned to ensure quality, coded and recorded in Microsoft Excel spreadsheets and was subjected to statistical analysis using Statistical Package for Social Sciences (SPSS) version 20 software.

Results

Perceived and observed climate change

Farmers' perception

The results indicated that it is widely perceived that climate is changing in terms of rainfall and temperature, respectively. Findings indicate that the majority (84.7%) of respondents perceived an increase in temperature above average. Likewise, a larger proportion (97.3%) of respondents perceived a decrease in the amount of rainfall below average (Fig. 11.2). Thus, the perception clearly shows that the amount of rainfall is decreasing and temperature is increasing in the study area (see also rainfall and temperature trends shown in Figs 11.3–11.6).

During focus group discussions, farmers recounted that there is noticeable climate variability in the form of changing rainfall patterns, late onset and early offset of the cropping/rainy seasons and occurrences of off-season rainfall. They also realized the annual temperature variability, as it becomes extremely cold in cold seasons and extremely hot during the hot seasons.

Observed changes

The study area is characterized by well-defined bimodal rainfall pattern with short (March–May) and long (June–September) rainy seasons, which have different rainfall intensity (Dereje, 2011). The variation of rainfall over time showed a weak correlation coefficient (R = 0.054), which is close to zero (Fig. 11.3). However, the weak correlation could be as a result of the influence of extreme variation over average data. For example, extremely low rainfall was recorded in 1984 and 1989 (the former was a devastating drought in Ethiopian history) and extremely high rainfall was recorded in 1964 and 2010 (see Fig. 11.3).

Observed records indicated the incidence of dry spells occurred with an interval of 2–5 years. As indicated in Fig. 11.4, the long rainy season (June–September) had smaller rainfall amounts of 483.1 mm in 1966, 472.1 mm in 1972, 507.6 mm in 1976, 245.9 mm in 1983 and 285.1 mm in 1984 compared with the average amount of 706.5 mm. Other dry spells were recorded in 1987, 1989 and 1993.

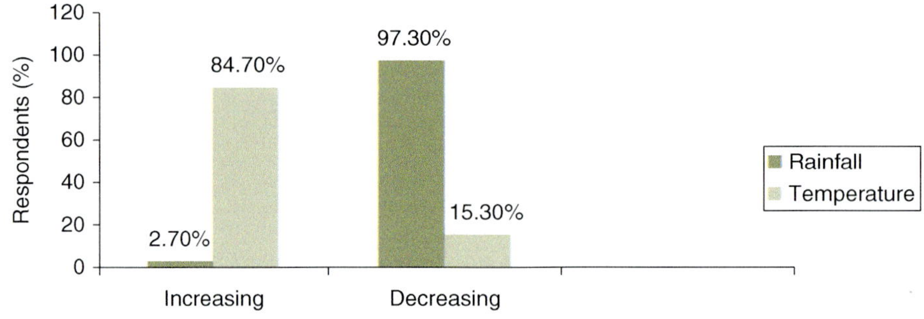

Fig. 11.2. Farmers' perceptions of rainfall and temperature trends.

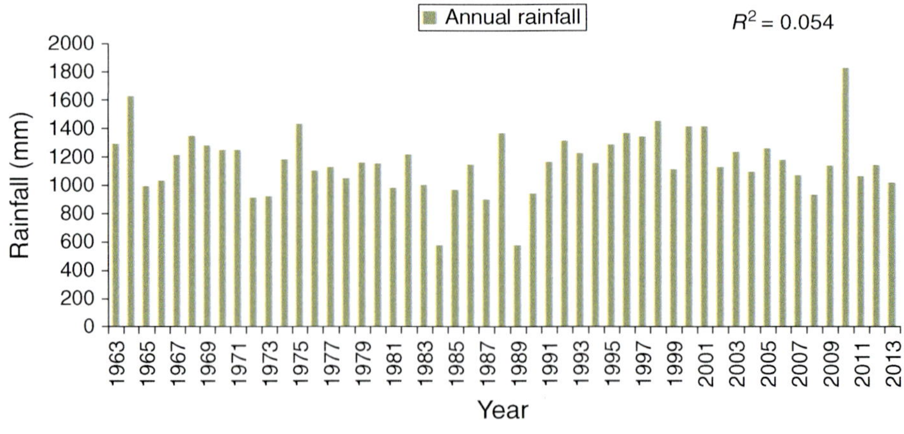

Fig. 11.3. Rainfall trend over the period 1963–2013.

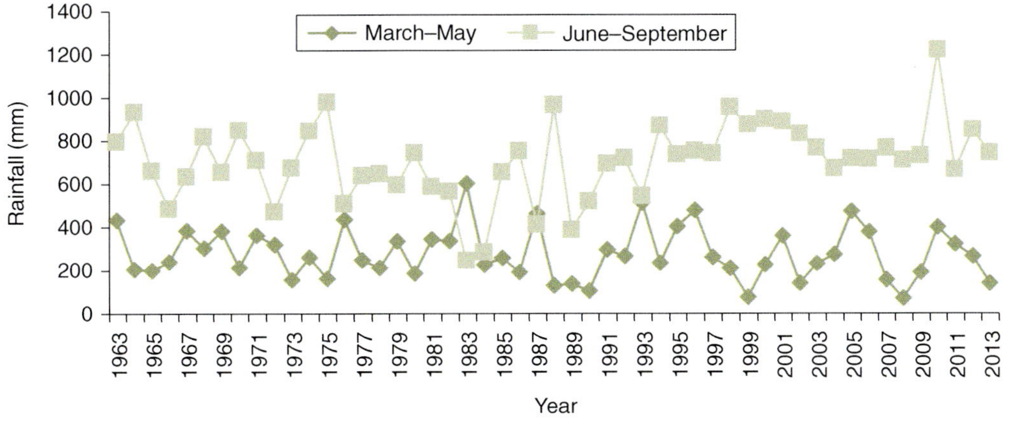

Fig. 11.4. Mean annual rainfall for the two rainy seasons.

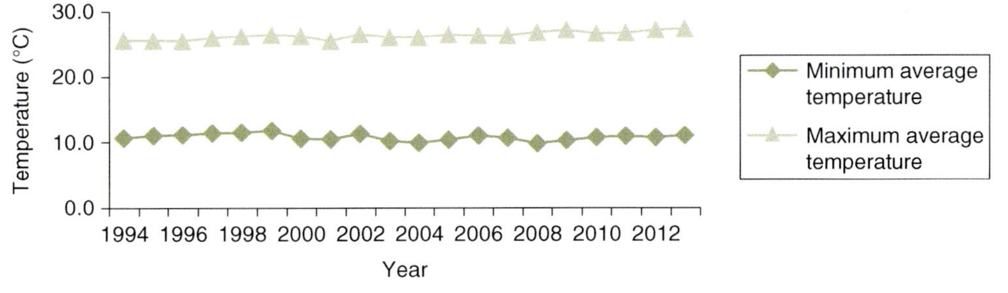

Fig. 11.5. Maximum and minimum average temperature (1994–2013).

The temperature variability is one of the key components of climate change. Findings from analysis of the household survey, focus group discussions and meteorological data revealed that there was a slight increasing trend in annual maximum average temperature and an even trend in annual minimum average temperature (Fig. 11.5). Despite the fact that most of the respondents perceived that there was an increasing trend of temperature, the analysis of meteorological data indicated that the overall trend of the temperature for the last 20 years was decreasing at a rate of 0.2 per decade (Fig. 11.6).

CSA practices

The understanding and perceptions about existing CSA practices was profiled. Accordingly, the respondents were asked to indicate their perceptions about CSA, after the concept and practices of CSA had been explained to them. They were asked if they believed that climate change can be adapted to through CSA practices and if they have adopted any CSA practices. Most of the respondents (73.7%) perceived that it is possible to respond to climate change impacts using CSA practices. Similarly, 74% of the respondents recounted that they practise CSA. A relatively small proportion (19.3%) perceived that it is not possible to effectively respond to climate change and variability using CSA practices (Fig. 11.7). In the same line, only 11% said that they were not practising CSA.

Regarding CSA practices that were adopted in Tehuledere District, findings indicated that the majority of respondents (95.3%) practised soil and water conservation, and most respondents (83.3%) used organic fertilizer. Fewer respondents in the study area were using improved crop varieties

(44.7%) or diversified the crops they grew (43.3%) (Table 11.1).

It was observed that the purpose of using improved crop varieties varied from farmer to farmer. Accordingly, the majority (82.7% and 70%) were adopting improved crop varieties for the purpose of high yield and early maturity, respectively (Fig. 11.8).

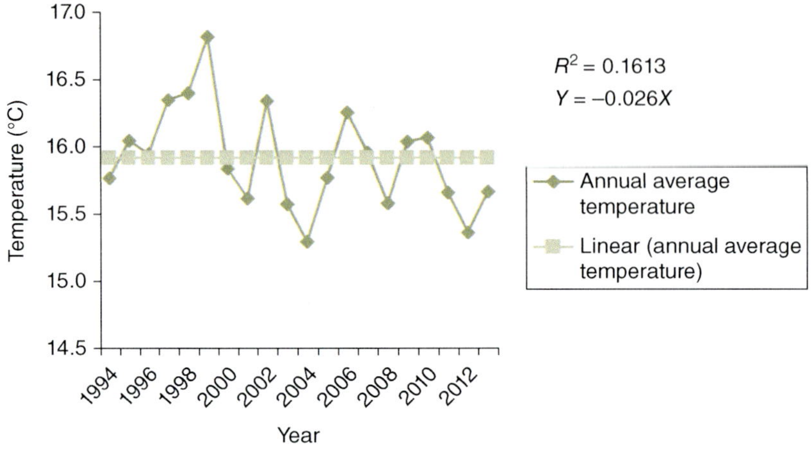

Fig. 11.6. Mean annual temperature (1994–2013).

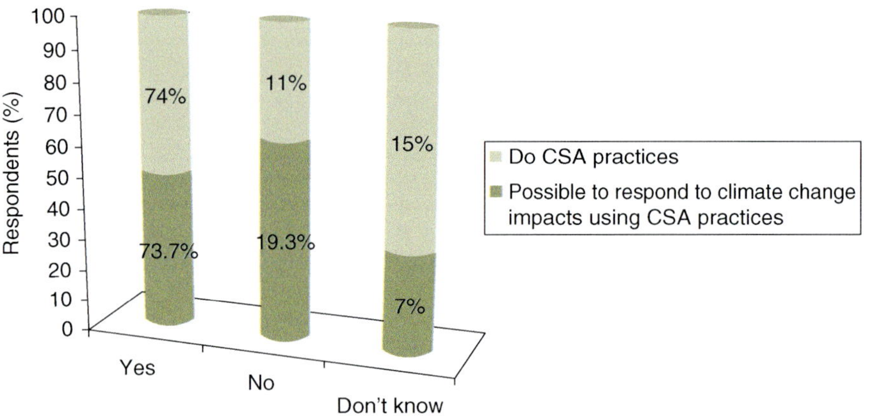

Fig. 11.7. Farmers' perceptions of climate smart agriculture (CSA) practices.

Table 11.1. Climate smart agriculture (CSA) practices adopted in Tehuledere District.

	Respondents who had adopted the practice[a]	
CSA practices	Frequency	Percentage[b]
Soil and water conservation	143	95.3
Organic fertilizer utilization	125	83.3
Using improved crop varieties	67	44.7
Crop diversification	65	43.3
Weather forecast	39	26

[a]Out of a total of 150 respondents who were asked.

[b]The total percentage exceeds 100% because multiple responses were allowed.

Drought tolerance (46.7%) and resistance to diseases and pests (47.3%) were other reasons for adopting improved crop varieties.

Soil and water conservation practices

There are a number of soil and water management practices that can be considered as CSA approaches. Findings revealed that 92.7% of respondents used terracing as a soil and water conservation technique. Fewer respondents used agroforestry (41.3%) and minimum tillage (33.7%). About 21.3% of the respondents replied that they were using cover crops and the remaining 18.7% used other soil and water conservation techniques like water harvesting and mulching (Fig. 11.9). During the focus group discussions, farmers reported that the green development programme facilitated by the government was a means by which they could adapt to climate change impacts and prevent future climate-related problems. They reported that this is part of an afforestation programme and it is implemented in all rural areas.

Types of organic fertilizers

Common organic fertilizers were listed and respondents were asked if they had any experience of using any of them. Results

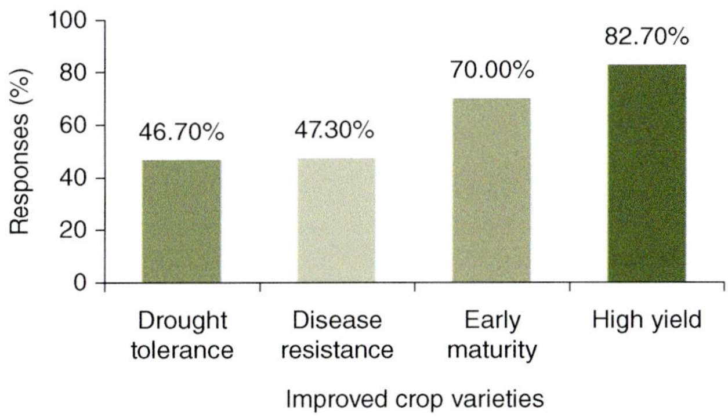

Fig. 11.8. Purpose of using improved crop varieties in the study area as a CSA practice. Respondents may have grown more than one type of improved crop variety.

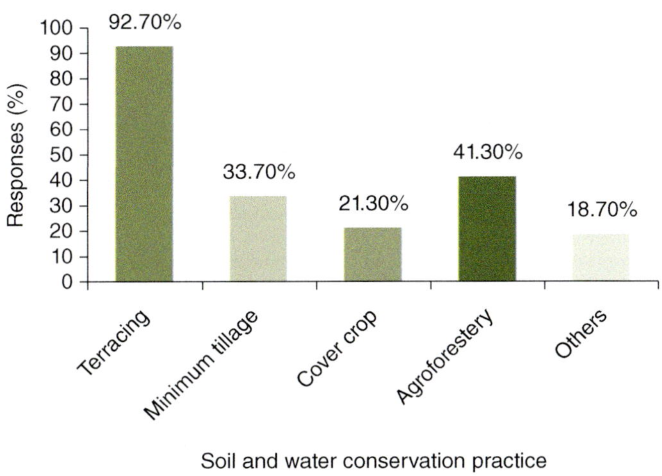

Fig. 11.9. Common soil and water conservation practices adopted in the study area. Respondents may have adopted more than one type of conservation practice.

showed that compost was the most commonly used organic fertilizer as reported by 81.3% of respondents. This was followed by farmyard manure (63.3%) and green manure (20.7%) (Fig. 11.10). Some respondents used other kinds of organic fertilizers, including crop residues, and some had not applied organic fertilizer to farmland. During focus group discussions, participants showed that they were aware of the importance of using organic fertilizers, which was believed to be a sustainable agriculture practice. Almost all participants agreed that compost was a common organic fertilizer, since it can be easily prepared from any waste material and plant residue even outside the farm, compared with farmyard manure which is also used as a source of energy for rural households. The participants also reported that green manure is used rarely because it requires a lot of land, and farmers only have small farms for cultivation.

Information from the key informants, and zonal and district agriculture office documentation indicated that farmers in Tehuledere District were using organic fertilizers, especially compost, to replenish soil fertility and this was supplemented with synthetic fertilizers in order to improve agricultural productivity. Results from the district agriculture office showed that about 299,678 m^3 of compost had been prepared every year since 2012.

Discussion

Traditionally, farmers in the study area have been practising different agricultural techniques to tackle water and soil fertility problems that are compounded by climate change and variability. Agricultural practices like terracing, stone bunds, contour ploughing, agroforestry, mulching, intercropping and rotation, crop diversification and adopting a mixed farming system have been practised in Ethiopia for quite a long period of time as part of the livelihood of the smallholder farmers (Jirata *et al.*, 2016). By definition, all these activities are CSA practices and thus can build adaptation/resilience, reduce greenhouse gas emissions and bring about sustainable agriculture productivity. For example, soil conservation practices improve the soil's carbon storing capacity that mitigates greenhouse gas emissions and enhances soil fertility increasing its productivity (Branca *et al.*, 2011; Jirata *et al.*, 2016). Most of the farmers were not aware that these practices played an important role in adapting to climate change and variability. They practised these activities because they perceived these practices were feasible and could help in increasing yield and food security.

The current agricultural development activities that are supported by policies and strategies focus on soil and water management, which play a paramount role in tackling

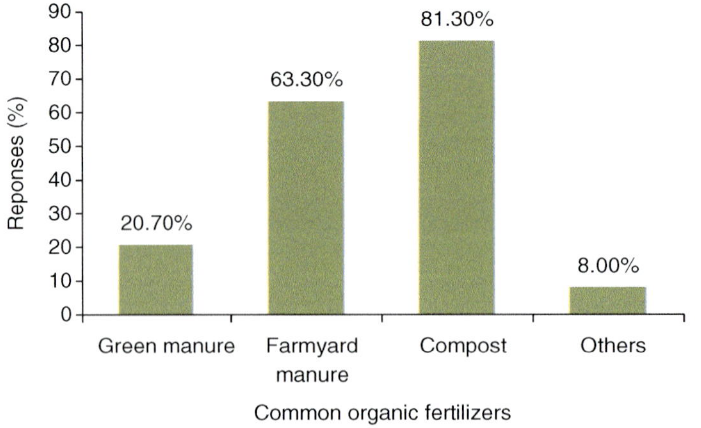

Fig. 11.10. Respondents' experience of using common organic fertilizers in the study area. Respondents may have used more than one type of organic fertilizer.

climate change and variability. Soil and water management programmes in Ethiopia comprise integrated soil fertility and watershed management projects run by the Ethiopian government, non-governmental organizations (NGOs) and other stakeholders (Jirata *et al.*, 2016). The setting up of a green development strategy by the government to ensure sustainable agricultural development can contribute to reducing greenhouse gas emissions and carbon sequestration which is the core aim of CSA. The strategy helps the farming community to plant trees on eroded and vulnerable areas to regenerate the soil and improve its productivity while fighting climate change and variability (FDRE, 2007).

The findings indicate that soil and water conservation is the most commonly used CSA practice. Similarly, Jirata *et al.* (2016) and Gebre Hadgu *et al.* (2015) found that soil and water conservation, with other practices, has been used as a common climate change adaptation option in Ethiopia. Additionally, Gadisa Chimdesa (2016) reported that after the famine of 1984/85, Ethiopia focused on environmental management in the form of watershed protection such as soil and water conservation to reduce soil erosion and land degradation. This implies that for more than three decades Ethiopia has been promoting the use of various CSA-related practices in addressing climatic stresses through different programmes, leading to wide of adoption of some of them, such as soil and water conservation practices.

Various techniques of soil and water conservation are used by farmers, terracing being the most widely used, followed by agroforestry and minimum tillage. This result is in agreement with Gadisa Chimdesa (2016) who reported that terracing with other erosion control structures like stone bunds reduced the total soil loss by 12.6%. Jirata *et al.* (2016) also report that agroforestry is a deep-rooted agricultural practice in many parts of Ethiopia. The technique involves the integration of trees and shrubs into farmland to enhance soil fertility and provision of shade for some other crops like coffee (Jirata *et al.*, 2016).

Application of organic fertilizers was reported to improve soil fertility and boost crop production, thus most farmers used this CSA practice. This result is in agreement with Branca *et al.* (2011) who reported that application of compost in Tigray Province, Ethiopia showed an increase in yield compared with crops that were chemically fertilized. For example, yields of barley increased by 9%, wheat by 20%, maize by 7%, tef by 10.7%, and finger millet by 3%. The interest of the farmers to integrate organic fertilizers with commercial ones increased following the concept of green development and climate change adaptation which became a concern of agriculture extension service providers (FDRE, 2011a).

Despite the wide adoption of the CSA practice of using organic fertilizers, it was found that farmers in the study area often only used organic fertilizers to supplement use of synthetic fertilizers. The limited use of organic fertilizers is due to the fact that it is labour intensive and slow to increase the productivity of the soil compared with synthetic fertilizers. The latter is used for immediate food security because of its immediate action (Niggli, *et al.*, 2009). Lack of technical support from extension service workers to prepare standard compost was mentioned as one of constraints for limited use of organic fertilizers in the study area. Branca *et al.* (2011) and Lemmi Legesse Kotu (2013) reported that lack of technical knowledge and skilled manpower to demonstrate climate beneficial practices are among the constraints limiting quick adoption of agricultural technologies.

Despite the importance of weather forecast information as a key CSA practice (WMO, 2002), few farmers were using it. The reason might be due to unreliability and limited source of weather forecast information. A similar result was found by Feleke (2015) who reported that farmers in the Central Rift Valley of Ethiopia neither rely on nor trust the weather forecast. The author also suggests that the forecast given by the National Meteorology Agency is unreliable and it targets the entire region and hence it lacks location specification. Okonya *et al.* (2013) also report that unreliable weather forecasts are among the factors that hinder adaptation to climate change.

The survey further revealed that planting improved crop varieties and engaging in crop diversification were important CSA practices in the study area. However, attaining high yield and escaping from risk of crop failure in case of a dry spell during the cropping season by planting early-maturing crop varieties were the major purposes of adopting improved crop varieties. This finding is in agreement with Okonya *et al.* (2013) who indicated that crop varieties which are able to mature early and are high yielders are farmer's preferred varieties in Uganda and Kenya. Resistance to disease and pests and tolerance of drought were other reasons for the farmers adopting improved crop varieties. Jirata *et al.* (2016) suggest that crop varieties that are resistant to pests and diseases will reduce the need for pesticides. Research conducted in the Nile Basin of Ethiopia also indicated that farmers use soil conservation, different crop varieties, changing planting dates and irrigation to reduce the impact of climate change and variability (Deressa *et al.*, 2009). Similar work by Kinde Gebru (2014) reported that improved crop varieties were farmers' highly preferred climate change adaptation strategy followed by soil and water conservation in Adwa District, northern Ethiopia.

Generally, a combination of a green development programme and traditional soil and water management practices like terracing, stone bunds, cover crops, agroforestry, minimum tillage and other agricultural technology practices enhance productivity in a sustainable manner and can build a climate-resilient agricultural system. This result was supported by Jirata *et al.* (2016) who reported that the aim of CSA is promoting existing technologies, bringing them to the hands of farmers and integrating them with newly developed technologies such as drought-tolerant crops and advanced weather forecasts to cope with the changing climate.

Results showed that indicators of the effectiveness of the CSA practices are visible in the study area (Table 11.2). These include eroded areas that have been rehabilitated, tree planting on the hillsides, improved crop production and continuous yield despite rainfall that is spatially and temporally varied, and improved livestock production due to the use of a cut-and-carry system. Other indicators were minimized soil erosion and efficient water use as demonstrated by the fact that with water harvesting technology even small amounts of rainfall became productive. Furthermore, the participants of the study agreed that agricultural inputs supplied through extension services have much improved the productivity of agriculture. Improved varieties have been utilized for drought tolerance, disease resistance, and for early maturity that help crops to give substantial yield by escaping the dry spell in the growing season.

Soil erosion prevention and sustainable land use programmes run by the Ministry of Agriculture indicate the effort made to fight climate-change-related problems. The green development programme has mobilized the rural community to plant trees and practise conservation agriculture. All the farmers contribute 60 days labour for soil and water conservation each year as a safety net programme and this has improved the adaptation and mitigation capacity of the farmers towards climate change and its variability. The outcomes from these efforts are an indicator of the effectiveness of CSA practice. Mass promotion of compost utilization might have improved soil carbon sequestration which in turn contributes to greenhouse gas emission reduction (FAO, 2010). Application of compost improves the chemical, physical and biological characteristics of soils (Niggli *et al.*, 2009). It is one of the means of reducing external inputs like synthetic fertilizers, which are a main source of greenhouse gas emissions during their production, transportation and application, so reducing its use will thereby reduce the adverse effect it has on global warming (FDRE, 2011a, b).

The enclosure of an erosion-prone area for rehabilitation also improved the land cover with trees and residue that might have created an environment conducive to microorganisms and the ability to store soil carbon. Diversification of crop types and use of a mixed farming (crop-livestock) agricultural system plays an important role in

Table 11.2. Indicators of the effectiveness of climate smart agriculture (CSA) practices in Tehuledere District. (From focus group discussions, key informant interviews and field observation.)

CSA practice in the district	Indicators of CSA	Effectiveness of CSA
Soil conservation	• Constructed stone bunds, terraces • Improved soil productivity	✓ Erosion is minimized ✓ Improves physical properties of the soil
Water harvesting technology	• Reduced risk of crop failure • Reduced water shortage for humans and animals	✓ Improves water use efficiency
Organic fertilizer (compost)	• Farmers practise compost preparation and use • Improved soil condition	✓ Improves soil carbon sequestration ✓ Improves soil organic matter ✓ Reduces use of synthetic fertilizer hence reduces greenhouse gas emissions
Area of enclosure for rehabilitation	• Rehabilitation of enclosed area	✓ Avoids overgrazing ✓ Supplies feed for cattle ✓ Improves adaptive capacity
Tree planting	• Number of trees planted • Land cover change	✓ Modifies temperature and creates a microclimate ✓ Serves as a sink of CO_2 ✓ Increases resilience
Adoption of improved crop varieties	• Productivity is increased • Improved adaptability	✓ Risk of drought and disease is minimized ✓ Improves yield and income ✓ Improves adaptability and resilience
Crop diversification	• Number of crops planted per growing season • Reduced risk of total crop failure • Improved resource use efficiency (e.g. nutrients and water)	✓ Improves food security ✓ Increases resilience and adaptation to climate variability
Mixed farming	• Presence of crop and livestock production	✓ Increases income ✓ Improves food security
Scheduling of different planting date	• The change of harvesting date of crop following the rainfall pattern	✓ Reduces risk of crop failure ✓ Improves adaptive capacity
Cut-and-carry livestock production system	• Improved livestock production	✓ Avoids overgrazing ✓ Reduces greenhouse gas emission ✓ Improves physio-chemical condition of the soil

improving the income of the farmers and enhances food security, which is one of the aims of CSA practices (FAO, 2011). The risks of crop failure due to recurrent drought and other climate-related stress have been minimized through crop diversification by increasing the probability of getting a yield (Branca *et al.*, 2011).

We found that CSA practices have proved to be effective measures in addressing climate change stresses, due to the fact that they strike a 'triple-win' by ensuring mitigation of greenhouse gases, improving the adaptive capacity and increased agricultural production. Various scholars within and outside Ethiopia have affirmed that utilization of green manure, compost, farmyard manure and other organic fertilizers increase carbon sequestration in the soil which has a pre-eminent role in countering the greenhouse gas concentration in the atmosphere and improving soil fertility and crop production in a sustainable manner (IPCC, 2007; Niggli *et al.*, 2009; Branca *et al.*, 2011; Teka *et al.*, 2012).

Conclusion

The study was conducted in Tehuledere District, north-eastern Ethiopia. Most of the respondents perceived that climate change and variability in terms of temperature and rainfall variability is a reality. Farmers observed different indicators of climate change and variability including drought, off-season rainfall, too little and too much rainfall and high temperatures. Climate change and variability had adverse impacts on livelihoods and food security. Climate-change-related impacts include: (i) decreasing crop yields and livestock production; (ii) increasing pests and disease; and (iii) decreasing water quantity and quality. In response to the impacts of climate change and variability, farmers were engaging in various adaptation measures such as changing crop varieties, integrating livestock and crop production as a mixed farming system, adopting soil and water conservation practices and adjusting the planting date of crops to follow the pattern of the weather system. The study found that existing CSA practices, such as integrated soil and water management techniques (e.g. terracing, stone bunds); application of organic fertilizers on the soil (e.g. compost and farmyard manure); adoption of improved crop varieties for their drought-tolerant, disease-resistance and high-yielding ability and diversification of crop production, play a significant role in adapting to climate change. We found that those existing and newly adopted CSA practices are effective in curbing climate change impacts. Some of the indicators of the success of these CSA practices include the observed constructed stone bunds and terraces, compost preparation and use, rehabilitation of enclosed areas, number of trees planted, land cover change, improved soil productivity and reduced risk of crop failure. Moreover, strengthening the capacity of farmers, through provision of quality extension and weather services, multiple sources of information and incentives for the efforts put into ensuring sustainable agriculture systems are important prerequisites for the wide adoption of affordable, local and effective response measures (i.e. CSA practices).

References

Adhikari, U., Pouyan Nejadhashemi, A. and Woznicki, S.A. (2015) Climate change and eastern Africa: a review of impact on major crops. *Food and Energy Security* 4(2), 110–132.

Branca, G., McCarthy, N., Lipper, L. and Jolejole, M.C. (2011) Climate smart agriculture: a synthesis of empirical evidence of food security and mitigation benefits from improved cropland management. *Mitigation of Climate Change in Agriculture Series* 3. Food and Agriculture Organization of the United Nations (FAO), Rome.

Brohan, P., Kennedy, J.J., Harris, I., Tett, S.F.B. and Jones, P.D. (2006) Uncertainty estimates in regional and globalobserved temperature changes: A new data set from 1850. *Journal of Geophycal Research* 111, 1–21.

Central Statistical Authority (2007) Population and Housing Census of Ethiopia: Results for Amhara Region, Vol. 1, part 1. Central Statistical Authority, Addis Ababa.

Deressa, T.T., Hassan, R.M. and Ringler, C. (2009) Perception of and adaptation to climate change by farmers in the Nile basin of Ethiopia. Journal of Agricultural Science 149, 23–31.

Federal Democratic Republic of Ethiopia (FDRE) (2007) Climate Change National Adaptation Programme of Action of Ethiopia. FDRE, Addis Ababa.

Federal Democratic Republic of Ethiopia (FDRE) (2011a) Ethiopia's Climate-Resilient Green Economy Strategy (ECRGES). FDRE, Environmental Protection Agency, Adis Ababa.

Federal Democratic Republic of Ethiopia (FDRE) (2011b) Agriculture Sector Programme of Plan on Adaptation to Climate Change. FDRE, Addis Ababa.

Feleke, H.G. (2015) Assessing weather forecasting needs of smallholder farmers for climate change adaptation in the Central Rift Valley of Ethiopia. *Journal of Earth Science and Climate Change* 6(10), 1–8.

Food and Agriculture Organization of the United Nations (FAO) (2010) *'Climate-Smart' Agriculture: Policies, Practices and Financing for Food Security, Adaptation and Mitigation*. The Hague Conference on Agriculture, Food Security and Climate Change. FAO, Rome.

Food and Agriculture Organization of the United Nations (FAO) (2011) *Proven Technologies for Smallholder Farmers*. FAO, Rome.

Gadisa Chimdesa (2016) Climate change impacts and adaptation actions in Central Rift Valley of Ethiopia. *Journal of Natural Sciences Research* 6(3), 84–93.

Gebre Hadgu, Kindie Tesfaye, Girma Mamo and Belay Kassa (2015) Farmers' climate change adaptation options and their determinants in Tigray Region, northern Ethiopia. *African Journal of Agricultural Research* 10(9), 956–964.

Intergovernmental Panel on Climate Change (IPCC) (2007) *Climate Change 2007: Impacts, Adaptation and Vulnerability. Contribution of Working Group II to the Fourth Assessment Report of the Intergovernmental Panel on Climate Change.* [Parry, M.L., Canziani, O.F., Palutikof, J.P., van der Linden, P.J. and Hanson, C.E. (eds)]. Cambridge University Press, Cambridge.

International Food Policy Research Institute (IFPRI) (2014) Farmers preference for climate smart agriculture: an assessment in the Indo-Gangetic Plain. IFPRI discussion paper. IFPRI, New Delhi.

Jirata, M., Grey, S. and Kilawe, E. (2016) *Ethiopia Climate-Smart Agriculture Scoping Study*. Food and Agriculture Organization of the United Nations (FAO), Addis Ababa.

Kinde Gebru (2014) Smallholder farmers' adaptation strategies to climate changes in Ethiopia: a case of Adwa District, Tigray Region. MSc. thesis, Mekelle University, Mekelle, Ethiopia.

Lemmi Legesse Kotu (2013) Climate change perception and smallholder farmers' adaptation strategies: the case of Tole District, Southwest Showa Zone, Oromiya Regional state, Ethiopia. MSc. thesis, Haramaya University, Haramaya, Ethiopia.

Niang, I., Ruppel, O.C., Abdrabo, M.A., Essel, A., Lennard, C., et al. (2014) Africa. In: Barros, V.R., Field, C.B. Dokken, D.J., Mastrandrea, M.D. Mach, K.J. et al. (eds) Climate Change 2014: Impacts, Adaptation, and Vulnerability. Part B: Regional Aspects. Contribution of Working Group II to the Fifth Assessment Report of the Intergovernmental Panel on Climate Change Cambridge University Press, Cambridge, pp. 1199–1265.

Niggli, U., Fließbach, A., Hepperly, P. and Scialabba, N. (2009) Low Greenhouse Gas Agriculture: Mitigation and Adaptation Potential of Sustainable Farming Systems. FAO, April 2009, Rev. 2 – 2009.

Okonya, J.S., Syndikus, K. and Kroschel, J. (2013) Farmers' perception of and coping strategies to climate change: evidence from six agro-ecological zones of Uganda. *Journal of Agricultural Science* 5(8), 252–263.

Stabinsky, D. (2014) Climate-smart agriculture: myths and problems. The Green Political Foundation, E-Paper. Heinrich Böll Foundation, Brazil.

Teka, K., Van Rompaey, A., Poesen, J., Welday, Y. and Deckers, J. (2012) Impact of climate change on small-holder farming: a case of eastern Tigray, northern Ethiopia. *African Crop Science Journal* 20(2), 337–343.

Verhagen, J., Vellinga, T., Neijenhuis, F., Jarvis, T., Jackson, L. *et al.* (2014) *Climate Smart Agriculture: Scientists' Perspectives*. CGIAR Research Program on Climate Change, Agriculture and Food Security (CCAFS), Copenhagen.

Wassie, A.S. and Pauline, N.M. (2018) Evaluating smallholder farmers' preferences of climate smart agricultural practices in Tehuledere District, northeastern Ethiopia. *Singapore Journal of Tropical Geography* 39(2), 300–316. doi:10.1111/sjtg.12240

World Meteorological Organization (WMO) (2002) *Application of Climate Forecasts for Agriculture: Proceedings of an Expert Group Meeting for Regional Association I (Africa)*, 9–13 December 2002, Banjul, Gambia. WMO, Geneva, Switzerland.

World Wide Fund for Nature (WWF) (2006) *Living Planet Reports*. WWF, Gland, Switzerland.

Yanda, P.Z. and Mubaya, C.P. (2011) *Managing a Changing Climate in Africa: Local Level Vulnerabilities and Adaptation Experiences*. African Books Collective, Oxford.

12 Community Livelihoods and Ecosystem Integrity in Makere Forest Reserve, Western Tanzania

Pius Z. Yanda,[1] Edmund B. Mabhuye,[1]* Anselm R. Mwajombe[2] and Namkunda Johnson[1]

[1]*Institute of Resource Assessment, University of Dar es Salaam, Dar es Salaam, Tanzania;* [2]*Centre for Climate Change Studies, University of Dar es Salaam, Dar es Salaam, Tanzania*

Abstract

Wealth creation and poverty alleviation processes in the forest-bounded areas entail the use of such forests to a greater extent. Studies elsewhere show that there is often a tendency to use such forests until they are depleted before technology comes in to improve livelihoods. In this chapter, we examine community livelihoods in relation to ecosystem integrity for communities surrounding the Makere Forest Reserve, particularly socio-economic characteristics of communities, their links to forest utilization and implications for ecosystem integrity. We used mixed methods to collect data: (i) a household questionnaire; (ii) focus group discussions; (iii) key informant interviews; and (iv) a literature review, backed up with satellite imagery. Quantitative and qualitative data collected were subjected to statistical and non-statistical tests, respectively, with the use of Statistical Package for Social Sciences (SPSS) computer software for quantitative data analysis. Livelihood activities in the area include shifting cultivation, livestock keeping, firewood fetching, charcoal making, harvesting timber and illegal logging. The motive for such forest exploitation is both for meeting household needs as well as for wealth accumulation. Forest users take part in such activities regardless of the distance they have to travel from their villages to come to the forest areas. We found education is an integral part of wealth status, but had nothing to do in terms of improving livelihood activities for ecosystem integrity. The absence of livelihood diversification of farm-related activities penetrates into weak forest governance strategies resulting in proliferation of deforestation and forest degradation. To maintain forest integrity, integrated approaches in forest management and alternative livelihood activities are needed such as beekeeping, fishing and modernized livestock keeping. These activities have the potential to increase household food and income and alleviate poverty levels without compromising ecosystem integrity.

Introduction

Forest ecosystems play a significant role in ecological and socio-economic functions. They provide environmental goods and services such as recreation and aesthetic values, non-timber forest products, timber products and fibre (Belcher, 2005). From an ecological perspective, apart from providing habitats for a variety of plants and animals and being hotspots for endemic species, forests also provide water catchment areas (Harrison, 2006). From a socio-economic perspective, forests are the source of livelihoods for communities

*edmund.mabhuye@gmail.com

reliant on those resources and provide water sources for hydro-electric power generation and irrigation. Forests are also used for spiritual matters (Harrison, 2006). In recent decades, forests have became renowned for the significant role they play in climate change mitigation through sequestering greenhouse gases (Indrarto *et al.*, 2012).

Ever since societies began to understand that human activities have an impact on forests, assessing the best way to manage them has been a topic of discussion (Castrein and Pillai, 2011). The discourse has changed over decades, but the interaction of human activities and the state of the forests has always been at the core (Castrein and Pillai, 2011). Such a discourse takes two strands, first is how forests contribute to economic development of those described as well off (Schmidt *et al.*, 2006; Agrawal *et al.*, 2013; Shepherd *et al.*, 2013), and second is poverty alleviation for poor societies at the forest edges (see Kahyarara *et al.*, 2002). These strands bestow the differentiated forms and magnitudes of forest resources utilization (Schmidt *et al.*, 2006).

The poverty perspective of forest utilization is mainly characterized by a shifting cultivation form of agriculture (Kahyarara *et al.*, 2002). Often forests attract the poor, since they provide new agricultural land and economic opportunities for people with limited options (FAO, 2006; Sunderlin *et al.*, 2007), and those who are politically weak or powerless (Murty, 2008). According to Boserup (1965), expansion of agriculture prevails until no more forests are available, before agricultural transformation takes on a new phase. Transformation is a function of several variables such as technology, capital and/or labour. Deforestation continues until the land loses productivity and the forest cover becomes depleted in a lose-lose relationship.

The wealth perspective may also entail win–lose ties on communities and forest cover, respectively. In order to achieve urban and industrial development, rural resources (agriculture and forestry) are capitalized as well as invested in for physical and technological infrastructure of developed areas (Davies and Johnson, 1995). Often, the investments are attracted by consumer demand that have changed the way in which firms must invest and market their products (Nacker, 2004). The poor receive low incomes from forest resources, while the wealthy are able to gain higher profits from the same resource (Schmidt *et al.*, 2006). The poor-wealthy ties with regard to forest resources utilization have often resulted in lose-lose overtime in different parts of the world, in particular, developing countries.

The wealthy–poor ties with respect to forest utilization are prevalent in Tanzania, a country highly endowed with forest resources with about 48.1 million ha, which is 55% of the total land area of mainland Tanzania (URT, 2015). However, both reserved and unreserved forest resources in many parts of Tanzania are under threat to deforestation and degradation by communities residing alongside these forests through activities such as agriculture and logging (URT, 2013). In particular, agriculture is mostly characterized by slashing, burning and shifting cultivation (Kahyarara *et al.*, 2002; Hosonuma *et al.*, 2012; Suich *et al.*, 2015). Similarly, harvesting timber, logging, charcoal making and firewood extraction, both legal and illegal, are also prevalent whereby the products are transported to town centres (Kahyarara *et al.*, 2002).

In western Tanzania, forest utilization is either for wealth creation or for poverty alleviation, particularly for the rural population who live adjacent to the forests. In due regard, participatory forest management (PFM) approaches implemented earlier (Blomley, 2006; Blomley *et al.*, 2008; CBD, 2009) have largely been unsuccessful partly due to such heavy dependence on forests. In this chapter, we examine the relationship between community livelihoods, and whether the poverty mentality or wealth creation mentality drives different forms of forest utilization. We examine ecosystem integrity by analysing the socio-economic characteristics of communities, their links to different forms of forest utilization and the resulting implications for forest integrity. Apart from policy implications, findings also contribute to the existing literature on the links between social stratification and integrity of forest ecosystems.

Conceptual Framework

The win–lose or lose–win relationship between human well-being and forest ecosystem health may be a common relationship in developing countries, synonymous with the history of agriculture and rural development (Sunderlin *et al.*, 2005). Agriculture through shifting cultivation has often been at the expense of natural forest cover and biodiversity. Current forest utilization may benefit the communities now, through increased agriculture yields due to inherently high soil fertility. However, over time the land may lose productivity as a result of overutilization and eventually result in a decline in livelihood for these communities. In the end, the expected results would be lose–lose due to lack of a resource base (Sunderlin *et al.*, 2005). The situation would be reversed by the use of technology (Boserup, 1965).

Thus the conceptual framework used in this study (Fig. 12.1) takes on board these aspirations which culminate in the resource use-degradation dynamics. This framework was adopted and modified from Cowie *et al.* (2018) which focuses on sustaining and improving the stocks of land-based natural capital and the associated flows of ecosystem services, to support the future prosperity of humankind. The interconnected elements such as drivers for change, pressures, state of the land-based natural capital, impacts and responses influence each other. The ultimate objectives are: (i) to maintain and improve ecosystem services through improving the productivity of the land; (ii) to increase food security; and (iii) to increase resilience of land systems and the people depending on them. Supported with other social, economic and environmental objectives through coherence between policies and measures that address separate environmental and development objectives, it also entails governance reinforcement (Cowie *et al.*, 2018).

In this framework, we narrow down Cowie *et al.*'s (2018) ideas to contextualize them to the grassroots where communities heavily depend on forests and the eventual outcomes on forest resources and communities'

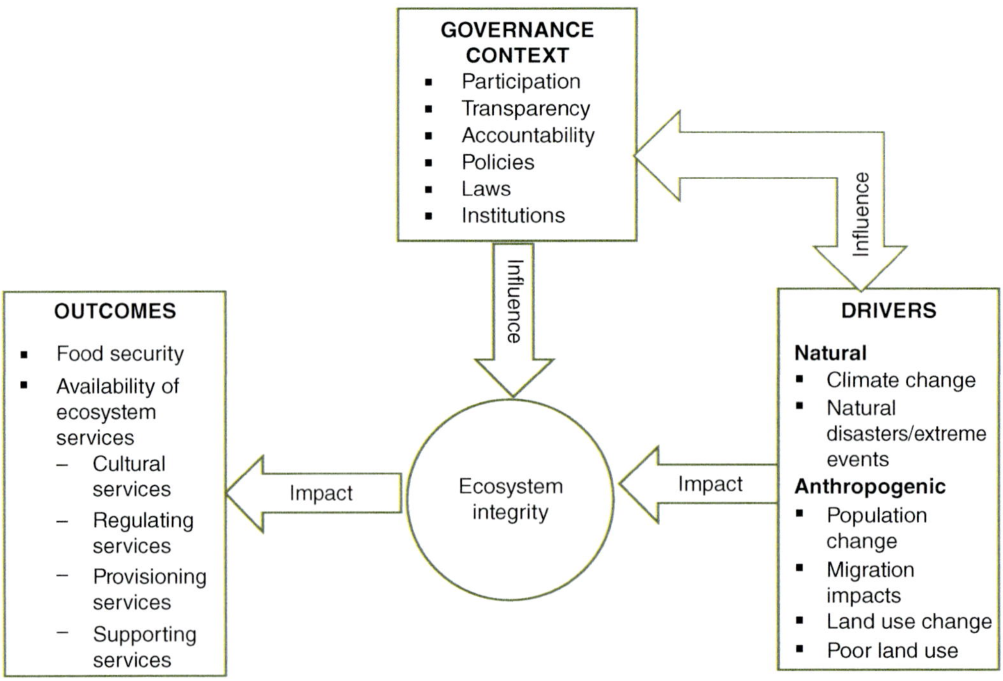

Fig. 12.1. Conceptual framework of interactions between human well-being and drivers of change. (Adopted and modified from Cowie *et al.*, 2018.)

livelihoods. We link the governance context, the drivers of ecosystem change as well the outcomes of livelihood activities from those drivers of change. In this regard, ecosystem integrity is affected by natural and anthropogenic factors coupled with the governance context. In the meantime communities may increase food production with a decrease in the availability of ecosystem services. In the long run, both food production and ecosystem services would eventually both become compromised (Sunderlin *et al.*, 2007). At that stage, injection of technologies, labour and capital become the only options (Boserup, 1965).

Methodology

Research design

The study design employed a mixed-method approach. According to Almalki (2017), the mixed method approach is applicable based on the following criteria: (i) when very little is known about a new concept, a qualitative approach can be used before a quantitative approach; (ii) when findings from a research approach can be better interpreted by using findings from the other approach; (iii) when a sole approach is not sufficient to generate meaningful findings; and (iv) when quantitative findings can be enriched by qualitative findings. Hence, the mixed methods were relevant to the existing study particularly when the information generated by one was to be complemented by the other approach.

Study area description

Makere Forest Reserve is located in Kasulu District in Kigoma Region (Fig.12.2). It is administratively divided into 'North Makere' and 'South Makere' Forest Reserves, and it covers a geographical area of 147,597,441 ha. The forest is currently facing acute threats of deforestation and forest degradation due to proliferation of livelihood activities particularly agriculture and livestock keeping.

Selection of study villages

This study was conducted in villages adjacent to South Makere Forest Reserve. The orientation and characteristics of this forest attract communities surrounding it and beyond. These entail easy soil workability and high soil fertility that leads to high crop yields. The natives ironically consider the forest as a mine, comparing abundant yield of maize due to high soil fertility to valuable minerals. The forest attracts a huge number of cultivators even from distant villages. We therefore sampled villages using non-probability and probability sampling. First, we identified villages located at a far distance from the forest area. To obtain a representative village from this category, random sampling was used whereby Buhoro was selected, located at a distance of approximately 70 km from the forest area.

A purposive sampling procedure was used to select specific study villages that were close to the forest area. In this case, Kagera Nkanda and Nyachenda villages were selected as the nearest villages to Makere Forest Reserve in the south. The specific location of each village is shown in Fig. 12.2. The purpose of selecting villages which varied in their distance (near/far) from the forest was to discern whether and to what extent influx into the forest was dependent on distance. It aimed also at analysing the forest features that attract the immigrants into the forest regardless of the distance from their respective villages to the forest.

Selection of subjects of the study

Similar to the selection of study villages, non-probability and probability sampling techniques[1] were also used to select the study respondents. The probability sampling technique employed simple random sampling under which questionnaire households were selected. On the other hand, non-probability (purposive) sampling was used to select heads of households for the questionnaire survey, and respondents for the key informant interviews and focus group discussions (FGDs).

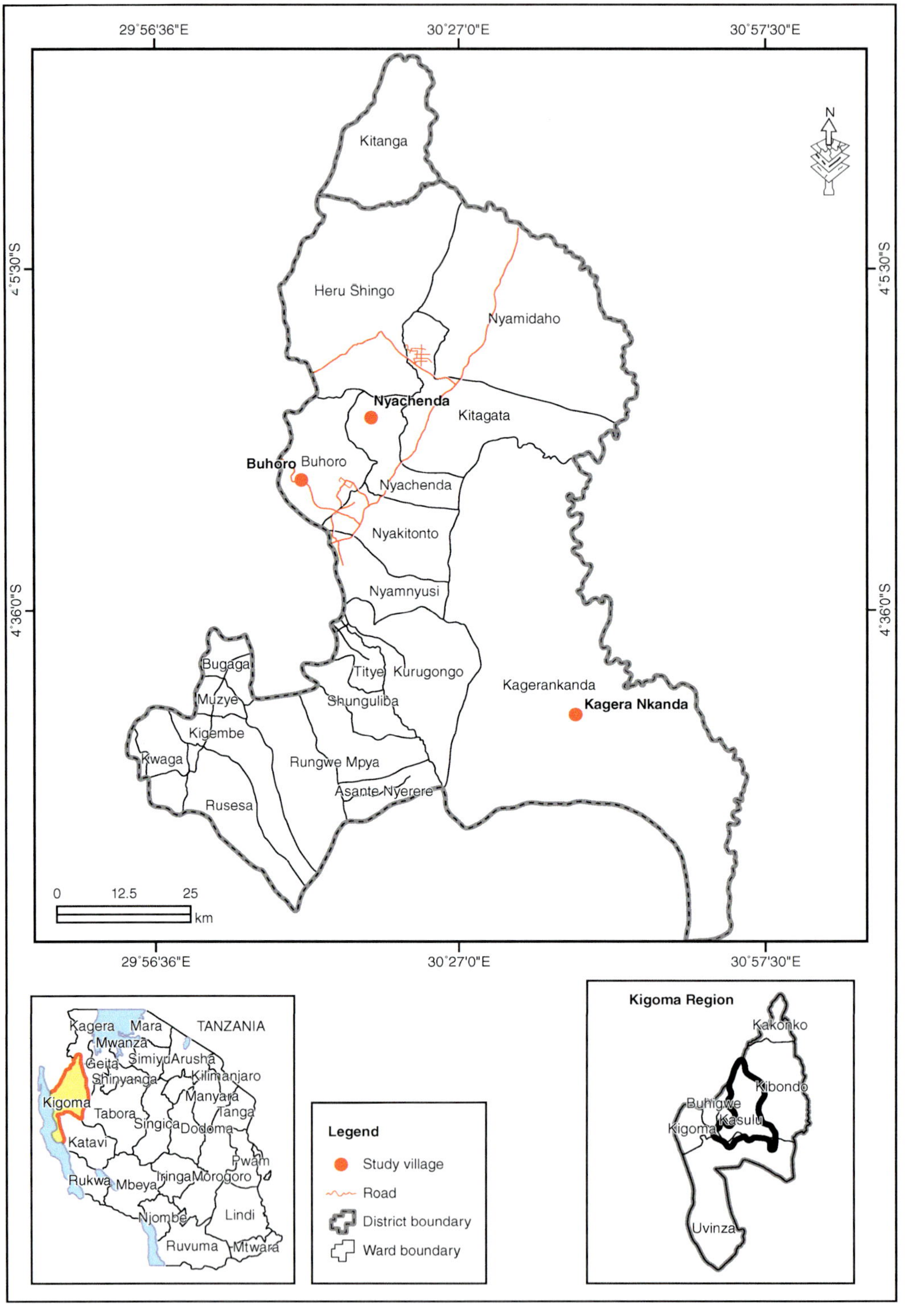

Fig. 12.2. Study villages. (From Geographic Information System (GIS) Laboratory, Institute of Resource Assessment, University of Dar es Salaam.)

Data collection methods

Participatory wealth ranking (PWR)

This tool aims at capturing community members' perceptions on poverty. According to Scoones (2017), PWR is particularly important when knowledge of the range of factors influencing households' asset ownership and income levels, as well as more intangible factors such as social or political standing, prestige and influence, are sought. PWR is more preferred in a local context because poverty indicators chosen by researchers may be an inadequate reflection of the complexity of rural settings. Unlike other tools of data collection in which the researcher chooses the criteria, the criteria for wealth ranking are decided by local people (Feulefack *et al.*, 2006). According to Feulefack *et al.* (2006), the PWR tool can portray less accuracy in the poor class than the middle one, hence facilitation is important.

For the purpose of this study, PWR aimed at uncovering underlying facts for Makere forest degradation, whether it was wealth creation or poverty alleviation. The wealth ranks were established in the FGDs as propounded by Scoones (2017). We postulated the link between wealth ranks and forest degradation relates to either wealth creation or short-term poverty alleviation, particularly for obtaining household income and food. There was a discussion about the meaning and understanding of poverty in the local context as suggested by Feulefack *et al.* (2006). Community members were conveyed by the facilitation team to form wealth groups based on their own criteria that considered assets, capabilities and activities performed by households. Three wealth groups were then formed: (i) the well off; (ii) the middle income; and (iii) the poor.

Key informant interviews

Key informant interviews are qualitative in-depth interviews with people who know what is going on in the community (Kumar, 1989). The purpose of key informant interviews is to collect information from a wide range of people, including community leaders, professionals or residents who have first-hand knowledge about the community (Boyce and Neale, 2006). These community experts, with their particular knowledge and understanding, can provide insights into the nature of problems and give recommendations for solutions. The study interviewed the following as key informants: (i) the District Environmental Officer; (ii) the District Wildlife Officer; (iii) the Community Development Officer; and (iv) the Village Executive Committee.

Focus group discussions (FGDs)

Focus groups consist of a moderator who supplies a topic and monitors the discussions. Focus groups provide an excellent opportunity to listen, observe the amount of interaction, explore issues in depth and obtain insights that might occur without the discussions (Tynan and Drayton, 1988; Almalki, 2017). In the existing study, FGDs were conducted at the village level. Selection of FGD participants was based on their long-term experience of Makere South Forest Reserve and those who noted changes in forest cover over time. Village leaders helped a research team to identify people with such characteristics.

Household survey

The household questionnaire was used during the household survey to solicit information on links between economic prosperity and poverty with Makere Forest Reserve degradation. The questionnaire was also used to cross-check information provided by key informants and FGDs. The surveyed households were purposively chosen based on wealth ranking groups (well off, middle income and poor). The surveyed households were drawn from the village household registry in accordance with wealth criteria provided during the wealth ranking activity in the FGD. A total of 245 respondents completed the household questionnaire as presented in Table 12.1.

Literature review

A review of the literature was also conducted to acquire further information relating to

Table 12.1. The sampling procedure of the study villages. (From field work, 2016.)

Village name	Ward	Hamlet	No. of households	Sample size	Sample (%)
Buhoro	Nyakitondo	Kahunga 'A' Kahunga 'B' Nyamisanze 'A' Nyamisanze 'B'	489	65	13.29
Nyachenda	Nyachenda	Fugwe Nyambogo Nyakasanda Bogo	1449	90	6.21
Kagera Nkanda	Kagera Nkanda	Katezi Kagunga Mkamba Nyazanza	1889	89	4.71
Total			3827	245	6.40

the subject matter and compare the results from this study with other research findings. Reviewed documents included journals, research papers, articles and websites.

Analysis of satellite imagery

Geographic information systems (GIS) and remote sensing are the most powerful techniques which employ multi-temporal satellite imagery in understanding landscape dynamics. In this study, the tools were employed to process Landsat TM of 1985 & 1995, Landsat ETM+ 2005 and Landsat 8 of 2016 with a resolution of 30 m by 30 m to quantify the changes in South Makere Forest.

The software was applied to perform various activities including image layer stacking to create a false colour composite (FCC) image subset to extract the study area by considering the geo-referenced outline boundary of the North and South Makere Forest, radiometric and geometric correction, image interpretation and classification and finally change detection was performed based on the classified images. Radiometric correction was carried out to reduce the influence of errors or inconsistencies on image brightness values that may hinder an ability to interpret or quantitatively process and analyse digital remotely sensed images. The process enhanced the quality of the images for better data extraction.

Geometric correction was also performed to correct errors caused by satellites or aircraft not staying at a constant altitude or due to sensor deviations from the primary focus plane. Thus the internal and external distortions were removed to identify the location of each pixel on the Earth. Images were compared to ground control points on accurate base maps and re-sampled so as to calculate the exact locations and appropriate values. Data analysis began after image correction, and then the image interpretation process was done using manual identification of features (i.e. creating signatures of each land cover) through visual observation. The process was based on image characteristics such as size, colour/tone, shape, texture, shadow, association and pattern.

Supervised classification method was applied to classify land cover use based on the spectral signature defined in the training set using Erdas Imagine 2011. The maximum likelihood algorithm was employed because of its capability to provide a consistent approach to parameter estimation problems. The algorithm has desirable mathematical and optimality properties (i.e. minimum variance and unbiased estimators as the sample size increases, so it provides approximate normal distributions and sample variance that can be used to generate confidence bounds and hypothesis tests for the parameters). Land-cover/use change detection was obtained based on pixel-based comparison using the post-classification method.

Data analysis plan

The collected data were analysed separately, depending on their nature. Qualitative data were sorted and arranged thematically in a matrix. Then they were subjected to content analysis. Quantitative data were sorted, arranged as well as coded and then analysed using descriptive statistics and inferential statistics. Such raw quantitative data were entered into Microsoft Excel and Statistical Package for Social Sciences (SPSS) software and results were expressed in frequencies, percentages and averages using tables, charts and graphs.

Results and Discussion

Socio-economic characteristics of communities

The study examined links between socio-economic characteristics of villagers surrounding Makere South Forest Reserve. The study considered education and wealth status of households as essential factors in influencing forest utilization through activities such as agriculture, livestock keeping, firewood and charcoal making as well as harvesting timber and logging.

Education levels of respondents

Education levels were classified as 'none', 'primary education', 'secondary education' and 'tertiary education'. Of the respondents from all three study villages, 3.7% had attained secondary education, whereas only 0.4% had attained tertiary education and 27.8% had received no formal education. The rest (68.2%) had attained only primary education level (see Table 12.2). These survey statistics depict the extent of illiteracy in the area.

Generally, the low levels of education of respondents may be reflected in the nature of livelihood activities they performed. They were mainly low skilled and agriculture was mostly practised. Kahyarara *et al.* (2002) indicated that low education levels are linked to limited non-farm livelihood diversification. In this study, as subsequently discussed, overdependence on agriculture was geared to attain good living standards.

Wealth ranks in the study area

Results in the previous section leads us to question whether or not there exists a relationship between wealth groups and education. Despite the majority of respondents having received low levels of education, there were links between education and wealth. Earlier, Phiri *et al.* (2004) showed that wealth differences do not prevent people from participating in conservation matters. In due regard, *t* tests showed a significant relationship ($P < 0.05$) between wealth groups and education at 95% confidence interval (Table 12.3). The upper and lower confidence intervals of the difference values (Table 12.3) imply more concentration of the well off in the upper and the middle classes.

We then characterized wealth groups to discern how and to what extent they related to the manner livelihood activities are performed. Consistent with other studies (e.g. Feulefack *et al.*, 2006; Kok and Lu,

Table 12.2. Levels of education of respondents. (From field data, 2016.)

	Village							
	Buhoro		Kagera Nkanda		Nyachenda		Total	
Education level	Frequency	%	Frequency	%	Frequency	%	Frequency	%
None	26	39.4	16	17.8	26	29.2	68	27.8
Primary	35	53.03	71	78.9	61	68.6	167	68.2
Secondary	4	6.06	3	3.3	2	2.2	9	3.7
Tertiary	1	1.51	–	–	–	–	1	0.4
Total	66	100	90	100	89	100	245	100

Table 12.3. Relationship between worth group and education of respondent. (Analysed by paired sample correlations from field survey, 2016.)

	Paired differences							
				95% CI of the difference[a]				
	Mean	Standard deviation	Standard error of the mean	Lower	Upper	t	Degrees of freedom	Significance (two-tailed)
Worth group–education level of respondent	0.64898	0.93171	0.05952	0.53173	0.76623	10.903	244	0.000

[a]CI, Confidence interval.

2016; Scoones, 2017), the wealth ranking exercise revealed that communities consider household assets accumulated including farm size, number of employees hired by a particular household, livelihood diversification and market networking as essential attributes that characterize wealth. Three major wealth groups were eventually categorized namely: (i) 'well off'; (ii) 'middle income'; and (iii) 'poor'. Table 12.4 shows the criteria used by respondents in FGDs to classify different households in the area according to wealth. Over 52.4% of households in the study area fall in the poor group, while the middle income group comprises 36.4% and the remaining 10.7% are the well-off group (Table 12.4).

The poor class

Of all the interviewed respondents, 52.4% were classified as poor (Table 12.4: 8.7% in Buhoro + 18.4% in Nyachenda + 25.3% in Kagera Nkanda). Their struggle was aimed at satisfying basic human needs: food, clothing and shelter and in the long run they had ambitions of attaining decent living standards like other successful people within and outside the village. They owned small farms ranging in size from 0.25 acres to 0.5 acres in the forest. Their housing was mainly mud or brick walled and roofed with grass. Most in this group could afford two meals a day.

The middle income class

The class constituted 36.4% of all surveyed households (Table 12.4: 15.1% in Buhoro + 13.1% in Nyachenda + 8.2% in Kagera Nkanda). They had living standards that are somehow better than the poor class. This group owned farms ranging in size from 2 to 14 acres in Makere Forest Reserve. Their housing could be brick walled with no floor. They could afford to hire a small amount of labour for farm activities. However, the assets they possess distinguish this class from the poor. The number of cows they owned ranged from one to four, and goats would range from four to five (Table 12.4).

The well-off class

Of all surveyed households, 10.7% were classified as well off. They possessed fairly large farms with sizes usually ranging from 10 to 30 acres in Makere Forest Reserve. They had greater potential to diversify their livelihood through collateral, owing to the valuable assets they owned. Such assets included business houses, motorized transport, decent housing and a good number of labourers. Farming was not mechanized implying that money generated due to sale of crops was invested in other sectors (Table 12.4).

Livelihood activities performed in the Makere Forest Reserve

The major ways in which the forest is utilized in the area include for agriculture, livestock keeping, firewood harvesting, charcoal making as well as harvesting timber and logging. The question was on whether the type of wealth group, and their differences in education level, had any bearing on the

Table 12.4. Wealth group classification in the study villages. (From field survey, 2016.)

	Characteristics of wealth group		
Village	Well off	Middle income	Poor
Buhoro	Farm 5–10 acres	Farm about 1 acre	Farm 0.25–0.5 acres
	Five maize bags/acre	Three to four bags of maize/acre	Harvest not quantified
	Livestock: 10–50 cows, goats, hens	Livestock: 1–4 cows, 4–5 goats	Can afford one/two meals/day
	Forest farm of 1 acre or more	Self-employed. Kiosk ownership	May have a large family
	Kiosk ownership	Ability to change food varieties	This group constituted 8.7% of all the interviewed respondents
	Capacity to hire manpower	This group made up 15.1% of all the interviewed respondents	
	Ability to change foods		
	Government employee. The well-off group in Buhoro made up 3.3% of all the interviewed respondents		
Nyachenda	A well-built house of bricks	A block house roofed with corrugated sheet with no floor	A mud house roofed with grass
	Ownership of a means of transport: cars, trucks and motorbike	Farm 2–5 acres	May have a bicycle or not
	Go down ownership with the capability of 20–30 t	May own means of transport like motorbike	Working on day worker
	Farm 50 and above acres	This group constitutes of 13.1% of all the interviewed respondents	Farm of not more than 2 acres
	Crop business		One meal/day
	Livestock: 15–20 cows		Unable to access health services.
	The well off in Nyachenda made up 4.1% of all the interviewed respondents		The poor group constituted 18.4% of all the interviewed respondents
Kagera Nkanda	Ownership of commercial houses like guest houses	A well-built house of blocks and roofing corrugated sheet	Has mud housed roofed with grass or rented a house
	A well-built house of blocks and corrugated sheet roofing	Ownership of means of transport like motorbike	Livestock keeping 2–5 goats, 2–10 poultry
	Ownership of means of transport e.g. cars	Livestock: 15–40 poultry	Farm 2–3 acres where yield is for subsistence or not having a farm
	Livestock: > 200 cows	Kiosk ownership	Is a day worker. This group made up 25.3% of all the interviewed respondents
	Ownership of milling (maize and cassava) machine	Milling machine ownership	
	Farm 30+ acres	Farm 10 acres	

Continued

Table 12.4. Continued.

Village	Characteristics of wealth group		
	Well off	Middle income	Poor
	Oxen ownership	Can hire day workers or self-employed. This group constituted 8.2% of all the interviewed respondents	
	TV show huts. This group made up 3.3% of all the interviewed respondents		

different forms of forest utilizations in the area. From chi-square tests on the data, we see that all forms of forest utilization were statistically significant ($P < 0.05$) at 95% confidence interval, except agriculture which was constant. This implies that most respondents interviewed were engaged in agriculture, with a small variation due to engaging in the other activities. The result means that the manner in which the forest was utilized was the same regardless of wealth group.

However, these forms of forest utilization were destructive and illegal, because the forest is protected by its reserve status under which any unspecified forest utilizations are to be avoided by law (e.g. URT, 2008). These livelihood activities were described as the main sources of income in the area and can be linked to the state of Makere Forest Reserve (Fig. 12.3). The manner in which the forest is utilized is discussed below.

Agricultural practice in the area

Agriculture was practised by 90% of respondents across the sample (Fig. 12.3). Crops grown include both food crops such as maize, beans and cassava, and cash crops such as tobacco. In particular, the repercussions of tobacco cultivation on the integrity of forests are well documented by Abdallah and Monela (2007) and Abdallah (2014). In this study we noted that most food crops grown (e.g. beans and maize) were for household consumption, some for sale (e.g. cassava), whereas cash crops were sold to buyers in the study villages as well as external buyers from within and outside Kasulu District. Farmers owned farms within and outside the villages where they were born and the sizes of farms were directly related to their wealth status. However, low soil fertility and high population size in the villages were constraints to increased agricultural yields to farms inside the villages. The majority of villagers who had less than 1 acre plots, were considered to be in the poor class (Table 12.5).

Despite variations, the relationship between wealth group and plot sizes was significant ($P = 0.001 < 0.05$). Dominance of small plots per household in the village depicts the extent to which land within village is highly occupied. There were claims among farmers that village farms were not productive. This mentality magnified the extent of migration to Makere Forest from their villages for agriculture (as discussed later). Moreover, rapid population growth in the studied villages was problematic in expanding farming areas to increase agricultural productivity. Those farmers' perceptions were confirmed by reports which show that while the Kigoma population is among the fastest growing at 2.4% per intercensal period (National Bureau of Statistics, 2013), that of Kasulu alone grew at a rate of 4.8% per intercensal period and these ranked among districts with the highest population growth rates (National Bureau of Statistics, 2013).

This result is consistent with those of Paavola (2004) who reported that, in attempts to secure livelihoods in Morogoro, communities extended cultivation, intensified agriculture, diversified livelihoods and migrated to gain access to land. Hence, due to land constraints for in-village farm expansion, cultivation inside Makere Forest

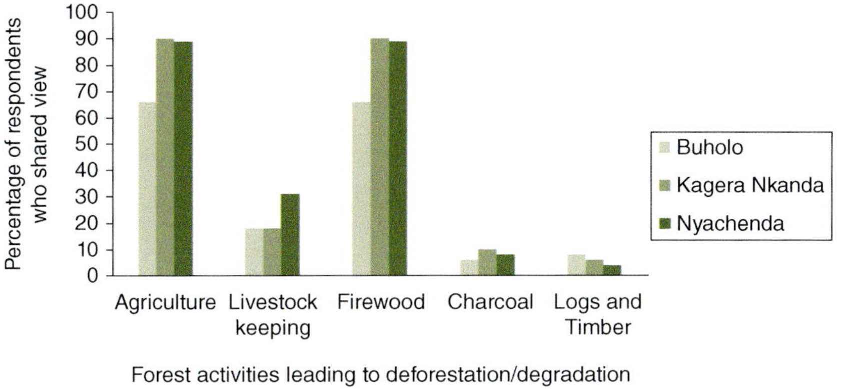

Fig. 12.3. Drivers of deforestation and forest degradation in Makere Forest Reserve. (From field data, 2016.)

Table 12.5. Farm ownership within the villages. (From field survey, 2018.)

		Respondents who owned farms (according to farm size) (%)						
Village	Wealth group	0–≤1 acres	>1–≤3 acres	>3–≤5 acres	>5–≤10 acres	>10–≤15 acres	N/A[a]	Total (%)
Buhoro	Well off	3.0	1.5	0	0	0	7.6	12.1
	Middle income	16.7	7.6	1.5	1.5	0	25.8	53.1
	Poor	22.7	0	1.5	0	0	7.6	31.8
	Total	**42.4**	**9.1**	**3.0**	**1.5**	**0**	**41**	**97**
Kagera Nkanda	Well off	0	0	0	1.0	0	7.8	7.8
	Middle income	2.2	3.3	0	0	0	16.7	22.2
	Poor	11.1	6.7	3.3	1.1	0	46.7	68.9
	Total	**13.3**	**10.0**	**3.3**	**2.1**	**0**	**71.2**	**98.9**
Nyachenda	Well off	2.2	2.2	1.1	1.1	4.5	0	11.1
	Middle income	16.9	7.9	4.5	4.5	2.2	2.2	38.2
	Poor	24.7	12.4	4.5	1.1	0	7.9	50.6
	Total	**43.8**	**22.5**	**10.1**	**6.7**	**6.7**	**10.1**	**99.9**

[a]N/A denotes farmers who didn't have farms within the villages.

was highly preferred because of the presence of fertile soils and easy workability, which assure increased harvests and incomes for farmers. These farm conditions in the forest have been an attraction to community members even from distant villages (such as 70 km away, Buhoro village being an example in this case) to conduct farming. Agricultural activities conducted in Makere Forest Reserve are anonymous and involved building temporary houses, which were occupied during planting and harvesting periods. It was observed that shifting cultivation was practised by all respondents across the sample regardless of wealth groups. The majority of farmers owned small plots of up to 5 acres and these farmers were mainly in the 'poor' wealth group (Table 12.6).

From Tables 12.5 and 12.6 we note the following trends. First, lack of farm ownership in village areas was high; and second, farms in the village were mainly small plots. Third, the majority of people who were well off had farms outside the village. In all the study villages, there was a significant ($P < 0.05$) relationship between wealth groups and plot sizes, implying that plot size increased with wealth. Hence, the well off acquired the large pieces of land compared with the poor who were able to acquire less land. Comparing Table 12.5 and Table 12.6 we note that farms inside the village are smaller than those

Table 12.6. Acreage of farms owned per wealth group in Makere Forest. (From field data, 2016.)

		Respondents who owned farms (according to farm size) (%)							
Village	Wealth group	0–≤5 acres	>5–≤10 acres	>10–≤15 acres	>15–≤20 acres	>20–≤25 acres	>25–≤30 acres	N/A[a]	Total (%)
Buhoro	Well off	0	1.5	1.5	0.0	6.0	3.0.	0.0	12.1
	Middle income	24.2	22.7	6.1	1.5	0.0	0.0	1.5	56.1
	Poor	28.8	3.0	0.0	0.0	0.0	0.0	0.0	31.8
	Total	**53.0**	**27.3**	**7.5**	**1.5**	**6.1**	**3.0**	**1.5**	**100**
Kagera Nkanda	Well off	1.1	0.0	2.2	1.1	1.1	3.3	0.0	8.8
	Middle income	13.5	0.0	7.8	0.9	0.0	0.0	1.1	23.3
	Poor	55.5	6.7	3.3	0.0	0.0	0.0	3.3	68.8
	Total	**70.1**	**6.7**	**13.3**	**2.0**	**1.1**	**3.3**	**4.4**	**100**
Nyachenda	Well off	0.0	0.0	0.0	1,1	4.5	3.4	2.2	11.2
	Middle income	4.3	9.0	7.9	0.0	0.0	0.0	15.9	38.2
	Poor	20.2	3.4	3.4	0.0	1.1	0.0	22.5	50.6
	Total	**24.7**	**12,4**	**11.2**	**1.1**	**5.6**	**3.4**	**41,5**	**100**

[a]N/A Denotes farmers who didn't have farms outside villages.

outside the village. This could be due to 'saturation' of land inside the village, and so farmers had to find farms outside their villages. Even villagers from the farthest villages (Buhoro in this case) had farms in the Makere Forest. In this regard, distance from the respective villages to the forest was not a determinant factor.

Due to their great financial capacity, the 'well offs' were capable of transporting their products from Makere Forest to urban areas where there was high demand and good prices. Relations between networking abilities, wealth group and forest degradation could be traced based on the fact that the abilities of the well-off group to acquire large farm sizes was particularly high. They were earning high yields and thus could influence external buyers on selling prices. The high crop yields produced by the well-off people motivated the 'middle income' and 'poor' groups to extend their farming areas. The aim of expanding farming areas was to produce high yields and obtain high income.

Firewood and charcoal making

The primary source of energy for households in the study area was mainly firewood followed by charcoal. Firewood was the most used source of energy for subsistence at the household level, whereas charcoal was seldom used for commercial purposes. About 87.0% (n = 227) of the respondents surveyed were firewood users compared with 2.6% (n = 6) who used charcoal. Households that used charcoal were considered to be wealthier than those who used firewood. A sack/bag of charcoal was sold at TSh8000. Charcoal production was associated with selective tree cutting. Jumbe *et al.* (2009) indicated that charcoal and firewood provide 70% of the country's energy needs.

Livestock keeping

Livestock kept by the native villagers in the area involved poultry/geese, goats/sheep, cows and pigs. With the exception of cows, goats and sheep, other minor livestock were home based. The number of cows kept differed per household whereas 8.2% (n = 20) had cows ranging from one to five animals. About 2% of households had between six and ten cows. Only 1.2% (n = 3) of households had the highest number of cows (21–50 cows). The majority of native villagers were not engaged in the livestock keeping activities at all (Table 12.7).

Table 12.7. Livestock kept in the study area. (From field survey, 2016.)

Type of livestock	Frequency	Percentage
Poultry/geese	57	23.3
Goats/sheep	40	16.2
Cows	32	13.1
Pigs	7	2.9
N/A[a]	109	44.5
Total	245	100

[a]Not applicable.

Results imply that livestock keeping by native villagers could not induce massive deforestation/forest degradation. Inherently it is deduced that when the management of a common pool resource is rendered difficult, such difficulty renders destruction due to overutilization (Ostrom *et al.*, 1999; Ostrom, 2000). The aftermath of resource degradation leads to improvement of production through the adoption of modern technology (Brogaard and Li, 2005). Interviews and discussion results indicate that the majority of herds of cows found in the Makere Forest were owned by immigrants who had no permanent residence in the studied villages or in any other village. Accordingly, these livestock keepers built temporary huts which they used only for sleeping. An influx of livestock keepers into the forest reserve was driven by their inherent belief that fodder in the forest reserve was highly nutritious.

Interview and discussion results indicated that maintaining herds of cattle in the Makere Forest entailed cutting trees for building cowsheds and creating resting places. Accordingly, as the number of livestock keepers migrating into the forest kept increasing, this has resulted in ecosystem degradation. The immigrant livestock keepers believed that fodder in the forest reserve was used for therapeutic and curative measures for some livestock disease attacks. The livestock keepers also indulge in cultivating the forest area thereby contributing to greater forest degradation. This concurs with work by Kissinger *et al.* (2012) who asserted that animal husbandry is among the drivers of forest degradation.

Timber and log harvesting

The study observed that selective harvesting of timber and illegal logging occurred in Makere Forest Reserve, and this may be due to demand in nearby regions and countries. The term illegal logging is often used as shorthand to describe illegal practices related to harvesting, processing and trade in timber, including timber products (Tacconi *et al.*, 2003). Illegal logging and the related trade occurs when national or sub-national laws are broken at any point along the supply chain (Tacconi *et al.*, 2003). In this study, not a single respondent across the sample admitted engaging in harvesting timber though discussion results revealed that communities accessed timber through the people who harvested in the Makere Forest Reserve.

Although selective logging was reported as the main cause of forest degradation (Schulze *et al.*, 2008; Matricardi *et al.*, 2010; Pearson *et al.*, 2014), logging with an illegally acquired licence in protected areas, harvesting over the allowed quotas, processing of logs without necessary licences, non-payment of taxes and exporting products without paying export duties all featured in this study. This is also consistent with Kissinger *et al.* (2012) who observed that logging and harvesting timber are among major factors of forest degradation. However, it is noticed from Fig. 12.2 that the number of respondents who stated that they were engaged in harvesting timber and logging were insignificant. As it may be expected, respondents would seldom admit to being timber harvesters/loggers.

Implications of community livelihoods on the ecosystem's integrity

Different forms of forest utilization discussed previously such as agriculture, livestock keeping, fuelwood collection as well as illegal harvesting of timber and logging, had differently impacted the forest cover. Results from GIS clearly depict the pace of deforestation/forest degradation over time from 1985 to 2015. Whereas the closed woodland decreased rapidly from 31.56% in 1985 to only 3.74% in 2015, the cultivated land increased from 0.67% in 1985 to 5.42% in 2015. Likewise, scattered cultivation increased from 1.43% to 26.00%, and open woodland increased from 30.09% to 36.08% during this period. This indicates that the forest has been undergoing degradation over time (Table 12.8 and Fig. 12.4).

As already stated (and shown in Table 12.8), cultivated land (or cropland) increased from 0.67% in 1985 to 5.42% in 2015 while scattered cultivation increased from 1.43% to 26.0% in the same time interval. The increase was partly associated with engagement in shifting cultivation within the forest reserve, which was largely attributed to income generation. Farmers did not mechanize the agricultural practice despite money generated by the sale of the harvested crops. Instead they used hand hoes, and when they needed to increase production, they expanded the farming areas by clearing the forest for new farms. Such a process entailed hiring additional labour (Table 12.9).

Despite some respondents paying for labour, buying a new farm and paying for farm inputs (Table 12.9), the money was also used for things other than modernizing the agricultural activities (e.g. buying household requirements and paying for social services). Buying a new farm partly meant that farmers were sure of the availability of some other land. Such land could eventually be used once the former land was exhausted. Hence shifting cultivation would continue until no more land is available. These findings confirm Boserup's theory which posits that population change drives the intensity of agricultural production (Boserup, 1965). From Table 12.10, it is further deduced that other direct drivers (e.g. livestock keeping, charcoal making, firewood extraction as well as timber harvesting and logging) changed some aspects of land cover. This includes change in closed woodland from 31.56% in 1985 to only 3.74% in 2015. Definition of the term 'agent' is adopted from Tejaswi (2007), referring to actually deforest the land. As such, agents of deforestation refer to individuals, corporations, government agencies or development projects that clear forests. Agents of deforestation and forest degradation discussed in this study are summarized in Table 12.10.

Forest management challenges

Results from interviews and discussions with forest managers revealed several constraints ans shortfalls inherent in the forest governance as discussed in the following subsections.

Inadequate finance for forest administration

Financial constraints led to failure to mark the forest borders. According to interview results, financial constraints caused failure

Table 12.8. Land use/land cover change, Makere Forest Reserve, 1985–2015. (From GIS Laboratory, Institute of Resource Assessment, University of Dar es Salaam, 2016.)

	Proportion of land under a particular use/cover (%) by year			
Land use/cover	1985	1995	2005	2015
Bushland	11.17	10.39	10.17	10.15
Closed woodland	31.56	24.95	13.53	3.74
Cultivated land	0.67	3.53	4.39	5.42
Grassland	16.29	16.29	13.07	9.76
Open woodland	30.09	29.74	36.49	36.08
Scattered cultivation	1.43	6.41	13.14	26.00
Settlement area	0.11	0.12	0.40	0.54
Swamp	8.51	8.48	8.70	8.19
Water	0.16	0.08	0.10	0.12

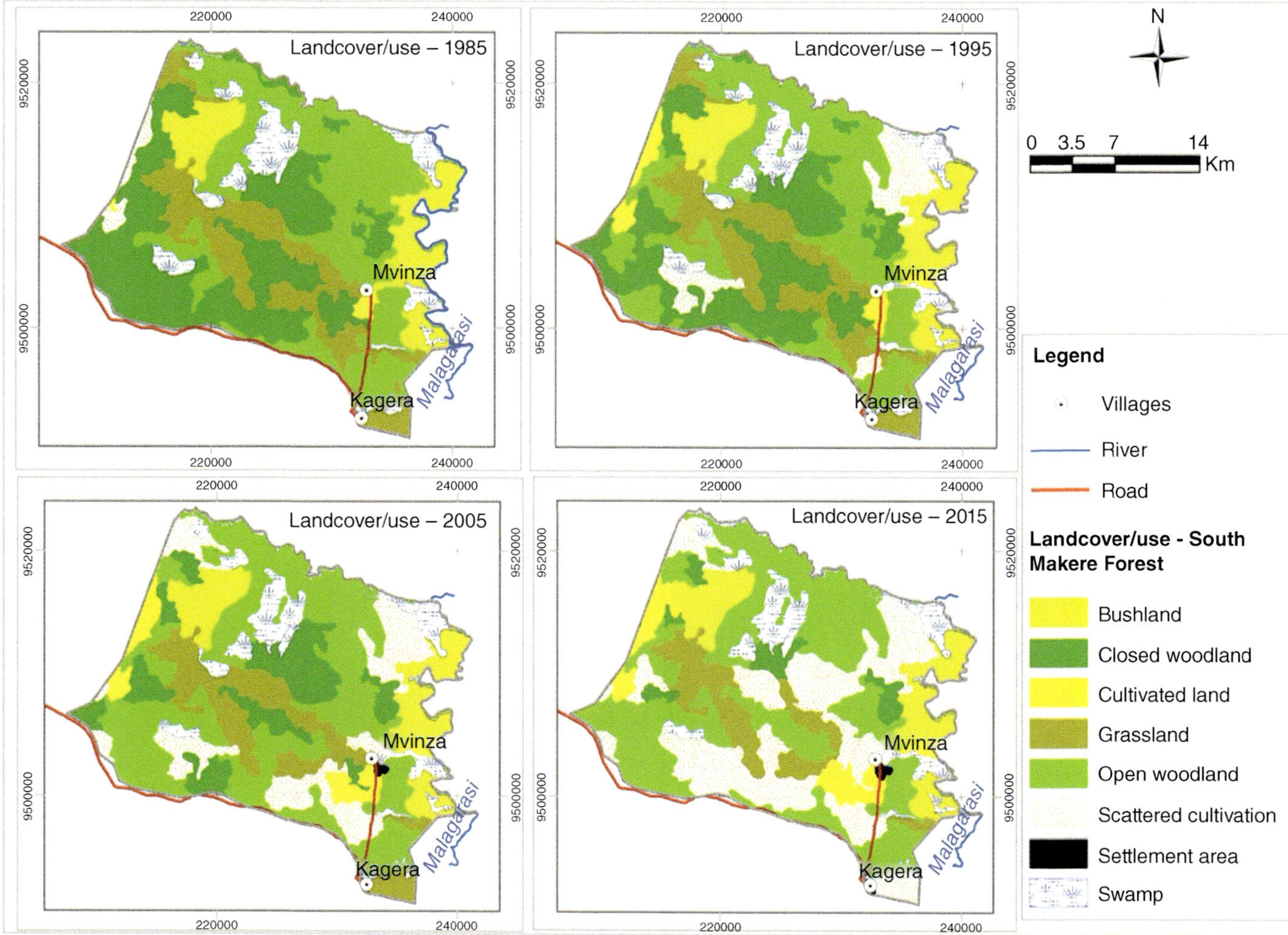

Fig. 12.4. Land use/land cover change in South Makere Forest Reserve, 1985–2015. (From GIS Laboratory, Institute of Resource Assessment, University of Dar es Salaam.)

Table 12.9. Use of money from the sale of crops from harvesting. (From field survey, 2015.)

Use of money generated from the crop sale	Responses	
	Frequency	Percentage
Used for paying labour	25	6.9
Used for buying household requirements	141	39.0
Bought means of transport	6	1.7
Business capital	5	1.4
Bought livestock	2	0.6
Used for house construction	95	26.2
Paying for social services	84	23.2
Bought farm inputs	3	0.8
Purchased a farm	1	0.3
Total	362	100

Table 12.10. Agents of deforestation and their role. (From field data, 2016.)

Agents	Their roles in causing deforestation and degradation
Cultivators	Farm management practices start with cultivating the land and this is followed by tree burning. Such practices kill all generations of trees
Livestock keepers	The dwelling process within the forest is often associated with felling of trees for the construction of homesteads (*boma*) as well as establishment of new informal settlements
Charcoal producers	Felling trees for charcoal production is often unsustainable with even rare and endangered species in jeopardy of extinction
Firewood collectors	Firewood is a source of energy which is accessible to all community members. The intensity of energy demand coupled with increasing populations puts forests at high risk of degradation
Logs and timber harvesting	The timber harvesting process is often selective whereby even endangered species are cleared and removed

to complete the forest reserve boundary identification exercise that would avoid forest encroachment as well as registering the villages in the forest reserve such as Kabhulanzwili, Chokanya and Katito hamlets. According to Kajembe *et al.* (2002) and Blomley and Iddi (2009), the Tanzanian Government's capacity to allocate sufficient budgets for this sort of work has declined progressively.

Institutional factors

Institutional challenges were also reported elsewhere as a reason for prolonged ecosystem degradation (e.g. Coleman, 2009; Rosendal and Andresen, 2011; Prokofieva and Gorriz, 2013). Insufficient financial and technical capacity combined with understaffing had rendered Tanzania Forest Services (TFS) less effective in monitoring the Makere Forest Reserve. Forest patrols were seldom conducted. To ameliorate this challenge, Kasulu District Council collaborated with TFS to manage and monitor the forest. However, these initiatives had not borne positive outcomes with regard to combating illegal forest practices. Moreover, the presence of multiple institutions overseeing management of the same resource poses challenges, particularly with respect of whose decision should be followed. The presence of more than one institution in monitoring the forest may attribute to poor accountability.

Other stakeholders who collaborate with TFS include the Community Environmental and Development Organisation (CEMDO). Recent collaborative efforts with central government encompassed a ban on agricultural activities in the forest. There has also been external support from Belgium Technical Cooperation (BTC) to assist with initiatives that support forest management such as fishing and beekeeping activities coupled with awareness-creation campaigns through workshops, training and forest reserve boundaries identification and marking.

Conclusion

Generally, results from this study indicate that wealth status and education levels have greater influence the improvement of livelihood particularly agriculture. However, their influence depends much on the availability of arable land. More investments by community members are directed towards livelihood extensification as opposed to intensification until no more land is available. This is noticeable for forests which are less monitored, and has serious repercussions on the state of forest ecosystems. Moreover, the distance from the forest to the study villages did not hinder its accessibility, as long as resource benefits were guaranteed. Further, poor forest management due to understaffing, poor resource mobilization, accountability and transparency provide entry points for illegal forest activities to continue. There is a need to promote and improve business conditions, beekeeping, fishing and modern livestock keeping and other alternative livelihood activities which are not directly related to forest utilization. These activities have the potential to increase household food levels and incomes and enhance the integrity of forest ecosystems.

Acknowledgements

For financial support for the research, sincere thanks go to the Building Capacity for REDD+ in East Africa for improved ecosystem health and for sustainable livelihoods in Eastern Africa (REDD-EA) project funded by NORAD through the Norwegian Programme for Capacity Development in Higher Education and Research for Development (NORHED). The project is implemented by Makerere University, Uganda, the Norwegian University of Life Sciences, Norway and the University of Dar es Salaam through the Centre for Climate Change Studies. Moreover, the authors extend their heartfelt recognition and thanks to the district officers, community members and respondents in the study villages. The invaluable information given to us was instrumental in accomplishing this task. We are not able to mention all those whose assistance helped us accomplish our goals but we really appreciate their support.

Note

[1] Under probability sampling, all people have an equal chance of being selected to form a sample, whereas non-probability sampling entails deliberate selection by a researcher (Kothari, 2004).

References

Abdallah, J.M. (2014) Overview of Miombo Woodlands in Tanzania, (January 2007).

Abdallah, J.M. and Monela, G.G. (2007) Overview of Miombo Woodlands in Tanzania. *Working papers of the Finish Forest Research Institute* 50: 9–23.

Agrawal, A., Cashore, B., Hardin, R., Shepherd, G., Benson, C. and Miller, D. (2013) Economic contributions of forests. Background paper prepared for the United Nations Forum on Forests. Tenth Session 8–19 April 2013, Istanbul, pp. 1-127.

Almalki, S. (2017) Integrating quantitative and qualitative data in mixed methods research – challenges and benefits. *Journal of Education and Learning* 26(3), 288–297. https://doi.org/10.5539/jel.v5n3p288

Belcher, B. (2005) Forest product markets, forests and poverty reduction. *International Forestry Review* 7(2), 82–89. https://doi.org/10.1505/ifor.2005.7.2.82

Blomley, T. (2006) Mainstreaming participatory forestry within the local government reform process. *Gatekeeper Series* 128. International Institute for Environment and Development, London.

Blomley, T. and Iddi, S. (2009) Participatory forest management in Tanzania: 1993- 2009. Lessons learned and experiences to date. Ministry of Natural Resources and Tourism, Dar es Salam, Tanzania.

Blomley, T., Pfliegner, K., Isango, J., Zahabu, E., Ahrends, A. and Burgess, N. (2008) Seeing the wood for the trees: an assessment of the impact of participatory forest management on forest condition in Tanzania. *Oryx* 42(3), 380–391. https://doi.org/10.1017/S0030605308071433

Boserup, E. (1965) *The Conditions of Agricultural Growth*. George Allen & Unwin, London.

Boyce, C. and Neale, P. (2006) Conducting in-depth interviews: a guide for designing and conducting in-depth interviews. *Evaluation* 2(May), 1–16. https://doi.org/10.1080/14616730210154225

Brogaard, S. and Li, X. (2005) Agricultural performance on marginal land in Eastern Inner Mongolia, China – development in the pre- and post-1978 reform periods. *GeoJournal* 64(3), 163–175. https://doi.org/10.1007/s10708-005-5645-z

Castrein, T. and Pillai, M. (2011) *Forest Govern-nanance 2.0. A Primer on ICTs and Governance*. Program on Forests (PROFOR), Washington, DC.

Coleman, E.A. (2009) Institutional factors affecting biophysical outcomes in forest management. *Journal of Policy Analysis and Management* 28(1), 122–146. https://doi.org/10.1002/pam.20405

Convention on Biological Diversity (CBD) (2009) Sustainable Forest Management, Biodiversity and Livelihoods: a Good Practice Guide. Available at: http://www.cbd.int/development/doc/cbd-good-practice-guide-forestry-booklet-web-en.pdf (accessed 16 December 2019).

Cowie, A.L., Orr, B.J., Castillo Sanchez, V.M., Chasek, P., Crossman, N.D., *et al.* (2018) Land in balance: the scientific conceptual framework for Land Degradation Neutrality. *Environmental Science and Policy* 79, 25–35. https://doi.org/10.1016/j.envsci.2017.10.011

Davies, P. and Johnson, J. (1995) Rural development forestry network. Spring. https://doi.org/10.1177/030913339802200306

Feulefack, J.F., Zeller, M. and Schwarze, S. (2006) *Accuracy of Participatory Wealth Ranking*. University of Goettingen, Germany.

Food and Agriculture Organization of the United Nations (FAO) (2006) *Better Forestry, Less Poverty. A Practitioner's Guide*. FAO, Rome.

Harrison, P. (2006) *Socio-economic Study of Forest-adjacent Communities from Nyanganje Forest to Udzungwa Scarp: a Potential Wildlife Corridor*. Incorporating Livelihood Assessments and Options for Future Management of Udzungwa Forests. World Wide Fund for Nature (WWF)-Tanzania Programme Office, Dar es Salaam, Tanzania.

Hosonuma, N., Herold, M., De Sy, V., De Fries, R.S., Brockhaus, M., *et al.* (2012) An assessment of deforestation and forest degradation drivers in developing countries. *Environmental Research Letters* 7(4). https://doi.org/10.1088/1748-9326/7/4/044009

Indrarto, G.B., Murharjanti, P., Khatarina, J., Pulungan, I., Ivalerina, F., *et al.* (2012) The context of REDD+ in Indonesia: drivers, agents and institutions. *CIFOR Occasional Paper* 116. https://doi.org/10.1111/apv.12087

Jumbe, C.B.L., Bwalya, S.M. and Husselman, M. (2009) Contribution of dry forests to rural livelihoods and the national economy in Zambia. Paper presented at the XIII World Forestry Congress, 18–23 October, Buenos Aires, Argentina.

Kahyarara, G., Mbowe, W. and Kimweri, O. (2002) Poverty and deforestation around the gazetted forests of the coastal belt of Tanzania. *Research Report – Research on Poverty Alleviation* 02.3. REPOA, Dar es Salaam, Tanzania. Available at: https://opendocs.ids.ac.uk/opendocs/handle/20.500.12413/1799 (accessed 16 December 2019).

Kajembe, G.C., Monela, G.C. and Mvena, Z.S.K. (2002) Making community-based forest management work: a case study from Duru-Haitemba village forest reserve, Babati, Arusha, the United Republic of Tanzania. Paper presented at the Second International Workshop on Participatory Forestry in Africa. Defining the Way Forward: Sustainable Livelihoods and Sustainable Forest Management through Participatory Forestry, 18–22 May 2002, Arusha, Tanzania.

Kissinger, G., Herold, M. and De Sy, V. (2012) Drivers of deforestation and forest degradation: a synthesis report for REDD+ policymakers. Government of the UK and Norway, London.

Kok, M. and Lu, M. (2016) A new method for analysing socio-ecological patterns of vulnerability. *Regional Environmental Change* 16(1), 229–243. https://doi.org/10.1007/s10113-014-0746-1

Kothari, C.R. (2004) *Research Methodology: Methods and Techniques*, 2nd edn. New Age International Publishers, New Delhi.

Kumar, K. (1989) Conducting Key Informant Interviews in Developing Countries. AID Program Design and Evaluation Methodology Report No. 13. United States Agency for International Development, Washington, DC, 1–33.

Matricardi, E.A.T., Skole, D.L., Pedlowski, M.A., Chomentowski, W. and Fernandes, L.C. (2010) Assessment of tropical forest degradation by selective logging and fire using Landsat imagery. *Remote Sensing of Environment* 114(5), 1117–1129. https://doi.org/10.1016/j.rse.2010.01.001

Murty, T.S. (2008) Forests source book practical guidance for sustaining forests in development cooperation World Bank–WWF alliance for forest conservation and sustainable use. *Natural Hazards* 48(2), 315–316. https://doi.org/10.1007/s11069-008-9278-7

Nacker, R.M. (2004) *Evaluation of Forest Products Industry Cluster in Wisconsin and Recommendations for Economic Development Actions*. Wisconsin Economic Development Institute, Madison, Wisconsin.

National Bureau of Statistics (2013) *2013 Population and Housing Census; Population Distribution by Adminstrative Areas*. National Bureau of Statistics, Dar es Salaam, Tanzania.

Ostrom, E. (2000) Collective action and the evolution of social norms. *Journal of Economic Perspectives* 14(3), 137–158. https://doi.org/10.1257/jep.14.3.137

Ostrom, E., Burger, J., Field, C.B., Norgaard, R.B. and Policansky, D. (1999) Revisiting the commons: local lessons, global challenges. *Science*, New Series, 284(5412), 278–282.

Paavola, J. (2004) Livelihoods, vulnerability and adaptation to climate change in Morogoro, Tanzania. *Environmental Science & Policy* 11(7), 642–654. https://doi.org/10.1016/j.envsci.2008.06.002

Pearson, T.R.H., Brown, S. and Casarim, F.M. (2014) Carbon emissions from tropical forest degradation caused by logging. *Environmental Research Letters* 9(3). https://doi.org/10.1088/1748-9326/9/3/034017

Phiri, D., Franzel, S., Mafongoya, P., Jere, I., Katanga, R. and Phiri, S. (2004) Who is using the new technology? The association of wealth status and gender with the planting of improved tree fallows in Eastern Province, Zambia. *Agricultural Systems* 79(2), 131–144. https://doi.org/10.1016/S0308-521X(03)00055-6

Prokofieva, I. and Gorriz, E. (2013) Institutional analysis of incentives for the provision of forest goods and services: an assessment of incentive schemes in Catalonia (north-east Spain). *Forest Policy and Economics* 37, 104–114. https://doi.org/10.1016/j.forpol.2013.09.005

Rosendal, G.K. and Andresen, S. (2011) Institutional design for improved forest governance through REDD: lessons from the global environment facility. *Ecological Economics* 70(11), 1908–1915. https://doi.org/10.1016/j.ecolecon.2011.04.001

Schmidt, S., Altanchimeg, C., Tungalagtuya, K., Narangerel, Y., Ganchimeg, D., *et al.* (2006) Depleting natural wealth – perpetuating poverty: rural livelihoods and access to forest resources in Mongolia. Food and Agriculture Organizaiton of the United Nations (FAO) Livelihood Support Programme. Available at: http://www.fao.org/forestry/19630-0251966a898321c31cf2193325df2762e.pdf (accessed 6 December 2019).

Schreckenberg, K. (1998) Rural development forestry. *Progress in Physical Geography* 22(3), 389–397. https://doi.org/10.1177/030913339802200306

Schulze, M., Grogan, J., Uhl, C., Lentini, M. and Vidal, E. (2008) Evaluating ipê (Tabebuia, Bignoniaceae) logging in Amazonia: sustainable management or catalyst for forest degradation? *Biological Conservation* 141(8), 2071–2085. https://doi.org/10.1016/j.biocon.2008.06.003

Scoones, I. (2017) Investigating difference : applications of wealth ranking and household survey approaches among farming households in Southern Zimbabwe. *Development and Change* 26(1), 67–88. https://doi.org/10.1111/j.1467-7660.1995.tb00543.x

Shepherd, G. and Kazoora, C. with Mueller, D. (2013) Forests, livelihoods and poverty alleviation : the case of Uganda. *Forestry Policy and Institutions Working Paper* No. 32. Food and Agriculture Organization of the United Nations (FAO), Rome.

Suich, H., Howe, C. and Mace, G. (2015) Ecosystem services and poverty alleviation: a review of the empirical links. *Ecosystem Services* 12, 137–147. https://doi.org/10.1016/j.ecoser.2015.02.005

Sunderlin, W.D., Belcher, B., Santoso, L., Angelsen, A., Burgers, P., *et al.* (2005) Livelihoods, forests, and conservation in developing countries: an overview. *World Development* 33(9 Special issue), 1383–1402. https://doi.org/10.1016/j.worlddev.2004.10.004

Sunderlin, W.D., Dewi, S. and Puntodewo, A. (2007) *Poverty and Forests: Multi-country Analysis of Spatial Association and Proposed Policy Solutions*. Centre for International Forestry Research, Bogor, Indonesia.

Tacconi, L., Boscolo, M. and Brack, D. (2003) National and international policies to control illegal forest activities. Report for the Ministry of Foreign Affairs, Government of Japan. CIFOR, Bogor, Indonesia. Available at: https://www.researchgate.net/publication/230676885_National_and_International_Policies_to_Control_Illegal_Forest_Activities (accessed 17 December 2019).

Tejaswi, G. (2007) *Manual on Deforestation, Degradation, and Fragmentation Using Remote Sensing and GIS*. Forestry Department, Food and Agriculture Organization of the United Nations (FAO), Rome.

Tynan, C.A. and Drayton, J.L. (1988) Conducting focus groups – a guide for first-time users. *Marketing Intelligence & Planning* 6(1), 5–9. https://doi.org/10.1108/eb045757

United Republic of Tanzania (URT) (2008) *National Forest Policy*. Ministry of Natural Resources and Tourism, Dar-es-Salaam, Tanzania.

URT (2013) National Strategy for Reduced Emissions from Deforestation and Forest Degradation (REDD+). Vice President's Office, Dar-es-Salaam, Tanzania.d

URT (2015) *National Forest Resources Monitoring and Assessment of Tanzania Mainland (NAFORMA): Main Results*. Ministry of Natural Resources and Tourism, Dar-es-Salaam, Tanzania

Part III

Knowledge Systems and Climate Change

13 Weather Forecasting and Communication in the Upper Great Ruaha Catchment Area

Anselm R. Mwajombe[1]* and Godwin A. Lema[2]

[1]*Centre for Climate Change Studies, University of Dar es Salaam, Dar es Salaam, Tanzania;* [2]*College of Social Sciences, University of Dar es Salaam, Dar es Salaam, Tanzania*

Abstract

Effective weather forecast dissemination depends on how effective dissemination channels are in informing decision making for improved management of water resources and livelihood activities, which depend on water resources in catchment areas. In this chapter, the effectiveness of the channels for weather forecast dissemination is assessed in terms of magnitude of awareness creation and versatility to end users. Our findings show that both traditional and conventional channels of weather forecasting and communication can be used to create awareness to end users in various parts of the country. For local communities, traditional weather forecasting and communicating were contingent on indigenous knowledge acquired through interaction with the local environment. Such information was accessed through indicators or signs that entail plant phenology, astronomical and meteorological events as well as mammals' behaviour. Conventional forecasting is communicated via modern communication technologies including radio, television, the Internet and posted letters. Communication of traditional weather forecasting is mainly through oral traditions. Results from our respondents revealed that 40% received weather forecasts through traditional channels, 11% through modern channels and 49% through modern and traditional channels. The majority of respondents said that weather forecasts from modern sources were not reliable to inform the decision-making process when compared with traditional sources. The study recommends synchronizing modern and traditional channels for effective weather forecast delivery.

Introduction

Background

Communicating climate information, particularly weather forecasts, to local communities around catchments is important for improving the management and infrastructure of water resources and farming practices, with a view to increase agricultural productivity, and developing an outreach programme to improve the quality of water management (Munang *et al.*, 2010). Countries with effective weather forecast services greatly contribute to reducing the risks and maximizing the opportunities associated with climate (Blench, 1999; Amissah-Arthur, 2003; WMO, 2006).

Forecasts originate from traditional and conventional sources. Traditional forecasts are rooted in indigenous knowledge acquired from interaction with the local environment. Whereas traditional forecasts are mainly used to predict seasonal weather patterns (Kijazi *et al.*, 2013), conventional weather forecasts are of three types and levels, namely, present,

*mwajombea@gmail.com

seasonal and climate change at deterministic (1–7 days), probabilistic (1–3 months) and projections (scenarios) (10–500 years) degrees of certainties, respectively (Ziervogel *et al.*, 2010; Gubbels, 2013).

Weather forecasts from traditional sources are mainly communicated through oral traditions. Such a form of communication has for years enabled the transmission of knowledge that makes use of verbal messages from past generations (Madzingira, 2001; Muguti and Maposa, 2012; Kijazi *et al.*, 2013). Weather forecasts from conventional sources are communicated through information communication technology (ICT) via writing, print, radio, telephones, telegraph, photography, film, disk, tape recording, television, radio, computers and mobile phone.

There have been claims that there are mismatches between the weather forecasts generated by meteorological services and those needed by smallholder farmers and other users (Blench, 1999). The mismatches may be due to shortcomings inherent in the reliability of the forecast that depends on three basic aspects: (i) consistency; (ii) quality; and (iii) value of the information. Whereas consistency is contingent on the forecasters' judgements and their forecasts, the quality depends on correspondence between forecasts and the matching observations (Murphy, 1993). To decision makers the value of the forecast depends on incremental economic and/or other benefits realized through the use of the forecasts (Murphy, 1993; Thornes *et al.*, 2001; Tall *et al.*, 2012).

Challenges of communicating climate information at regional and catchment scale

The quality of the forecast from conventional sources is contingent on the abilities to downscale the Atmosphere Ocean General Circulation Models (AOGCMs). It is argued that current models are still of the order of a few hundred kilometres. Such a spatial scale is too large to capture the effects of local forces such as complex topography and land surface characteristics. As such, coarse resolution prevents global models from providing accurate descriptions of the extreme events which are of fundamental importance to users of weather forecasts with respect to regional and local impacts of climate variability (Giorgi *et al.*, 2009).

Intra-basin communication of climate information at the local level makes use of education campaigns, public awareness campaigns as well as interpersonal communications (e.g. TMA, 2008; GWP and INBO, 2009). On the one hand public campaigns use newspapers, television, radio, printed materials, state of the basin reports, basin scorecards and the Internet. On the other hand, workshops, meetings, websites and communication tools such as libraries, mobile phone services and community-based information services are also used (GWP and INBO, 2009; Gubbels, 2013). However, digital divides between rural and urban areas as well as socio-economic differences limit the abilities to access climate information. Against this background, forging a relationship between communicating climate information and its application in agriculture and water resources management in the Great Ruaha Catchment Area (GRCA) is necessary because of the importance of the catchment area to Tanzania's economy.

An overview of the GRCA

The availability of water resources and soil fertility in the GRCA has attracted intensification of economic activities such as small-scale and large-scale agriculture which, to a great extent, depend on water from rivers for irrigation. Also, the richness in biodiversity of the GRCA attracts tourism, and waters from the Great Ruaha River are used for generation of hydroelectric power (WWF-TCO, 2010). The catchment experiences a semi-arid climate with total annual rainfall ranging from 450 mm to 750 mm (Malley, 2011; Mwakalila, 2011). In addition, the livelihood strategies are supported by rivers which pass through the catchment area such as the Kimani, Chimala and the Great Ruaha. These rivers have given rise to increased competing needs for water resources due to diversification of human activities such as agriculture, livestock keeping and generation of hydroelectric power at Mtera and Kidatu dams. Thus effective water resources management practices are important in order to increase efficiency and productivity.

As explained in the paragraph above, the GRCA is important for the economic development of the country and livelihoods of households in and around the catchment, and the rich biodiversity means it is important for tourism and a wide range of other economic activities such as irrigation agriculture. However, the catchment has been impacted by climate change. In addition, land use/land cover changes have impacted hydrological regimes and fresh water resources by declining volume and runoff. Also, resources and livelihoods depend on reliable availability of water which depends on rainfall. Since climate change has altered the rainfall patterns, the use of climate information becomes an important intervention for these climate dynamics.

A wide range of studies pertaining to weather forecasts conducted in GRCA hardly focus on communicating and using climate information in water resources management. Devisscher *et al.* (2010) proposed a water resources governance system in the basin which incorporates the use of climate information. Although the GRCA receives weather forecasts from Tanzania Meteorological Agency (TMA), the modes under which they are sent and received by wider communities at the local level are questionable. It is also not clear whether the weather forecasts already in place reflect the users' needs. Furthermore, it is less clear how effective the channels of communicating climate information from TMA are among information users at the local level. It is on these grounds that this chapter was constituted. The general objective was to examine the channels for communicating weather forecasts and the inherent awareness of the wider community in the Upper GRCA to such forecasts. Findings from the chapter add to the body of knowledge on climate services, dissemination and delivery. They may also be useful for practitioners in water resources management, the agriculture sector, local communities and the meteorological services department to improve forecast dissemination and delivery.

Conceptual framework

The conceptual framework which was adopted in this study (Fig. 13.1) was modified from the Global Framework for Climate Services (GFCS) developed by the World Meteorological Organization (WMO, 2009). The original GFCS version was intended for use at global, regional and national levels and it has the following components: (i) observations and monitoring; (ii) research, modelling and prediction; (iii) climate services information systems; and (iv) user interface and users in government, the private sector, research, agriculture, water, health, construction, disaster reduction, environment, tourism and transport.

The framework has five components: (i) sources of climate information both traditional (using indigenous knowledge[1]) and modern sources (TMA); (ii) channels for communicating climate information; (iii) awareness of actors in water resources management; (iv) application or non-application of such forecasts; and (v) the associated positive or negative impacts of application/non-application, respectively.

Channels of communicating climate information include public awareness programmes and education campaigns (GWP and INBO, 2009). Public awareness campaigns include the use of mass media such as newspapers, television, radio, printed materials, state of the basin reports and basin scorecards (GWP and INBO, 2009; Ziervogel *et al.*, 2010). Education campaigns include workshops, meetings and communication

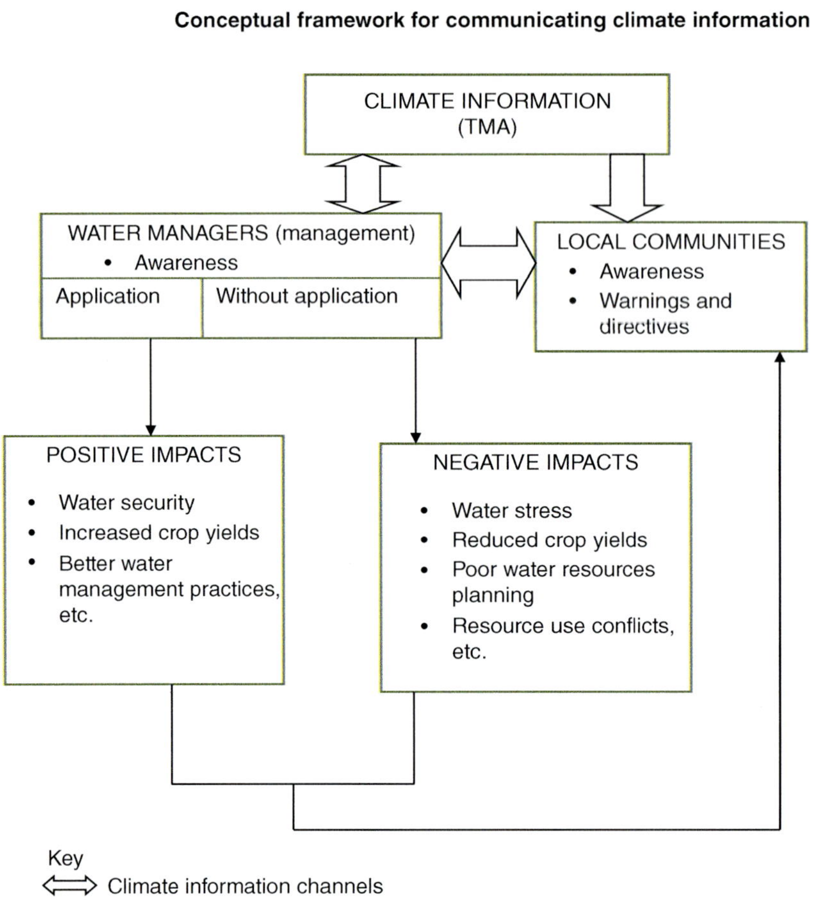

Fig. 13.1. Sources and impacts of communicating climate information. (Modified from WMO, 2009.)

tools such as websites, libraries and district forums (GWP and INBO, 2009). Interpersonal communication is also often used (Madzingira, 2001).

Awareness of actors in water resources management includes knowledge and understanding of water resource managers and users at the local or community level. This awareness is enhanced by exposure to educational programmes and the accessibility of various communication tools such as rural radio, television broadcasts, printed materials, mobile phone text messages and the Internet (Gubbels, 2013). Socio-economic status determines the kinds of communication tools users can access.

Another component is the application of climate information. This depends on the extent the water resources managers and other users understand the information (Rayner *et al.*, 2005; Ziervogel *et al.*, 2010). Application of climate information will have positive impacts on a wide range of water resource operations (Rayner *et al.*, 2005) and vice versa. On the one hand, positive impacts manifest themselves through good water services delivery (Ziervogel *et al.*, 2010), increased crop yields (Podesta *et al.*, 2001) and alleviation of resource-use related conflicts. On the other hand, negative impacts of poor water resource management practices include resource use conflicts, water stress and decreased crop yields. These dynamics have repercussions on livelihoods at the local communities' level.

Material and Methods

Description of the study area

The study was conducted at Chimala and Mengele villages which are found in the Mkoji sub-catchment of the Upper GRCA in Mbarali District (Fig. 13.2). The area was chosen purposively because it has been experiencing a serious decrease in rainfall and prolonged drought (Rayner *et al.*, 2005; Lusuva, 2009). Such climate variability is magnified by the land use changes that impacts the hydrological regime of rivers such as the Great Ruaha, Kimani, Mbarali, Chimala and Ndembera (WWF-TCO, 2010).

Climatologically, the Upper GRCA is semi-arid and its climate is influenced by the Intertropical Convergence Zone whereby rainfall is highly seasonal with a single rainy season from November to April. The area is characterized by high intensity rainfall events accompanied by thunderstorms (WWF-TCO, 2010). Highland areas receive between 1000 mm and 2000 mm of rain annually, whereas the plains receive around 700–800 mm. The climate of the study area is reflected in its hydrology, with rivers showing extremely peak-flow patterns that distinguish the dry and wet season. The Great Ruaha River downstream of Usangu Wetland receives most of its annual runoff over about 4 months of the year (WWF-TCO, 2010).

Data collection methods

The study employed a descriptive and explanatory approach in order to generate a wide range of information. Purposive sampling was used to select the study villages, focus group discussants and key informants. Random sampling was used to select 100 households for the questionnaire survey from a total of 1662 respondents, equivalent to 6.0%. Selection of key informants was done from households and governmental institutions, particularly Rufiji Basin Water Office (RBWO), Mbarali District Council (MDC) and TMA. Primary quantitative data were collected using semi-structured questionnaires whereas qualitative data was collected using key informant interviews and focus group discussions (FGDs). A literature review was used to collect secondary data.

The focus groups in the study consisted of at least 25% women, as suggested by the village executive officers. Such a rate was consistent with the gender-aspect threshold in the village committee that requires at least 25% women.[2] Four FGDs were conducted, with respondents ranging from four to ten, two groups from each village. According to Saunders *et al.* (2007), a typical FGD involves four to 12 participants depending on the interviewer's skill and subject matter.

A total of 20 key informants were interviewed. Key informant interviews[3] included five Mbarali District Council officials from agriculture and water departments, three Rufiji Basin Water Office officials, nine Water User Committee leaders and three TMA officials. According to Kumar (1989), key informant numbers ranging from 15 to 25 are sufficient to represent the study. To complement in-depth interviews, semi-structured questionnaires were prepared and administered face to face.[4] They were used to obtain responses from household heads and any person aged 18 years and above when the household head was absent.

A total of 100 questionnaires were administered, and the open-ended questions enabled a researcher to capture detailed and varied responses. Efforts were made to ensure that all aspects of the specific objectives of this study were captured. In cases where some aspects were not captured by these questionnaires, information was collected using other tools. Secondary data was obtained through documentary review from both published and unpublished (grey) literature. The Internet was also a very useful resource for this study.

The data collected was cleaned, coded, entered and appropriately analysed. Quantitative data from questionnaires were entered into Statistical Package for Social Sciences (SPSS) version 16.0 for descriptive statistics whereas Microsoft Excel was used to generate figures. Qualitative data was analysed following a thematic analysis

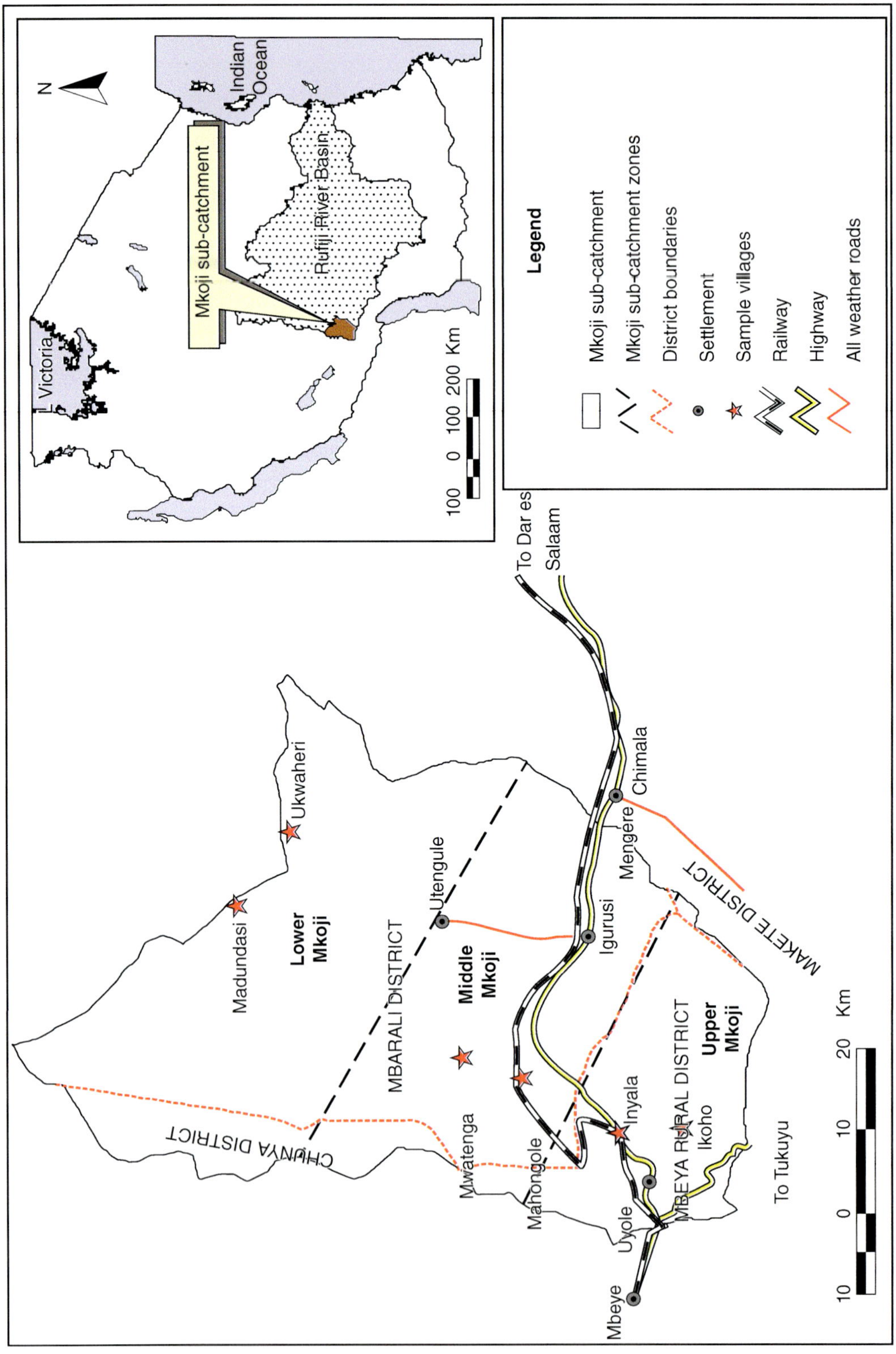

Fig. 13.2. Upper Great Ruaha Catchment Area (GRCA), Mbarali. (Adapted from Lusava, 2009.)

framework as well as narrative analysis.[5] The qualitative data was used to cross-check the quantitative information for novelty of findings and outcomes of the study.

Results and Discussion

Communicating weather forecasts from traditional sources in practice

Respondents were asked about the channels through which they are informed about the upcoming weather. About 40% of the weather forecast sources were from traditional forecasters, which were mainly communicated to them through interpersonal means (using word of mouth). Interpersonal communication was in the form of one to one or from one to many. Forecasting of weather by traditional forecasters was mainly through religious rituals, and dissemination was through the oral tradition. Following forecasting, the climate information so generated was conveyed to people in traditional forecasters' cycles face to face, who subsequently disseminated them to their neighbourhood. Thus, weather forecasts were disseminated to many people within and outside the village.

Apart from traditional forecasters, traditional ecological knowledge (TEK)[6] was used by individuals to forecast weather using indicators/signs. Traditional weather indicators were abundantly available within the local environment. Furthermore, TEK had contributed significantly to the provision of weather forecasts to farmers as end users of climate information. The information is useful to inform farming households on appropriate timing for farm preparations prior to the onset of rainfall. Traditional weather forecasts are based on the observation of mainly three classes of indicators, namely, plant phenology, animal behavioural changes within their habitat and astronomical and meteorological events.

TEK has been used from time immemorial and is transmitted from one generation to another through narrations.[7] In spite of the presence of traditional forecasters in the study area, local people were able to use indigenous knowledge to predict the beginning and ending of the rainfall onset on their own accord.[8] They acknowledged traditional sources as accurate and it was mainly used to predict the beginning and ending of rainfall onset. They also indicated that intensity cannot be predicted by use of TEK. Other aspects of weather such as temperature, wind and cloud cover are usually used as rainfall indicators, but cannot be predicted by traditional means.

Moreover, 49% of the respondents received weather forecasts from traditional forecasters via interpersonal means. As such, they also reported their ability to use TEK in predicting the beginning of rainfall onset based on local indicators. Appearance of large swarms of red, white and black ants from September to November, also the occurrence of large swarms of flying ants (*kumbikumbi*) and *Spodoptera exempta* (*upanga uhavi*) indicated that the rainy season was near. These results are similar to Chang'a *et al.*'s (2010) findings whereby black and white ants were observed moving together to the trees. The appearance of some reptiles such as the *Ndumilakuwili*, which has two heads, is believed to signify the onset of the rainy season.

More often, they also obtained weather forecasts from modern environmental knowledge or sources as shown in Table 13.1 and channelled through conventional means, as discussed in the next section. The mixing of the two approaches was mainly due to being skeptical about the accuracy and reliability of weather forecasts from either source.[9]

Another important weather indicator is the behaviour of birds. It was found during this study that changing types of singing for birds like the caucal (*Centropus cauca*) and the *Ngolikyaka* is indicative of weather conditions. For example, when the *Ngolikyaka* sings only once during a dry season, it means that rains are near; on the other hand, when it sings more frequently and randomly, it is an indicator of the probability of drought. When the caucal sing in pairs by alternation responding to each other it means the rainy season is near. But when only one bird sings without a response from the others, it means the dry season could possibly be prolonged.

Table 13.1. List of documented indicators and their application in rainfall forecasts. (From field data, 2014.)

Indicators	Swahili/local name	English name	Scientific name	Signs for rainfall forecast
Plant phenology	*Mipogoro*		*Acacia* spp.	When their leaves sprout indicates the approach of the rainy season
	Miembe	Mango tree	*Mangifera indica*	Significant flowering of mango trees indicate a potential drought season
	Mpombep-ombe	Marula tree	*Sclerocarya birrea*	Leaves become greener and sprouting signifies closure of rainy seasons
	Mitono	Acacia	*Acacia* spp.	Used for rainfall prediction where they sprout during dry conditions and do not grow during the rainy seasons
	Mkuyu	Mulberry	*Morus* spp.	When their leaves sprout indicates the approach of the rainy season
	Mzambarau pori	Wild plum tree	*Prunus americana*	When their leaves sprout indicates the approach of the rainy season
	Mkrismasi	Christmas tree	*Pinus* spp.	When their leaves sprout indicates the approach of the rainy season
	Mtewele			When their leaves sprout and flowering occurs indicates the approach of the rainy season
	Mibuyu	Baobab tree	*Adansonia* spp.	Their flowering signifies the onset of upcoming rainfall
Birds	*Dudumizi*	Caucal	*Centropus* spp.	Its singing, especially in the early hours of the morning, in October and November signifies the onset of imminent rainfall and as such if such songs progress over long periods then the rains are found to be profound
		Sun birds	*Cinnyris* spp.	Their singing signifies the early onset of rainfall
Insects	*Mapanga uhavi*	African army worms	*Spodoptera exempta*	Their presence signifies closure of the rainy season or more precisely its onset
	Filetandonya			Locally the presence of such insects signifies upcoming rainfall and as such their local name means rain bringers
Amphibians and reptiles	*Chura*	Toads	*Bufo* spp.	They emerge out from muddy soil, signifying the onset of the rainy season
	Ndumilakuwili			The two-headed reptile is believed to signify the onset of the rainy season

Table 13.2 shows local indicators of rainfall based on astronomical and meteorological events/signs.

Communicating weather forecasts from modern sources

Table 13.3 shows that 11% of the respondents interviewed used modern channels to obtain weather forecasts from conventional sources.

Generation of climate information using modern sources was evident at national and sub-national levels of government institutions.[10] At the national level, TMA generates daily weather forecasts, prospects/weather forecast for every 10 days/monthly/seasonal and early warnings of severe weather events. At the sub-national level, despite receiving climate information from TMA, government institutions also generate rainfall-related information. In particular, RBWO receives climate information from

Table 13.2. Local indicators of rainfall based on astronomical and meteorological events/signs. (From field data, 2014.)

Indicator	Description	Signifies
Stars	Constellation of stars which are locally known as *Ndimila*	Onset of the rainy season
Cloud	Formation of thick dark cloud cover locally known as *Nyamande*	Heavy rainfall, and as such the time to start preparation on farms for water storage
Wind	Strong wind during the months of September–October; heat/hot in mountainous land areas and shedding of leaves by trees	Onset of rainy seasons
Temperature	The rise of temperature at the end of July–August	Early rain fall
	If temperature rise is late (end of September–November)	Rain will be small in amount and will be late

Table 13.3. Channels of communicating weather forecasts. (From field data, 2014.)

Channel of weather forecast	Frequency	Percentage
Traditional channels	40	40
Modern (conventional) channels	11	11
Traditional and modern	49	49
Total	100	100

TMA through the Ministry of Water and also generates rainfall-related information. Actors at these government institutions do not send queries to TMA as a feedback to climate information sent to them. Table 13.4 shows institutions responsible for generating climate and meteorological data and information at the national and sub-national levels, types of climate/meteorological services generated and the sources of information.

Regarding types of weather forecasts generated from modern sources we observed that climate information generated by TMA includes weather forecasts, early warnings and advisories. Weather forecasts exist in two forms: (i) daily/weekly forecasts; and (ii) seasonal forecasts. Table 13.5 shows the types of weather forecasts and parameters that are often tested. The information is generated from meteorological stations, radar observations, satellite pictures and climate models.

We also investigated climate information delivery pathways from modern sources. The results show that after generation at TMA, forecasts are issued via e-mails (to registered users), mobile phone text messages (to registered users), newspapers, the TMA website and blog. Weather forecast dissemination also makes use of mass media, central government ministries, government parastatals and agencies through stakeholder meetings. Madzingira (2001) noted that modern forms of communication make use of writing, print, radio, telephones, telegraph, photography, film, television and computers. The local government system used posted letters and e-mails. Through interviews it was established that the content of the information is normally the same; however, analyses are performed by individual institutions to suit the needs of their respective sectors. Besides, individual users can request climate information relevant to their sectors from TMA. The seasonal forecasts are associated with advice whereas daily/weekly forecasts are not.

The outlook reports of rainfall patterns for autumn (October–December 2014) and the long rains (March–May 2015) provide important extracts from interviews with TMA officicals about the long rains. The same information was also posted on the TMA website (http://www.meteo.go.tz/).

In the local government system, the Prime Minister's Office Disaster Management Department (PMO-DMD) and Ministry of Agriculture, Food Security and Cooperatives (MAFSC) receive climate information from TMA and it is taken through local government's authorities via the Regional Commissioner's office where it is reanalysed. Then it

Table 13.4. Institutions responsible for generating climate information in the study area. (From field data, 2014.)

Institutions	Type of climate/ meteorological services	Sources of information[a]
Tanzania Meteorological Agency (TMA)	Climate information (trends, models, extremes, inter-annual variability, higher order statistics)	Own-generated data (weather stations)
	Weather forecasts	Global and regional data (WMO, IPCC, UNFCCC secretariat, EAC)
	Early warnings	Satellite data
	Advisories	Community data (feedback on seasonal rainfall performance and research)
Rufiji Basin Water Office (RBWO)	Rainfall data	Own-generated data (installed automatic and manual weather stations)
	Hydrological data	Hydrological survey

[a]EAC, East African Community; IPCC, Intergovernmental Panel on Climate Change; UNFCCC, United Nations Framework Convention on Climate Change; WMO, World Meteorological Organization.

Table 13.5. Types of weather forecasts generated. (From field data, 2015.)

Type of information	Weather parameters tested
Seasonal forecasts	Sea surface temperature (SST)
	Wind direction
Daily/weekly weather forecasts	Atmospheric pressure
	Wind
	Humidity
	Cloud cover
	Temperature
Early warnings	Extremes of the above weather parameters

is channelled down to district-level authorities via posted letters where it is reanalysed further by the Agriculture Department. It is then distributed to users at the local level in the form of advice from experts (i.e. extension services by extension officers).

The outlook could informally be translated as follows:

> The Regional Commissioner's office has received a report on rainfall patterns for the period from March to May 2015 from Tanzania Meteorological Agency. The report shows that there will be a shortage of rains in Morogoro Region, especially the district councils of Kilosa, Mvomero and Gairo. You are therefore requested to consult farmers on planting drought-resistant crops such as sesame, sunflower, cassava, sorghum and short-grown seeds, such as maize and rice.

Climate information is delivered to RBWO from TMA through the Ministry of Water via e-mails. The e-mails and stakeholder meetings are often used by RBWO to inform large-scale users within the basin such as Tanzania National Electric Supply Compan (TANESCO) to take the necessary precautions such as adjusting the volume of water uptake and thus adapting to changes in hydrological regimes in case of heavy rains or drought.[11] The existing model for communicating climate information from modern source to end users is presented in Fig. 13.3.

Flow of climate information and institutional limitations

There were breaks and inconsistencies in the lines of communication among government institutions as opposed to the 2002 Water Policy requirements, sub-section 4.10 (URT, 2002). Inter-institutional communication links with respect to water resources management was non-existent. For example, RBWO generates climate information but this is not shared with MDC.[12] Contrary to policy provisions, an inherent imposition on water resources management by MDC officials

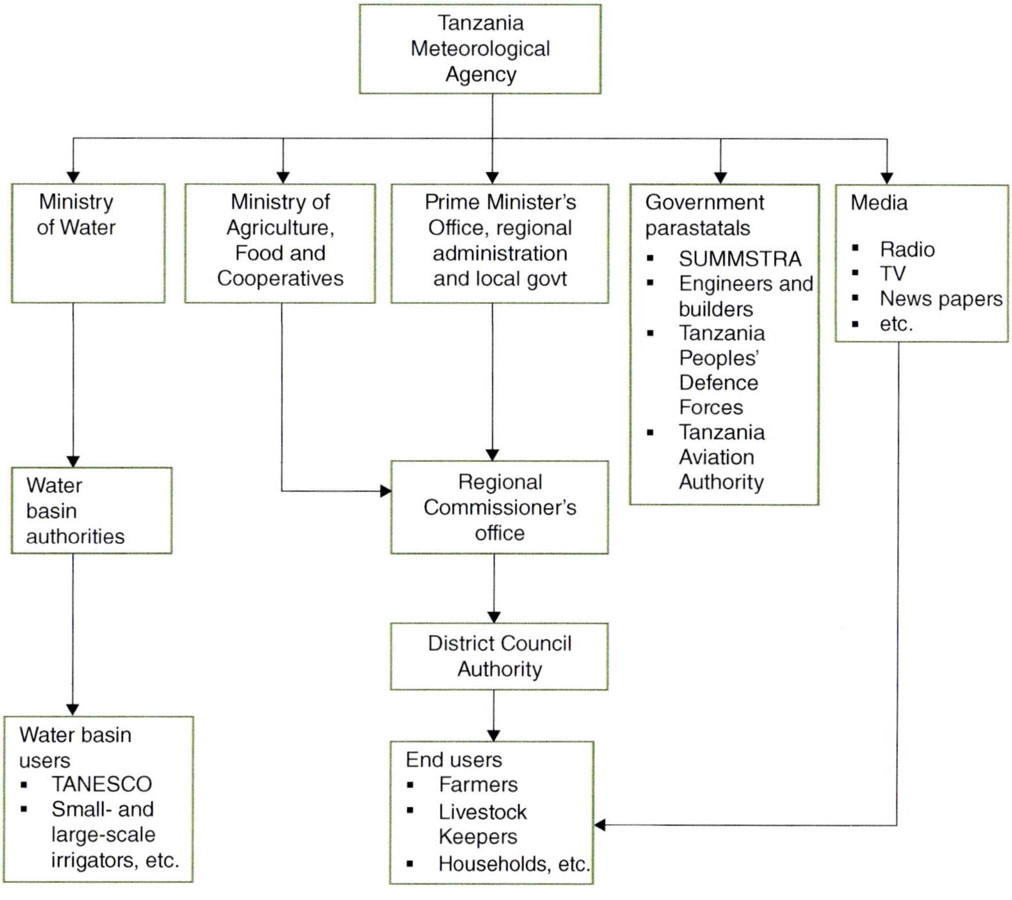

Fig. 13.3. Existing institutional model for communicating climate information. (From field data, 2014.)

to RBWO was evident. Likewise, seasonal forecasts received from TMA by the Agriculture Department were not received by farmers as advice. RBWO and MDC only collaborated in formation of village water committees. At district council level, the communication disconnect was also noted across departments in the district council itself (i.e. water and agriculture departments). The intercommunication link between the Ministry of Water, RBWO and TMA with respect to sharing climate information was also acknowledged. The World Bank (2016) indicated the role institutions should play in enhancing adaptation to climate extremes, is mainly through: (i) enhancing information gathering and dissemination; (ii) resource mobilization and allocation; (iii) skills development and capacity building; (iv) providing leadership; and (v) networking with other decision makers.

TMA plays a central role in generating and communicating climate information at the national level. However, its capacity to discharge climate information and services to decision-making institutions at the national, district and local levels is hampered by its weak institutional position within the national planning framework and its legal mandate. TMA is located in the Ministry of Transport, whereas implementation tools are in the PMO-DMD. TMA is mandated to provide technical services (information, advice and warnings) while the PMO-DMD deals with operational aspects of climate and weather extremes like floods, mud flow and droughts. All operational issues relating to climate- and weather-induced disasters and extremes are handled by the PMO-DMD. However, key informant interviews at MDC showed that the PMO is resource constrained, thus issuing climate information to end users is faced with long and frequent delays. TMA only provides advice, which can either be taken seriously or ignored by the institutions responsible for implementation.

Sub-national level institutions, MDC, RBWO and water committees in particular, are mandated to collect climate information and hydrological data (Section 4.10 of Water Policy 2002, p. 27). However, the link between MDC and RBWO was disjointed. TMA and RBWO share climate data; however, the widespread distribution of climate information is centralized to TMA, which has no capacity to enforce its implementation under the existing institutional landscape. That there was poor dissemination of seasonal forecasts among government institutions was evident through delays in seasonal forecast delivery to end users of often up to 1 month. The delay was due to excessive bureaucratic processes that impeded its flow. This was described as a result of the long channel from TMA through the PMO, to regional offices and to district council departments.

Awareness of actors on climate information

Reflection from actors at the community level revealed three types of weather forecasts that were communicated. Few respondents were aware of daily/weekly forecasts while more of the respondents were aware of warnings about extreme events and seasonal forecasts (Fig. 13.4).

Daily weather forecasts are issued throughout the year, whereas seasonal forecasts are usually issued from September to October. Warnings of extreme events are usually issued whenever abnormal climate events are observed by the Meteorological Department, normally forecasted with special insistence. Those respondents who did not have access to weather forecasts, or were uncertain of such information through conventional channels used TEK from local indicators, as pointed out earlier in this chapter. Respondents described this dependence on both conventional weather forecasting and TEK indicators was because of the limitations in both sources.

Access to weather forecasts and their inherent reliability

Although the TMA-generated weather forecasts are communicated through a wide range of channels, the study noted flaws associated with the communication system for climate information. The flaws that were noted were that access to forecasts could be impeded by the time that the forecast was delivered and the forecast's duration, and also the language used to communicate the forecast may not have been understood by

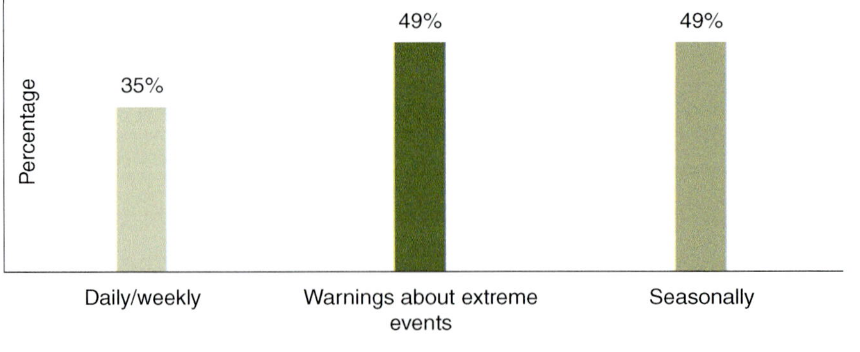

Fig. 13.4. Awareness of the respondents of weather forecasts. (From field data, 2014.)

all the recipients. Thus, all the respondents across the sample complemented such inadequacies of the modern sources through TEKs. From conventional sources, 22% of the respondents did not receive climate information, 28% used radio, 2% used television and 47% used both radio and television (Table 13.6).

The aspect of the timing of forecast delivery had a significant impact on access to weather forecasts. In this regard 58% of respondents reported that weather forecasts were not timely. Usually, weather forecasts are aired briefly once a day, at the end of news bulletin between 7:00 and 10:00 p.m. through either radio or television. Forecast presentation was usually too fast for both viewers and listeners to concentrate and capture the messages.

The question of reliability of weather forecasts raised concerns during the study whereby 77% of respondents acknowledged unreliability of weather forecasts from TMA when compared with those from traditional sources. The degree of reliability of weather forecasts from traditional sources varied significantly, with 55% of respondents acknowledging their reliability whereas 31% reported they were somewhat reliable, 13% somewhat unreliable and 1% unreliable (Fig. 13.5).

The scale 'reliable', 'somewhat reliable', 'somewhat unreliable' depicted the various levels of occurrence of a forecasted weather event in line with observed indicators. Thus those who mentioned 'somewhat unreliable' and 'unreliable' had put less trust in traditional forecasts over recent years. Risiro *et al.* (2012) and Kijazi *et al.* (2013) reported that indigenous weather forecasts are losing their strength due to the elderly people who believed in them passing away and changes in the environment and climate change.

Table 13.6. Respondents' access to climate information. (From field data, 2014.)

Tool for weather forecast delivery	Respondents (%)
Radio	28
Television	2
Radio and television	47
Face to face	1
N/A[a]	22
Total	100

[a]N/A, Not applicable as respondent did not receive climate information.

Weather-related advice and contents of forecasts

The advice that forecasts were supposedly accompanied by was acknowledged as an important aspect of forecasts. People's understanding of which form of weather forecast was accompanied by advice varied significantly whereby 35% of respondents perceived that they received daily/weekly advice, 49% received seasonal advice and 49% received warnings of extreme events. In addition, 39% of respondents had received weather advice more than 5 years ago (Fig. 13.6).

These perceptions were against the expert opinions whereby according to interviews held with TMA officials, only seasonal forecasts are associated with advice. This indicated that people's understanding of advice varies. There was no evidence of delivery of extension services in line with the seasonal forecasts through the local government system to smallholder farmers and other climate service users as may be expected (Daniel, 2013). Perceptions of time

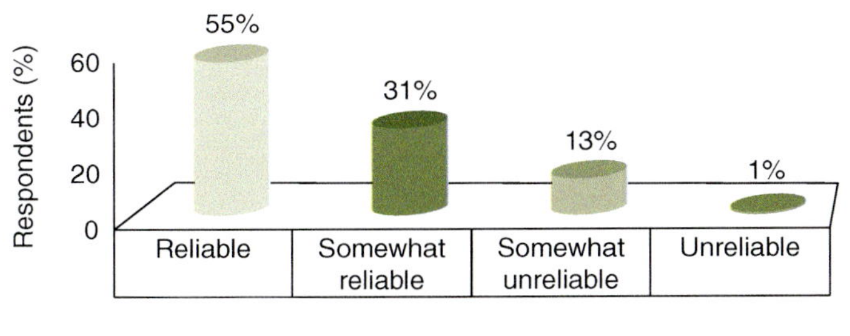

Fig. 13.5. Respondents' views (%) on reliability of climate information from traditional sources. (From field data, 2014.)

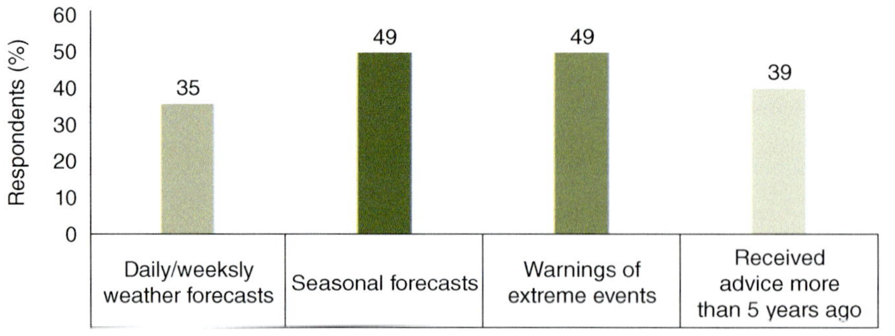

Fig. 13.6. Respondents' (%) receipt of weather-related advice from Tanzania Meteorological Agency (TMA). (From field data, 2014.)

intervals when such advisories were issued indicated that 16% of respondents perceived they received them weekly, 2% every few months, 5% several times a year, 3% once a year and 13% less frequently (Table 13.7).

There were concerns to devise mechanisms under which stakeholders may convene and share experiences, particularly the Meteorological Department, traditional forecasters and the district departments that have a stake in climate services. This is because TMA is not part of the local government system. Placing TMA services at the district level would equip farmers with agricultural knowledge and skills to improve farming practices. With regard to language, 51% of respondents reported on the clarity of the language of communication as weather forecasts are issued in Kiswahili language. However, 49% reported that the language used was not clear because it was difficult to comprehend the content of the information (Fig. 13.7).

Respondents indicated that although language used is familiar to them, it contained phrases which were not easily understood. For example, a weather outlook that reads:

> Southwestern highlands (Mbeya, Iringa, and Njombe regions and southern Morogoro areas): The ongoing rains are *expected to be normal to below* normal over much of the region. However Njombe, parts of southern Morogoro (Mahenge) and Mbeya (Tukuyu) regions are expected to experience *normal to above normal rains*. Rains are expected to end during the fourth week of April 2015.
> (TMA, 2015)

Table 13.7. Respondents' perception of frequency of receiving weather-related advice. (From field data, 2014.)

Frequency of receiving weather advice	Respondents (%)
After every week	16
After every few months	2
Several times a year	5
Once per year	3
Less frequently	13
N/A[a]	61
Total	100

[a]N/A, Not applicable as respondent did not receive climate information.

This text and the italicized words raised concerns during interviews about the abilities of end users to understand such information easily. With regard to adequacy, 34% of respondents were satisfied whereas 66% were not (Fig. 13.8).

The climate information was described as inadequate because it failed to address the information needs of specific users, lacked feedback loops and was not location specific. For example, when travelling, one would like to know about the weather of the place one is heading to, such as mist or fog so that precautions can be taken. Likewise, a water resource manager would like to know precisely which side of the catchment it is where it ought to rain, the possibility of flooding due to rainfall intensity and its implications on dams and precautions to be taken. Local people regarded climate information from

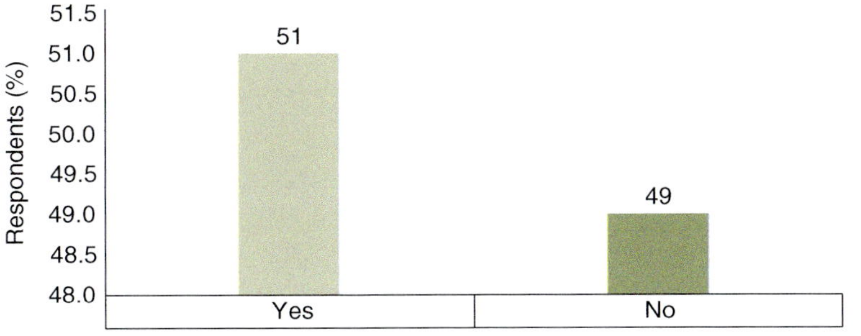

Fig. 13.7. Respondents' response (%) to question 'Was the language of communicating weather forecasts clear?' (From field data, 2014.)

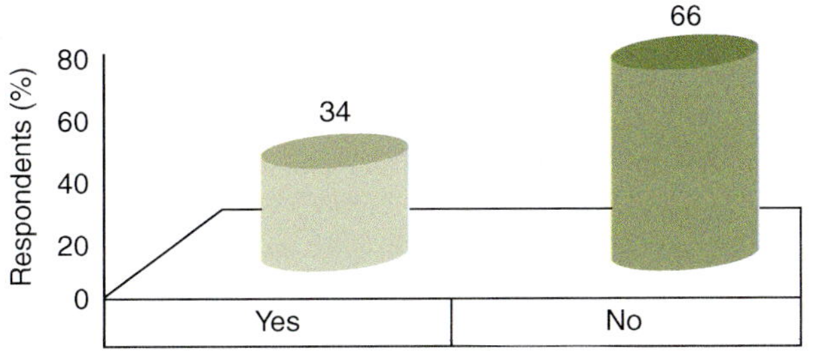

Fig. 13.8. Respondents' response (%) to question 'Are the weather forests provided adequate?' (From field data, 2014.)

modern sources inadequate as it was a one-way communication since there was no room for feedback.[13] According to Munang *et al.* (2010), it is through adequate climate information that rational decision making regarding natural resources management can be enabled.

Awareness of actors of water resource management institutions

Water committee members, who are principally water resource managers at the community level, acknowledged their inability to link climate information with water management practices. Lack of capacity building to acquaint them with knowledge and skills to enable them to apply climate information in water resource management operations was described as a major limitation. Although non-governmental organizations (NGOs) such as SONGAMBELE and WWF (World Wide Fund for Nature) have been providing environmental education, integrating climate services into livelihood systems had not been advocated. With regard to actors' awareness at the district level, it was revealed that actors in the Agriculture Department were aware of the importance of climate information (in the form of advice) to farmers at the community level. However, limited personnel as well as financial constraints prevented extension officers from discharging extension services to farmers. Moreover, despite the Water Department acknowledging the importance of weather forecasts in water supply, the inapplicability of such forecasts was due to lack of user specificity. Discussions revealed absence of capacity building to newly selected water committee members, despite their responsibilities under the 2002 water policy to collect various data and information.

Conclusion

The effectiveness of the channels of communicating weather forecasts to water resource managers and other users of climate services in the Upper GRCA was examined, with regard to the impact generated to end users in terms of awareness creation and usefulness in decision making. Two broad sources of weather forecasts were observed, namely, traditional and conventional. Whereas forecasts from traditional sources were communicated face to face via oral traditions, forecasts from conventional sources were communicated through ICT using two broad routes, on the one hand the local government system through posted letters and e-mails and on the other hand via mass media (radio or television or both).

These communication channels are used to create awareness to end users at the community level as well as in government institutions. At the community level, people are aware of weather forecasts from modern sources through mass media (mainly radio and television) despite the fact that TMA disseminates climate information through ICT. Overreliance on radio and television indicates that many people are not familiar with ICT. Likewise, socioeconomic characteristics, such as gender, education and livelihood strategies, tend to limit access to climate information from modern sources. Despite awareness, the reliability of forecasts is affected by space and time limitations. Comparatively, weather forecasts from traditional sources were found to be more reliable than those from modern sources.

Awareness is also linked to language and the adequacy of climate information. Language issues are implicated in the content of forecasts, whereas adequacy issues are implicated in that communication is a one-way traffic (i.e. no feedback from recipients) as well as the inability of climate information to address place and time requirements, which are necessary for decision making. Water resources managers are aware of the importance of climate information, however, with their varied expertise, experience and seriousness among them implies that issues relating to climate services have to great extent not been their focus. Water resource managers at the local level are not knowledgeable on the use of weather forecasts from modern sources. This may be linked to the poor placing of climate information in planning at local levels.

Notes

[1] Indigenous knowledge is also referred to as traditional ecological knowledge (TEK) (Usher, 2000).
[2] Accounts from village executive officers, Chimala and Mengele.
[3] According to Kumar (1989), key informant interviews involve interviewing a selected group of individuals who are likely to provide needed information, ideas and insights on a particular subject.
[4] Face-to-face interviews allow for complex questions to be explained, if necessary, to the interviewee. There is also scope to ask open questions as well as use visual aids (Phellas *et al*., 2011).
[5] A narrative is a 'story with a chronological order i.e. with a beginning, middle and an end' that summarizes an unambiguous set of phenomena which has achieved the 'status of conventional wisdom within the development arena' (Adger *et al*., 2001, p. 685).
[6] According to Usher (2000) TEK refers specifically to all types of knowledge about the environment derived from the experience and traditions of a particular group of people. TEK and indigenous knowledge can be used interchangeably.
[7] Narrated during FGDs in Chimala and Mengele villages; see also Ruiz-Mallen and Corbera (2013).
[8] Narrated during FGDs in Mengele village.
[9] Accounts from FGDs in Chimala and Mengele villages.
[10] Key informants at MDC and RBWO and FGDs in Chimala and Mengele villages.
[11] Informant interview with head of RBWO.
[12] Interview with RBWO Iringa office.
[13] Accounts from FGDs in Chimala and Mengele.

References

Adger, W. N., Benjaminsen, T. A., Brown, K. and Svarstad, H. (2001) Advancing a Political Ecology of Global Environmental Discourses. Development and Change 32(4), 618–715.

Amissah-Arthur, A. (2003) Targeting climate forecasts for agricultural applications in sub-Saharan Africa: situating farmers in user-space. Climatic Change 58(1–2), 73–92. https://doi.org/10.1023/A:1023462613213

Blench, R. (1999) Seasonal climate forecasting: who can use it and how should it be? Natural Resources Perspectives 47, 1–4.

Chang'a, L.B., Yanda, P.Z. and Ngana, J. (2010) Indigenous knowledge in seasonal rainfall prediction in Tanzania: A case of the south-western Highland of Tanzania. Journal of Geography and Regional Planning 3, 66–72.

Daniel, E. (2013) Assessment of agricultural extension services in Tanzania. A Case Study of Kyela, Songea Rural and Morogoro Rural Districts [Report]. Africa Rice, Dar es Salaam, Tanzania. Available at: http://www.parasite-project.org/wp-content/uploads/2013/12/Elifadhili-2013-Internship-report-final.pdf (accessed 13 January 2020).

Devisscher, T., Taylor, A., Bood, N. and Jeans, H. (2010) Ecosystems, development, and climate adaptation: improving the knowledge base for planning, policy and management. Final Report: Belize Study. SEI, Oxford. Available at: https://mediamanager.sei.org/documents/Publications/Climate/wwf-sei-2010-belize-final-report.pdf (accessed 6 January 2020).

Giorgi, F., Jones, C. and Asrar, G.R. (2009) Addressing climate information needs at the regional level: the CORDEX framework. World Meteorological Organization Bulletin 58(3), 175–183.

Global Water Partnership (GWP) and International Network of Basin Organizations (INBO) (2009) A Handbook for Integrated Water Resources Management in Basins. Global Water Partnership and International Network of Basin Organizations, Sweden.

Gubbels, P. (2013) West Africa Learning Event Community based Adaptation [Conference proceedings]. Groundswell International, Cotonou, Benin

Kijazi, A.L., Chang'a, L.B., Liwenga, E.T., Kanemba, A. and Nindi, S.J. (2013) The use of indigenous knowledge in weather and climate prediction in Mahenge and Ismani wards, Tanzania. Journal of Geography and Regional Planning 6(7), 274–279. https://doi.org/10.5897/JGRP2013.0386

Kumar, K. (1989) Conducting key informant interviews in developing countries. A.I.D. Program Design and Evaluation Methodology Report No. 13(13), 1–33. https://doi.org/10.1017/CBO9781107415324.004

Lusuva, E.A. (2009) An assessment of gender mainstreaming in water resources management: a case study of Mkoji sub-catchment in Usangu Plains, Tanzania. MSc. dissertation in Integrated Water Resources Management, Faculty of Engineering, University of Zimbabwe, Harare.

Madzingira, N. (2001) Culture, Communication and Development in Africa. Available at: https://pdfs.semanticscholar.org/e873/26de105fa9bcfcc8071e81a478cd6feb80c1.pdf?_ga=2.160330749.1734296298.1575922303-1812651368.1573034300 (accessed 28 September 2014).

Malley, Z. (2011) Climate change and water resources for energy generation in Tanzania. Climate Change Issues 2, 705–711. Available at: http://www.ep.liu.se/ecp/057/vol2/018/ecp57vol2_018.pdf?origin=publication_detail (accessed 25 September 2013).

Muguti, T. and Maposa, R.S. (2012) Indigenous weather forecasting: a phenomenological study engaging the Shona of Zimbabwe. Journal of Pan African Studies 4(9), 102–112A. Available at: http://www.jpanafrican.org/docs/vol4no9/4.9Indigenous.pdf (accessed 18 October 2017).

Munang, R., Rivington, M., Takle, E.S., Mackey, B., Thiaw, I. and Liu, J. (2010) Climate information and capacity needs for ecosystem management under a changing climate. Procedia Environmental Sciences 1(1), 206–227. https://doi.org/10.1016/j.proenv.2010.09.014

Murphy, A.H. (1993) What is a good forecast? An essay on the nature of goodness in weather forecasting. American Meteorological Society 8, 281–293. https://doi.org/10.1175/1520-0434(1993)008<0281:WIAGFA>2.0.CO;2

Mwakalila, S. (2011) Vulnerability of people's livelihoods to water resources availability in semi arid areas of Tanzania. Journal of Water Resource and Protection 3(9), 678–685. https://doi.org/10.4236/jwarp.2011.39078

Phellas, C., Bloch, A., and Seale, C. (2011) Structured methods: interviews, questionnaires and observation. In: Seale, C (ed) Researching Society and Culture, Sage Publications Ltd, London, pp. 181–205.

Podesta, G., Letsona, D., Messina, C., Royce, F. R., Ferreyra, A. et al. (2002) Use of ENSO-related climate information in agricultural decision making in Argentina: a pilot experience. Agricultural Systems 74, 371–392.

Rayner, S., Lach, D. and Ngram, H. (2005) Weather forecasts are for the wimps: why water resource managers do not use climate forecasts. Climate Change, 63, 2–9. https://doi.org/10.1007/s10584-005-3148-z

Risiro, J, Mashoko, D., Tshuma, D, T. and Rurinda, E. (2012) Weather Forecasting and Indigenous Knowledge Systems in Chimanimani District of Manicaland, Zimbabwe, Journal of Emerging Trends in Educational Research and Policy Studies 3(4), 561–566.

Ruiz-Mallen, I. and Corbera, E. (2013) Community-based conservation and traditional ecological knowledge: implications for social-ecological resilience. Ecology and Society 18(4), article 12.

Saunders, M., Lewis, P. and Thornhill, A. (2007) Research Methods for Business Students,

4th Edition. FT Prentice Hall, Harlow-England. p. 624.

Tall, A., Mason, S.J., Van Aalst, M., Suarez, P., Ait-Chellouche, Y., et al. (2012) Using seasonal climate forecasts to guide disaster management: the Red Cross experience during the 2008 West Africa floods. International Journal of Geophysics 2012. https://doi.org/10.1155/2012/986016

Tanzania Meteorological Agency (TMA) (2015) Seasonal Forecast, Press Release Climate Outlook for Tanzania March–May 2015 Rainfall Season.

Thornes, J.E., Sciences, E. and Stephenson, D.B. (2001) How to judge the quality and value of weather forecast products. Meteorological Applications 8(3), 307–314.

TMA, (2008) Client Service charter, Tanzania Meteorological Agency, Dar es Salaam, Tanzania.

United Republic of Tanzania (URT) (2002) The National Water Policy. Ministry of Water (July), Government of Tanzania, Dar es Salaam, Tanzania. https://doi.org/10.1002/ejoc.201200111

Usher, P. J. (2000). Traditional Ecological Knowledge in Environmental Assessment and Management. Arctic 53(2), 183–193. https://doi.org/10.14430/arctic849

World Meteorological Organisation (WMO) (2009) Climate Knowledge for Action: a Global Framework for Climate Services. Available at: https://gfcs.wmo.int//sites/default/files/FAQ/HLT/HLT_FAQ_en.pdf (accessed 27 October 2017).

World Wild Fund for Nature, Tanzania Country Office (WWF-TCO) (2010) Environmental flow assessment. The Great Ruaha River and Ihefu Wetlands, Tanzania, and options for the restoration of dry season flows. (July) 293. Available at: http://www.ourwater.vic.ghttp//www.waterandnature.org/sites/default/files/grr_flow_assessmentfinal_reportwwf2010.pdf (accessed 13 December 2013).

Ziervogel, G., Johnston, P., Matthew, M. and Mukheibir, P. (2010) Using climate information for supporting climate change adaptation in water resource management in South Africa. Climatic Change 103(3), 537–554. https://doi.org/10.1007/s10584-009-9771-3

Part IV

Conclusion

14 Lessons Learnt and the Way Forward for Research on Climate Change in Tanzania

Pius Z. Yanda,* Claude G. Mung'ong'o and Edmund B. Mabhuye*

Institute of Resource Assessment, University of Dar es Salaam, Dar es Salaam, Tanzania

Introduction

This book set out to collate and disseminate the findings of selected research projects done by researchers and faculty members at the Centre of Climate Change Studies of the University of Dar es Salaam for the period ending 2016. Historically, conventional approaches to understanding climate change have been limited to identifying and quantifying the potential long-term climate impacts on different ecosystems and economic sectors. While useful in depicting general trends and dynamic interactions between the atmosphere, biosphere, land, oceans and ice, this top-down, science-driven approach failed to address the regional and local impacts of climate change and the local abilities to adapt to climate-induced changes (Parry and Carter, 1998; Burton *et al.*, 2002).

This impact-driven approach has since given way to a new generation of scholarship, which utilizes bottom-up or vulnerability-driven approaches that assess past and present current vulnerability, existing adaptation strategies, and how these might be modified with climate change. Hence, the research themes in this volume have focused on impacts of climate change, vulnerability and resilience of ecosystems and communities to climate change. Other key sub-themes have been community access, control, and rights to climate information, importance of integration of indigenous and scientific knowledge in weather forecasting, participation of communities in decision making, and gender roles in response to impacts of climate change in various areas of mainland Tanzania. This last chapter of the book synthesizes the lessons learnt from the reported case studies and tries to chart out a road map for future research on climate change in Tanzania.

Vulnerability and Resilience to Climate Change

Climate is changing worldwide and adaptation results not only from the magnitude of the change in a particular area but in the existing vulnerability and resilience embedded in each community (Chishakwe, *et al.*, 2012). According to the Intergovernmental Panel on Climate Change (IPCC) (2001) 'vulnerability' is a function of the character, magnitude and rate of climate variation to which a system is exposed; its sensitivity, and adaptive capacity. All three factors form the equation: degree of climate change → rate of exposure → rate of sensitivity and → rate of adaptive capacity (Cinner *et al.*, 2011; Arndt *et al.*, 2012).

The history of the resilience perspective is dominated by empirical observations of ecosystem dynamics interpreted in

*pyanda@gmail.com and edmund.mabhuye@gmail.com

mathematical models, developing into the adaptive management approach for responding to ecosystem change (Folke, 2006). Serious attempts to integrate the social dimension is currently taking place in resilience work reflected in the large number of sciences involved in explorative studies and new discoveries of linked social–ecological systems. Recent advances have included understanding of social processes such as 'social learning and social memory ... agents and actor groups, social networks, institutional and organizational inertia and change, adaptive capacity, etc. ... that allow for management of essential ecosystem services' (Folke, 2006).

In this collection these important factors have been dealt with, each in their own way, by Shirima and Mung'ong'o in **Chapter 4**, and Richard Katondo and Agnes Nyomora in **Chapter 10**. Katondo and Nyomora assessed the role ecosystem services play in enhancing climate change resilience to local communities in the Ngarambe-Tapika Wildlife Management Area (WMA) in Rufiji District, while Shirima and Mung'ong'o attempted to construct a social–ecological vulnerability index to climate change for the agroecosystems of Mount Kilimanjaro.

The Ngarambe-Tapika case study has largely confirmed earlier studies elsewhere (e.g. Igoe and Croucher, 2007; Yang *et al.*, 2018) by showing that benefit sharing was still a problem in many WMAs. The little revenue that was realized did not outrightly benefit individual households. The results show, for example, that income obtained from Ngarambe-Tapika ecosystems by the communities were mainly invested in material welfare and livelihoods that largely enhanced resilience to climate change at the WMA level. Social services such as construction of dispensaries, rehabilitation of primary schools and maintenance of electricity supply to the community for lighting and operating the water pumps benefited the community as a whole, but could not increase the adaptive capacity of individual members of households who had foregone many benefits for wildlife conservation.

The research reported in **Chapter 4** was done to assess the maize-coffee-banana agroecosystems' resilience to climate change on the southern slope of Mount Kilimanjaro. The study assessed the agronomic practices and the socio-economic status of farmers to compute a social–ecological vulnerability index. The study depicted variations in agronomic practices with altitude due to microclimatic differences, terrain and soil characteristics. Climatic shocks like frequencies of drought, floods and below average rains were found to have implications on agricultural yield. Socio-economic indicators such as size of households and dependency ratios, social safety networks, off-farm contribution, possession of land titles, wood usage as cooking energy, and access to extension services also showed a significant implication on household vulnerability to a changing climate which could later affect the agroecosystem productivity due to association of these parameters with the natural environment. Indicators chosen for the vulnerability index depicted light variation of vulnerability with altitude, except the mid-lower zone which was indicated to be the most vulnerable. More such studies need to be replicated in other similar geographical areas to facilitate the construction of a nationwide vulnerability index to climate change.

From a theoretical viewpoint, while seminal papers like Smit *et al.* (2000) and Adger *et al.* (2005) have mainly analysed crucial components of adaptation to climate change, the research field has continued to gain complexity by increasingly including a consideration of adaptive capacity. Authors like Brooks (2003) are careful to distinguish between actual adaptation and adaptive capacity (which they conceptualize as the potential for adaptation that does not necessarily become real). Hence, they link adaptive capacity to the discourses on vulnerability and resilience (Eisenack and Stecker, 2012). Researchers at the Centre for Climate Change Studies (CCCS) at the University of Dar es Salaam are yet to address these nuanced conceptual relationships succinctly. Elsewhere, a wide canvas of theoretical literature has tried to grapple with such relationships. Examples from which the CCCS researchers could draw reference from include, among others, studies by Smit and Wandel (2006), O'Brien *et al.* (2007), Nelson *et al.* (2007) and Few *et al.* (2015).

Baker (2016) reacting to ecologists such as Smit and Wandel (2006), O'Brien *et al.* (2007), Nelson *et al.* (2007) and Few *et al.* (2015), for example, wonders whether resilience in social systems is as positive an attribute as it is in the more complex natural systems. In fact, Baker (2016) points out, for example, that many social systems are corrupt, non-democratic and promote brutal inequalities. Furthermore, many social practices are unsound when judged by sustainable development principles. Thus, from a social science perspective, rather than asking whether a system is or is not resilient, under what circumstances and how best to promote that resilience, a different question should be proposed, and this is 'For whom is the system resilient?'

In answering this question Baker (2016) suggests involving research on the distributional consequences of social arrangements and processes, including their impact on inter- and intra-generational equity, and ultimately involves making value judgements about social processes, collective norms, and how best to promote societal well-being. The material from the Ngarambe-Tapika WMA case study is instructive.

Impacts of Climate Variability and Change

This theme has been represented in this book by the work done by Cyrilo and Mung'ong'o in **Chapter 8** and Yamat and Mung'ong'o in **Chapter 6**. Yamat and Mung'ong'o carried out a comparative cost–benefit analysis of mobile pastoral and sedentary production systems using cases from Makame and Dofa villages in Kiteto and Karatu districts, respectively, to come up with reliable evidence-based data to analyse the rationality of claims that sedentary livestock raising is more productive, and utilizes fewer resources and less space than mobile pastoral systems, especially in the context of climate change. The work by Cyrilo and Mung'ong'o embarked on assessing the socio-ecological resilience of agropastoralists to climate change and variability impacts in Bariadi District. The study specifically focused on assessing the impacts of climate change and variability to livelihood activities as well as assessment of vulnerability of different livelihood options to climate change and variability. Lastly, resilience status of the agropastoralists was assessed by looking at the significance of climate change and variability impacts, the level of vulnerability, as well as assessment of the efficiency and effectiveness of coping and adaptation strategies.

The study by Yamat and Mung'ong'o used the cost–benefit analysis framework to compare the cost and benefits of the mobile pastoral system and the sedentary one in the two villages. The results showed that the cost of maintaining a mobile pastoral production system was lower compared with that of the sedentary one. The cost–benefit ratios were 1:0.5 for mobile pastoral production, while it was 1:0.25 for the sedentary production system. The benefits accrued from the mobile production system were also found to be higher. Net present values at a discount rate of 10% for a period of 10 years revealed the mobile pastoral production system gives more returns to those engaged in it, by at least four times that of the sedentary production system, but both exhibiting positive signs. The study recommended enacting of policies that would sustain the pastoral production system by mitigating the impacts of climate change, especially droughts, which had shown to be a big challenge that reduced the economic potential through high livestock mortality at local and national levels.

These findings confirm the observations made by earlier studies elsewhere, such as Hesse and MacGregor (2006) on the subject. Some useful recommendations have been made to strengthen the capacity of livestock keepers to adapt to climate change, including integrating modern climate services with available indigenous knowledge (IK) as proposed by Matowanyika (2016) and Mafongoya and Ajayi (2017). However, very few practical suggestions have been made so far on the issues of governance and organization of groups such as the livestock keepers forming into viable modern associations to support one another, with emphasis being placed on the four cross-cutting research themes of the role of power as suggested by Nightingale (2017) or of knowledge, norms and scale as suggested by Schulz and Siriwardane (2015).

Response Measures and Adaptation Strategies

Adaptation to climate change refers to adjustments in the human–environment system in response to actual and/or anticipated different climatic conditions in order to avoid or to mitigate the associated risks or to realize potential opportunities (Chishakwe *et al.*, 2012; Wheeler *et al.*, 2013). Adaptation is a key strategy that can alleviate the severity of climate change impacts on agriculture and food production. Adaptation strategies are unlikely to be effective without an understanding of the communities' perceptions of climate change. Climate change affects countries, regions and communities in different ways and thus they differ in terms of their adaptation strategies. The factors responsible for the variation in adaptive responses across regions are the agroecological system, socio-economics, climatic impact, and existing infrastructure and capacity (Monirul-Alam *et al.*, 2017). Much of the early work at the CCCS has hinged itself on this strand of research.

In this collection, this research is represented by the work of Mbwambo and Liwenga in **Chapter 3** where they analyse the role of cassava in adaptation to current climate variability and change in Mkuranga District in Tanzania. The findings revealed a decline in crop production which resulted in food shortages and livelihood insecurity in the study villages. The respondents in both villages reported cassava to be the crop that was least affected by climate and environmental extremes, hence ensured food availability and security in households. Hence growing cassava was a viable adaptation strategy to climate change and variability now and in the future. Improving cassava production, processing, marketing and value chain infrastructures were very crucial for enhancing this sustainable adaptation mechanism. **Chapter 7** by Mabhuye and Yanda also provides key adpative trends in the Maasai landscapes in northern Tanzania. They present key challenges induced by climate change that affect traditional livestock keeping in the form of pastoralism. The chapter emphasizes that climatic conditions are changing, with key climate variables such as rainfall becoming erratic and temperature steadily increasing. The erratic pattern of rainfall and increasing trends of temperature eventually have been affecting availability of pasture and water for livestock. Eventually, the effects compel local communities to respond through traditional knowledge systems, with modern knowledge initiatives complementing their efforts.

And finally, in **Chapter 5** Josephine Zimba and Emma Liwenga explore the effects of conservation agriculture (CA) as an adaptation option for farmers' livelihoods in the face of climate change in Balaka District of Malawi. CA has been highly promoted due to its potential to ensure high crop yields even in the face of changes in climatic factors. However, the actual benefits associated with CA are not only primarily focused on food security but are also site specific. Therefore, this study seeks to understand the benefits of CA in improving livelihoods in a changing climate in the Hanjahanja and Sawali sections of Bazale Extension Planning Area. Specifically, it analysed CA's contribution to farmers' livelihoods and challenges and opportunities of CA in climate change adaptation.

The study found out that due to adoption of CA, the majority of the farmers in both Hanjahanja and Sawali sections had realized positive livelihood outcomes, mainly through improved food security and increased incomes. Despite the similarities, Hanjahanja farmers reported a decrease in yields in seasons marred by floods. However, farmers faced several challenges due to CA adoption, which included high labour demands, rainfall variability and lack of inputs. Even so, improvement in soil moisture, soil erosion control, improved food security, the presence of several institutions and an enabling environment offered more opportunities for CA as an adaptation option for coping with climate change. CA, therefore, improves the livelihoods of the farmers except in times of floods. Hence, deliberate policies by the government to promote adoption of CA are required to take advantage of the benefits of CA. Further, research should also be done on how best to reduce the negative effects of CA on farmers' livelihoods.

A general conclusion that can be made from these case studies is that the respondents' perceptions of changes in the climate and of extreme climatic events are similar to the observations made elsewhere (e.g. Monirul-Alam *et al.*, 2017 for Bangladesh). The results indicate that respondents recognized the impacts on their livelihood and resources, resulting in an increased sense of vulnerability. To build resilience, households have undertaken a number of farming and non-farming adaptation strategies, which varied significantly among the farming and non-farming groups. The important adaptation strategies used included expansion of farms, modifying agricultural practices (including adoption of CA) and fishing activities, and engaging in different income-generating activities such as casual labour and petty businesses. However, it is also observed that not everyone in the studied communities managed to adapt to all of these strategies. The question for further research then is why did many smallholder farmers fail to adopt what appear to be relatively simple adaptation strategies that could help them cope with the climate-induced stressors they experienced (cf. Mulwa *et al.*, 2017 for Malawi)? An answer is suggested in Ubisi *et al.* (2017) whereby limited adaptive capacity to withstand climate change for the South African case study was due to compromised social, human, physical, natural and financial assets, the dynamics of which are, however, not well unpacked.

Second, what is also lacking in the studies done at CCCS so far is the aspect of theory building as a contribution to the already established theoretical literature on adaptation, such as developed by Smit *et al.* (2000), Adger *et al.* (2005, 2009) and others. As early contributions to the discourse on adaptation, a shared culture and related issues, these seminal writings provide a sound basis for understanding adaptation and propose crucial variables to characterize adaptations. However, the current studies at CCCS are not adequately linking to these theoretical texts and hence they fail to comprehensively draw useful conclusions about issues like barriers to adaptation and matters related to actors and decision-making processes involved in adaptation (Eisenack and Stecker, 2012).

Communicating Climate Change Information

Agriculture in East Africa, indeed all over sub-Saharan Africa, is very much dependent on the condition of climate at any particular time. The failure of crops to produce good yields is, therefore, more often than not attributable to bad weather and the incapability of farmers to utilize climate information provided by official weather forecasts to anticipate extreme weather events. Moreover, climate forecast information issued by official meteorology agencies like the Tanzania Meteorological Agency (TMA) is not informative enough due to the low level of understanding of its users. Hence, developing effective approaches to communicating climate information to the end users remains an imperative enterprise for ensuring resilience in a changing climate.

Efforts at CCCS to address issues of communicating climate change information to farmers have been an important aspect of its research activities. Historically and to date local communities in different parts of the world have continued to rely on IK to conserve the environment and deal with natural disasters. Communities, particularly those in drought- and flood-prone areas have generated a vast body of IK on disaster prevention and mitigation through early warning signs and preparedness (Rangoli *et al.*, 2002; Svotwa *et al.*, 2007; Anandaraja *et al.*, 2008).

The importance of integrating both the scientific and the indigenous climate forecast information for farm-level decision making is gaining momentum, as also documented in Mozambique and Kenya by Lucio

(1999) and Ngugi (1999), respectively. In Zimbabwe local communities have been coping with droughts by integrating scientific and indigenous climate forecasting techniques (Shumba, 1999). In many cases IK about natural hazards enables communities at risk to capitalize on this knowledge to protect themselves from natural disasters. This knowledge is still intact among indigenous communities in many parts of Africa and other regions of the world. However, it is not well documented and it stands in danger of being lost as custodians of IK are passing away.

Using IK in weather and climate prediction, local communities in different parts of Tanzania have been coping and adapting to increased climate variability normally manifested in the form of increased frequency and magnitude of various exigencies including droughts and floods, and outbreak of pests and diseases. Prediction of impending hazards has been an integral part of their adaptation strategies. Various environmental and astronomical indicators including plant phenology, behaviour and movement of birds, animals and insects are widely used in many parts of Tanzania to predict rainfall (Kihupi *et al.*, 2002; Mhita, 2006; Chang'a *et al.*, 2010). In spite of all the usefulness of IK in weather and climate prediction, the art is under threat of disappearing due to lack of systematic documentation of the knowledge and lack of coordinated research to investigate the accuracy and reliability of IK forecasts (Kijazi *et al.*, 2013).

In this volume this strand of research has been represented by the work of Anselm Mwajombe and Godwin Lema in **Chapter 13**, where they assess the effectiveness of the channels for weather forecast dissemination in terms of magnitude of awareness creation and versatility to end users. Their findings show that both traditional and conventional channels of weather forecasting and communication are used to create awareness to end users with different magnitudes. For local communities, traditional weather forecasting and communicating were contingent on the IK acquired through interaction with the local environment. They encompassed indicators or signs that covered plant phenology, astronomical and meteorological events as well as mammals' behaviour. The conventional forecasting is communicated via modern communication technologies including radio, television, the Internet and posted letters. Communication of traditional weather forecasting is mainly through oral traditions. Results from the respondents revealed that 40% received weather forecasts through traditional channels, 11% through modern channels and 49% through both modern and traditional channels. There were issues concerning the reliability and the content of the forecasts from modern sources as the majority of respondents said that weather forecasts from modern sources were not reliable enough to inform the decision-making process when compared with traditional sources. The study recommended synchronizing modern and traditional channels for effective weather forecast delivery.

Conclusion and Recommendations

This book has reviewed only 12 studies out of several impressive studies that have been conducted at the CCCS, University of Dar es Salaam, on various aspects of climate variability and change over the last decade. The aim of these studies has been to develop and test robust research methodologies in the field of climate change science and process the information in order to produce valuable findings for publications, informing policy and decision-making processes as well as engaging the wider general public through the media, community meetings and short courses for stakeholders across scales and levels. This volume is an attempt by the CCCS to systematically collate such studies into a coherent reference document.

As noted by Lukwale and Sife (2017), although the total number of publications is still generally low in Tanzania, despite the fact that climate change has been a topical area for over the last 10 years, these studies demonstrate an impressive growth of climate change literature not only at the University of Dar es Salaam, but also in the country as a whole. However, more effort needs to be made to:

- address issues of governance, with special emphasis being placed on the four cross-cutting research themes of the role of power, knowledge, norms and scale as suggested by Nightingale (2017) and Schulz and Siriwardane (2015);
- assess the potential and constraints of adopting innovative agricultural land use and farming systems for adaptation to and mitigation of climate change;
- assess and acknowledge the different roles, opportunities, perspectives and challenges that women and men have in the face of climate change; and
- investigate the security implications of climate change. Do changing weather patterns lead to future conflicts, demonstrations, riots, strikes and anti-government violence?

References

Adger, W.N., Arnell, N.W. and Tompkins, E.L. (2005) Successful adaptation to climate change across scales. *Global Environ Change* 15, 77–86.

Adger, W.N., Lorenzoni, I. and O'Brien, K. (eds) (2009) *Adapting to Climate Change: Thresholds, Values, Governance*. Cambridge University Press, Cambridge.

Arndt, C., Chinowsky, P., Robinson, S., Strzepek, K., Tarp, F. and Thurlow, J. (2012) Economic development under climate change. *Review of Development Economics* 16(3), 369–377. doi:10.1111/j.1467-9361.2012.00668.x

Anandaraja, N., Rathakrishnan, T., Ramasubramanian, M., Saravan, P. and Suganthi N.S. (2008). Indigenous weather and forecast practices of Coimbatore district farmers of Tamil Nadu. *Indian Journal of Traditional Knowledge*. 7(4): 630–633.

Baker, S. (2016) Nature in the Anthropocene: political science meets ecology debates. Paper presented to European Consortium for Political Research (ECPR) joint session of workshops, Workshop on Environmental Political Theory in the Anthropocene, directed by John Barry and Manuel Arias Maldonado, Pisa, Italy, 24-28 April 2016.

Brooks, N. (2003) Vulnerability, risk and adaptation: a conceptual framework. Working Paper 38. Tyndall Centre for Climate Change Research, University of East Anglia, Norwich. UK.

Burton, I., Huq, S., Lim, B., Pilifosova, O. and Schipper, E.L. (2002) From impacts assessment to adaptation priorities: the shaping of adaptation policy. *Climate Policy* 2, 145–159.

Chang'a L.B., Yanda P.Z. and Ngana, J. (2010) Indigenous Knowledge in seasonal rainfall prediction in Tanzania: A case of South-Western Highland of Tanzania. *Journal of Geography and Regional Planning* 3(4): 66–72.

Chishakwe, N., Murray, L. and Chambwera, M. (2012) *Building Climate Change Adaptation on Community Experiences: Lessons from Community-based Natural Resource Management in Southern Africa*. International Institute for Environment and Development, London.

Cinner, J.E., McClanahan, T.R., Graham, N.A.J., Daw, T.M., Maina, J., *et al*. (2011) Vulnerability of coastal communities to key impacts of climate change on coral reef fisheries. *Global Environmental Change* 2: 12–20. doi:10.1016/j.gloenvcha.2011.09.018

Eisenack, K. and Stecker, R. (2012) A framework for analyzing climate change adaptations as actions. *Mitigation and Adaptation Strategies for Global Change* 17(3), 243–260. doi:10.1007/s11027-011-9323-9

Few, R., Satyal, P., McGahey, D., Leavy, J., Budds, J., *et al*. (2015) Vulnerability and adaptation to climate change in semi-arid areas in East Africa. Adaptation at Scale in Semi-Arid Regions (ASSAR) Working Paper, ASSAR PMU, University of Capetown, Capetown, South Africa, 111 pp.

Folke, C. (2006) Resilience: the emergence of a perspective for social–ecological systems analyses. *Global Environmental Change* 16, 253–267.

Hesse, C. and MacGregor, J. (2006) *Valuing Pastoralism in East Africa*. International Institute for Environment and Development (IIED), London.

Igoe, J. and Croucher, B. (2007) Conservation, commerce and communities: the story of community-based wildlife management areas in Tanzania's northern tourist circuit. *Conservation and Society* 5(4), 534–561.

Inter-governmental Panel on Climate Change (IPCC) (2001) *Climate Change 2001: Impacts, Adaptation and Vulnerability. IPCC Third Assessment Report*, Cambridge University Press, Cambridge.

Kihupi, N.I., Kingamkono, R., Rwamugira, W., Kingamkono, M., Mhita, M. *et al*. (2002) Promotion and Integration of Indigenous Knowledge in Seasonal Climate Forecasts. Consultancy report submitted to Drought Monitoring Center, Harare, Zimbabwe.

Kijazi, A.L., Chang'a, L.B., Liwenga, E.T., Kanemba, A. and Nindi, S.J. (2013) The use of indigenous knowledge in weather and climate prediction in Mahenge and Ismani Wards, Tanzania. *Journal of Geography and Regional Planning* Vol. 6(7), pp. 274–280

Lucio FDF (1999) Use of contemporary and indigenous forecast information for farm level decision making in Mozambique. Consultancy report. UNDP/UNSO. p. 72.

Lukwale, S.R. and Sife, A.S. (2017) Climate change research trends in Tanzania: a bibliometric analysis. *International Journal of Biodiversity and Conservation* 9(6), 224–231.

Mafongoya, P.L. and Ajayi, O.C. (eds) (2017) *Indigenous Knowledge Systems and Climate Change Management in Africa*. CTA, Wageningen, The Netherlands, 316pp.

Matowanyika, J.Z.Z (2016) *Indigenous Knowledge Systems, Community Based Climate Observation Practices and Synergies with Climate Services and Adaptation in Zimbabwe. A Paper Presented at the Southern Africa Regional Climate Services Workshop: Toward Exploiting the Full Potential of Climate Services, 29 November to 2nd December 2016*, Victoria Falls, Zimbabwe.

Mhita, M.S. (2006) Training manual traditional knowledge for nature and environmental conservation, agriculture, food security and disaster management in Tanzania.

Monirul-Alam, G.M., Alam, K. and Mushtaq, S. (2017) Climate change perceptions and local adaptation strategies of hazard-prone rural households in Bangladesh. *Climate Risk Management* 17, 52–63.

Mulwa, C., Marenya, P., Rahut, D.B. and Kassie, M. (2017) Response to climate risks among smallholder farmers in Malawi: a multivariate probit assessment of the role of information, household demographics, and farm characteristics. *Climate Risk Management* 16, 208–221.

Nelson, D.R., Adger, W.N. and Brown, K. (2007) Adaptation to environmental change: contributions of a resilience framework. *Annual Review of Environmental Resources* 32, 395–419.

Ngugi RK (1999) Use of Indigenous and contemporary knowledge on climate and drought forecasting information in Mwinyi district, Kenya. Consultancy report. UNDP/UNSO. p. 28.

Nightingale, A.J. (2017) Power and politics in climate change adaptation efforts: struggles over authority and recognition in the context of political instability. *Geoforum* 84, 11–20.

O'Brien, K., Eriksen, S., Nygaard, L.P. and Schjolden, A. (2007) Why different interpretations of vulnerability matter in climate change discourses. *Climate Policy* 7, 73–88.

Parry, M. and Carter, T. (1998) *Climate Impact and Adaptation Assessment: a Guide to the IPCC Approach*. Earthscan, London.

Rangoli, C., Ingram, K. and Kirshen, P. (2002) Reading the Rains: Local Knowledge and rainfall forecasting in Burkina Faso. *Society and Natural Resources*. 15: 409–427.

Schulz, K. and Siriwardane, R. (2015) Depoliticised and technocratic? Normativity and the politics of transformative adaptation. Earth System Governance Working Paper No. 33. Earth System Governance Project, Lund, Sweden and Amsterdam.

Shumba O (1999) Coping with drought: Status of integrating contemporary and Indigenous climate/drought forecasting in communal areas of Zimbabwe. Consultancy report. UNDP/UNSO. p. 72.

Smit, B. and Wandel, J. (2006) Adaptation, adaptive capacity and vulnerability. *Global Environmental Change* 16, 282–292.

Smit, B., Burton, I., Klein, R.J.T. and Wandel, J. (2000) An anatomy of adaptation to climate change and variability. *Climatic Change* 45, 223–251.

Svotwa, E.J., Manyanhaire, I.O. and Makanyire (2007) Integrating Traditional Knowledge Systems with Agriculture and Disaster Management: A case for Chitora Communal Lands. *Journal of Sustainable Development in Africa*. 9(3): 50–63.

Ubisi, N.R., Mafongoya, P.L., Kolanisi, U. and Jiri, O. (2017) Smallholder farmer's perceived effects of climate change on crop production and household livelihoods in rural Limpopo province, South Africa. *Change Adaptation in Socioecological Systems* 3, 27–38.

Wheeler, S., Zuo, A. and Bjornlund, H. (2013) Climate change beliefs and irrigator adaptability in the Southern Murray-Darling Basin. *Global Environmental Change* 23, 537–547.

Yang, H., Yang, W., Zhang, J., Connor, T. and Liu, J. (2018) Revealing pathways from payments for ecosystem services to socioeconomic outcomes. *Science Advances* 4, eaao6652.

Index

abandoned farmland 161
action theory of adaptation to climate change 103
adaptation to climate change 240–241
  agropastoralist communities 142–148, 149–150
  cassava farmers 23–33, 240
  climate smart agriculture 180–192
  coffee and banana farmers 16–17
  conservation agriculture 44–56, 240–241
  pastoral communities 103–119, 240
adaptive capacity 150n, 238
  low 49, 181
African armyworm (*Spodoptera exempta*) 40–41, 224
age
  and perception of climate change 26, 135
  of studies' participants 81, 106–107, 132
agroforestry 38, 189
agroindustries 1–2
agropastoralism 71
  in climate smart agriculture 190–191
  comparison with mobile pastoralism 76–94, 239
  conflict with pastoralists 163
  resilience to climate change 123–150, 239
  switch from pastoralism 114–115
altitude, and farming practices 35–36, 38
Amorites 70
ants, appearance forecasts rain 223
armyworm (*Spodoptera exempta*) 40–41, 224
avocado 15

banana
  Kilimanjaro Region 39, 41
  Moshi Rural district 10–19
  pests and diseases 10, 14, 41
Bariadi District agropastoralists 125–150, 239
beekeeping 114
birds, forecasting rain 223, 224
building materials 175, 176
bush meat 174

carbon sequestration 181, 190, 191
carrying capacity 61–62
cash crops 16, 24, 35, 78, 127, 173, 204
  see also coffee
cassava 24–25, 29–32, 240
cattle *see* livestock
charcoal 206
chickens 142, 148
China 70
cholera 109
climate smart agriculture (CSA) 180–192
coffee
  Kilimanjaro Region 34, 41, 43
  Moshi Rural District 10–19
  pests and diseases 10, 14, 41
communication with farmers 38, 50, 111
  weather/climate forecasts 217–218, 219–220, 223–232, 241–242
compost 188, 189, 190
conflict 153–155, 243
  ethnic/political violence 65
  in the Kilombero and Kilosa Districts 156–167
    between various actors 162–164
    impacts of 164–166
    over land 157–162, 165
    over water 157, 160, 163, 165
  and pastoralism 74–75, 109, 116, 158–159, 162–163, 165–166
  resolution 159, 162, 166–167
conservation agriculture (CA) 4, 44–56, 240–241

conservation areas 75, 170
  conflict over 109, 158, 159, 161, 163, 166
corruption 161
cost–benefit analysis (CBA) 67–68
  mobile vs sedentary pastoralism 87–94, 239
cotton 127, 144
  pests 141
credit cooperatives 38
crop production
  adoption of improved varieties 17, 140, 143, 185–186, 190
  Bariadi District 127, 133–134, 138–141, 142–146
  cassava (Mkuranga District) 23–33, 240
  climate smart agriculture 185–192
  coffee and banana (Moshi District) 11–12, 14–19
  conservation agriculture 44–56, 240–241
  intercropping 17, 38–39, 54, 55, 144
  Kilimanjaro Region 36, 38–40, 238
  Makere Forest Reserve 204–206, 208
  Ngarambe-Tapika WMA 172–173
  rotation 144

deforestation 19, 40, 49, 195, 207–208
demographics
  Bariadi District 125–126, 132
  Makere Forest Reserve 204
  Monduli District 104–105
  pastoralists 74, 80–83
  population growth 1, 74, 75–76, 204
  Simanjiro District 105
diseases and pests
  banana 10, 14, 41
  coffee 10, 14, 41
  crops (in general) 109–110, 141
    pesticide use 146
    resistant varieties 190
  human 109, 176
  livestock 85, 86, 112–113, 141–142, 148
  maize 40–41
Dofa village (sedentary pastoralists) 76–94, 239
domestication of animals 70
drought
  Bariadi District 136, 139
  coffee and banana 12
  Ethiopia 184
  Kilimanjaro Region 40
  mobile pastoralism 74, 86–87
  Monduli/Ngorongoro/Simanjiro Districts 108, 109
drought-tolerant crop varieties 17, 140, 143

early maturing crop varieties 17, 143, 186, 190
ecosystem services 169–178
education
  affected by conflict 164
  Bariadi District 133
  Makere Forest Reserve 201
  Ngarambe-Tapika WMA 177–178
  pastoralists 83–84, 118
  and perception of climate change 135
electricity 173
enclosure of pastures 147
energy use 173–174, 206
erosion
  control 39, 39–40, 54–55, 187, 188–189, 190
  Kilimanjaro Region 38
Ethiopia (climate smart agriculture) 182–192
extension services 41, 49, 50, 148, 189
extreme events
  Kilimanjaro Region 39, 40
  Mkuranga District 26
  weather warnings 228
  see also drought

fertilizer use
  Bariadi District 140, 146
  Ethiopia 187–188, 189, 190
  Kilimanjaro Region 39
  Ngarambe-Tapika WMA 176
field observation 48, 80, 106, 130, 183
financial capital *see* wealth
firewood 174, 206
flooding
  and conservation agriculture 49, 53–54
  Kilimanjaro Region 39, 40
  Kilombero and Kilosa Districts 157
focus group discussions (FGDs) 47, 106, 129, 183, 199, 221
fodder
  crop residues 31, 64, 148
  tree branches 112, 116
food security 170
  affected by conflict 165
  Bariadi District 142–143, 170
  Malawi 51, 52, 55
  Mkuranga District 29
  Moshi Rural District 19
  Ngarambe-Tapika WMA 178
forests
  agroforestry 38, 189
  deforestation 19, 40, 49, 195, 207–208
  reforestation 49
  utilization 194–197
    Kilombero District 159
    Makere Forest Reserve 197–211
    Ngarambe-Tapika WMA 174–175
fuel 174, 206
future areas of research 242–243

game reserves 75, 109, 161
gender
    effect of land conflicts 164
    inequality 116–118
    profiles of studies' participants 81, 131–132
goats 84, 115, 147
government policy/governance
    conservation agriculture 55
    forests 208–211
    help for farmers in Bariadi District 148–149, 150
    land use planning/conflict 158, 160–161, 161–162, 167
        conflict resolution 166
    pastoralism 63, 66, 72–73, 86
    RRG paradigm 65
government-derived weather forecasts 218, 224–228
grazing 38, 84–85
    see also pastoralism
Great Ruaha Catchment Area (GRCA) 218–232
greenhouse gases 181, 190, 191

healthcare 138, 164–165, 176
herbicides 54
herd size and composition 84, 112–113, 116, 147–148
household size (Bariadi District) 126, 132–133
household surveys 48, 105–106, 130–131, 183, 199, 221
human capital 125
    see also education
hunting
    bush meat 174
    tourism 175–176, 178

impact of climate change 170, 239–240
    Bariadi District 138–141
    coffee and banana production 14–16
    Ethiopia 192
    Kilimanjaro Region 40
    Mkuranga District 29
    on pastoralists 108–110
indigenous knowledge
    mobile pastoralism 62
    weather forecasting 218, 223–224, 229, 241–242
infiltration ditches 40
information provision to farmers 38, 50, 111
    weather forecasts 217–232, 241–242
intercropping
    Bariadi District 144
    conservation agriculture 54, 55
    Kilimanjaro Region 38–39
    Moshi District 17
investors (from outside) 163–164
irrigation 17, 157

Kenya 74, 75
key informant interviews 106, 130, 183, 199, 221
Kilimanjaro Region 34–43, 238
Kilombero District 156–167
Kilosa District 156–167

labour
    increased demand 53
    paid 114, 134, 178
land use conflicts 74–75, 109, 116, 156–167
landholdings
    Kilimanjaro Region 38
    Kilombero and Kilosa Districts 157
    Makere Forest Reserve 205–206
    pastoralists 75, 114–115
    tenure issues 38, 75, 149
    undeveloped/abandoned farmland 161
livelihood assets 45–46, 124–125
livelihoods
    and conservation agriculture 47–54
    disrupted by conflict 164–165
    diversification
        agropastoralists 138, 145–146
        pastoralists 74, 75, 113–115, 119
    Makere Forest Reserve 202–207
livestock
    alternative fodders 31, 64, 112, 116, 148
    Bariadi District 127–128, 134, 141–142, 146–148
    distribution of herds (destocking) 112–113, 147
    forest areas 206–207
    Kilimanjaro Region 38
    mortality rates 85–87
    pests and diseases 85, 86, 112–113, 141–142, 148
    social importance of cattle 127–128, 147–148
    see also pastoralism/pastoralists
logging, illegal 207

Maasai ethnic group 83, 104, 105, 163
maize 24, 39, 41, 115, 144
    pests 40–41
Makame village (mobile pastoralists) 76–94, 239
Makere Forest Reserve 196–211
malaria 109, 176
Malawi (Balaka District) 47–56, 240–241
manure 16, 17, 39, 54, 146, 188
marital status 81–83, 133
mat-making 175
meteorology *see* weather forecasting
migration (human)
    of arable farmers 144
    due to political conflict 65
    government policy 160, 167
    into wetlands, causing conflict 160, 162
    of pastoralists 75, 111–112, 116, 146–147
    remittances from migrants 145

migratory routes for livestock 86, 119
mixed farming *see* agropastoralism
Mkuranga District cassava farmers 23–33, 240
mobile pastoralism *see* pastoralism/pastoralists
mobile phone ownership 38, 50
modernization paradigm in pastoralism 63–64
Monduli District (pastoral) 103–119, 240
mortality of livestock 85–87
Moshi Rural District (coffee and banana) 10–19
mulching 49, 53, 54

National Adaptation Programme of Action (NAPA) 73, 101
National Livestock Policy (NLP) 72
natural capital 125
net present value (NPV) 68
  mobile vs sedentary pastoralism 90–92, 94
Ngarambe-Tapika Wildlife Management Area (WMA) 170–178, 238
ngitiri pasture enclosures 147
Ngorongoro District (pastoral) 103–119, 240
NGOs (non-governmental organizations) 49, 55, 111, 148
nitrogen fertilizer *see* fertilizer use
nomadic pastoralism 71

ox-ploughs 146
Oxfam 111

participatory economic valuation (PEV) 67
participatory wealth ranking (PWR) 199
pastoral mobility paradigm 62–63
pastoralism/pastoralists 59–61, 94
  benefits 68–69, 89–90, 93–94
    estimation methods 66–68
  climate change 73–74, 76, 101–119
    adaptation to 103–119, 146–148, 240
  comparison of mobile and sedentary pastoralism 76–94, 239
  and conflict 65, 74–75, 109, 116, 158–167
  demographics 74, 81–83
  education 83–84, 118
  forms of 71–72
  in the future 75–76
  gender differences 117
  government policy 63, 66, 72–73, 86
  herd size and composition 84, 112–113, 116
  land holding 75, 114–115
  livelihood diversification 74, 75, 113–115, 119
  livestock mortality 85–87
  migration 75, 111–112, 116, 146–147
  origins 69–71
  residence 84
  sale of livestock 87, 113, 116
  sustainability theories 61–66

perception of climate change
  agropastoralists (Bariadi District) 135–137, 149
  coastal district 25–26, 31
  coffee and banana farmers 14–16
  Ethiopia 183–184
  pastoralists 107
pests and diseases
  banana 10, 14, 41
  coffee 10, 14, 41
  crops (in general) 109–110, 141
    pesticide use 146
    resistant varieties 190
  human 109, 176
  livestock 85, 86, 112–113, 141–142, 148
  maize 40–41
petty businesses 74, 114, 134, 145–146
phenology 140, 143
  rain forecasts 224
pigs 147, 148
planting season 143
ploughing 146, 148
population growth 1, 74, 75–76, 204
poultry 142, 148
poverty
  agropastoralists 138, 142–143, 149
  assessment 199
  forest ecosystems 195, 202, 203
  pastoralists 114
precipitation *see* rainfall

radio ownership 38, 50, 111, 229
rainfall
  Bariadi District 127, 136, 139, 140
  Great Ruaha Catchment Area 221
  Kilimanjaro Region 35, 40, 43
  Kilombero and Kilosa Districts 157
  Mkuranga District 24, 26, 28–29
  Monduli/Ngorongoro/Simanjiro Districts 103, 107–108, 109
  Moshi Rural District 14, 18–19
  Tehuledere District (Ethiopia) 184
  and vulnerability in conservation agriculture 49, 53–54
Ramsar sites 158, 163
remittances from relatives 145
resilience 34–35, 237–239
  agropastoralists (Bariadi District) 123–150, 239
  and cassava 24, 29–30, 32, 240
  climate smart agriculture (Ethiopia) 180–192
  ecosystem services in the Ngarambe-Tapika WMA 169–178, 238
  pastoralism 60
  see also vulnerability
Rift Valley Fever 109
rotation of crops 144
RRG paradigm of mobile pastoralism 64–66

Rufiji District (Ngarambe-Tapika WMA) 170–178, 238
rural–urban linkage 144–145

SAGCOT (Southern Agricultural Growth Corridor of Tanzania) 1–2
satellite data 200
Savings and Credit Cooperatives (SACCOs) 38
sedentary pastoralism, comparison with mobile pastoralism 76–94, 239
seed prices 116
sheep 84, 147
Simanjiro District (pastoral) 83, 103–119, 240
small businesses 74, 114, 134, 145–146
social cost–benefit analysis (SCBA) 67
social environments 50, 117–118, 125
soil
  and conservation agriculture 52, 54–55
  Ethiopia 187, 188–189, 190
  Kilimanjaro Region 38, 39–40
Sukuma ethnic group 126–127, 163
Sustainable Livelihoods Framework (SLF) 45–46, 124–125

Tanzania Forest Services (TFS) 210
Tanzania Livestock Modernization Initiative (TLMI) 73
Tanzania Meteorological Agency (TMA) 224–228
technical/physical capital 125
Tehuledere District (climate smart agriculture) 182–192
temperature
  Bariadi District 137
  effect on coffee and banana 12
  Kilimanjaro Region 35
  Mkuranga District 24, 26–28
  Monduli/Ngorongoro/Simanjiro Districts 103, 108
  Moshi Rural District 12, 14, 18
  predicted rises 2, 170, 180
  Tehuledere District (Ethiopia) 185
terracing 39, 189
tillage methods 39, 146, 148
timber harvesting 175, 207
total economic value (TEV) 66–67
tourism (hunting) 175–176, 178
transhumance 71

urban markets 144–145

vaccination of livestock 148
violent conflict 65, 162, 164
vulnerability 41, 123, 181, 237–239
  Bariadi District 137–138, 149
  coffee and banana farming 10–12
  conservation agriculture in Malawi 46, 49, 55–56
  Kilimanjaro region 36–37, 41–42, 43, 238
  pastoral communities 108–110
  vulnerability index calculation 36–37
  see also resilience

waged employment 114, 134, 178
water availability/conservation
  Bariadi District 136, 141, 142, 145
  conflict over 157, 160, 163, 165
  Ethiopia 187, 188–189, 190
  Great Ruaha Catchment Area 218–219, 220, 221
  Kilimanjaro Region 39, 40
  Mkuranga District 29
  Ngarambe-Tapika WMA 173
  pastoralism 65, 85, 141
wealth
  agropastoralists 137–138, 142–143, 149
  assessment 130, 199
  conservation agriculture adopters 50
  definition of financial capital 125
  forest ecosystems 195, 201–202
  pastoralists 114
weather forecasting 217–218, 241–242
  Ethiopia 189
  Great Ruaha Catchment Area 218–232
weeds 54, 109
wetlands
  conflict over natural resources 156–167
  newly cultivated 145
widows 81, 83, 133, 164
wildlife
  bush meat 174
  hunting by tourists 175–176, 178
  resource conflict and management 109, 157, 161, 170
wind 108
women
  effect of land conflicts 164
  gender inequality 116–118
  see also widows
woodlands *see* forests

yield
  Bariadi District 138–139
  coffee and banana 14
  conservation agriculture 45, 55
  decreased due to climate change 14, 18, 138–139
  fertilizer use 189
  higher-yield crop varieties 143, 186
  Ngarambe-Tapika WMA 174

Zambia 53
zebu cattle 127